LONDON MATHEMATICAL SOCIETY LECTURE NOTE SERIES

Managing Editor: Professor J.W.S. Cassels, Department of Pure Mathematics and Mathematical Statistics, University of Cambridge, 16 Mill Lane, Cambridge CB2 1SB, England

The books in the series listed below are available from booksellers, or, in case of difficulty, from Cambridge University Press.

4 Algebraic topology, J.F. ADAMS
5 Commutative algebra, J.T. KNIGHT
16 Topics in finite groups, T.M. GAGEN
17 Differential germs and catastrophes, Th. BROCKER & L. LANDER
18 A geometric approach to homology theory, S. BUONCRISTIANO, C.P. ROURKE & B.J. SANDERSON
20 Sheaf theory, B.R. TENNISON
21 Automatic continuity of linear operators, A.M. SINCLAIR
23 Parallelisms of complete designs, P.J. CAMERON
24 The topology of Stiefel manifolds, I.M. JAMES
25 Lie groups and compact groups, J.F. PRICE
27 Skew field constructions, P.M. COHN
29 Pontryagin duality and the structure of LCA groups, S.A. MORRIS
30 Interaction models, N.L. BIGGS
31 Continuous crossed products and type III von Neumann algebras, A. VAN DAELE
34 Representation theory of Lie groups, M.F. ATIYAH *et al*
36 Homological group theory, C.T.C. WALL (ed)
38 Surveys in combinatorics, B. BOLLOBAS (ed)
39 Affine sets and affine groups, D.G. NORTHCOTT
40 Introduction to H_p spaces, P.J. KOOSIS
41 Theory and applications of Hopf bifurcation, B.D. HASSARD, N.D. KAZARINOFF & Y-H. WAN
42 Topics in the theory of group presentations, D.L. JOHNSON
43 Graphs, codes and designs, P.J. CAMERON & J.H. VAN LINT
44 Z/2-homotopy theory, M.C. CRABB
45 Recursion theory: its generalisations and applications, F.R. DRAKE & S.S. WAINER (eds)
46 p-adic analysis: a short course on recent work, N. KOBLITZ
48 Low-dimensional topology, R. BROWN & T.L. THICKSTUN (eds)
49 Finite geometries and designs, P. CAMERON, J.W.P. HIRSCHFELD & D.R. HUGHES (eds)
50 Commutator calculus and groups of homotopy classes, H.J. BAUES
51 Synthetic differential geometry, A. KOCK
52 Combinatorics, H.N.V. TEMPERLEY (ed)
54 Markov processes and related problems of analysis, E.B. DYNKIN
55 Ordered permutation groups, A.M.W. GLASS
56 Journees arithmetiques, J.V. ARMITAGE (ed)
57 Techniques of geometric topology, R.A. FENN
58 Singularities of smooth functions and maps, J.A. MARTINET
59 Applicable differential geometry, M. CRAMPIN & F.A.E. PIRANI
60 Integrable systems, S.P. NOVIKOV *et al*
61 The core model, A. DODD
62 Economics for mathematicians, J.W.S. CASSELS
63 Continuous semigroups in Banach algebras, A.M. SINCLAIR
64 Basic concepts of enriched category theory, G.M. KELLY
65 Several complex variables and complex manifolds I, M.J. FIELD
66 Several complex variables and complex manifolds II, M.J. FIELD
67 Classification problems in ergodic theory, W. PARRY & S. TUNCEL
68 Complex algebraic surfaces, A. BEAUVILLE

69 Representation theory, I.M. GELFAND *et al*
70 Stochastic differential equations on manifolds, K.D. ELWORTHY
71 Groups - St Andrews 1981, C.M. CAMPBELL & E.F. ROBERTSON (eds)
72 Commutative algebra: Durham 1981, R.Y. SHARP (ed)
73 Riemann surfaces: a view towards several complex variables, A.T. HUCKLEBERRY
74 Symmetric designs: an algebraic approach, E.S. LANDER
75 New geometric splittings of classical knots, L. SIEBENMANN & F. BONAHON
76 Spectral theory of linear differential operators and comparison algebras, H.O. CORDES
77 Isolated singular points on complete intersections, E.J.N. LOOIJENGA
78 A primer on Riemann surfaces, A.F. BEARDON
79 Probability, statistics and analysis, J.F.C. KINGMAN & G.E.H. REUTER (eds)
80 Introduction to the representation theory of compact and locally compact groups, A. ROBERT
81 Skew fields, P.K. DRAXL
82 Surveys in combinatorics, E.K. LLOYD (ed)
83 Homogeneous structures on Riemannian manifolds, F. TRICERRI & L. VANHECKE
84 Finite group algebras and their modules, P. LANDROCK
85 Solitons, P.G. DRAZIN
86 Topological topics, I.M. JAMES (ed)
87 Surveys in set theory, A.R.D. MATHIAS (ed)
88 FPF ring theory, C. FAITH & S. PAGE
89 An F-space sampler, N.J. KALTON, N.T. PECK & J.W. ROBERTS
90 Polytopes and symmetry, S.A. ROBERTSON
91 Classgroups of group rings, M.J. TAYLOR
92 Representation of rings over skew fields, A.H. SCHOFIELD
93 Aspects of topology, I.M. JAMES & E.H. KRONHEIMER (eds)
94 Representations of general linear groups, G.D. JAMES
95 Low-dimensional topology 1982, R.A. FENN (ed)
96 Diophantine equations over function fields, R.C. MASON
97 Varieties of constructive mathematics, D.S BRIDGES & F. RICHMAN
98 Localization in Noetherian rings, A.V. JATEGAONKAR
99 Methods of differential geometry in algebraic topology, M. KAROUBI & C. LERUSTE
100 Stopping time techniques for analysts and probabilists, L. EGGHE
101 Groups and geometry, ROGER C. LYNDON
102 Topology of the automorphism group of a free group, S.M. GERSTEN
103 Surveys in combinatorics 1985, I. ANDERSON (ed)
104 Elliptic structures on 3-manifolds, C.B. THOMAS
105 A local spectral theory for closed operators, I. ERDELYI & WANG SHENGWANG
106 Syzygies, E.G. EVANS & P. GRIFFITH
107 Compactification of Siegel moduli schemes, C-L. CHAI
108 Some topics in graph theory, H.P. YAP
109 Diophantine Analysis, J. LOXTON & A. VAN DER POORTEN (eds)
110 An introduction to surreal numbers, H. GONSHOR
111 Analytical and geometric aspects of hyperbolic space, D.B.A.EPSTEIN (ed)
112 Low-dimensional topology and Kleinian groups, D.B.A. EPSTEIN (ed)
113 Lectures on the asymptotic theory of ideals, D. REES
114 Lectures on Bochner-Riesz means, K.M. DAVIS & Y-C. CHANG
115 An introduction to independence for analysts, H.G. DALES & W.H. WOODIN
116 Representations of algebras, P.J. WEBB (ed)
117 Homotopy theory, E. REES & J.D.S. JONES (eds)
118 Skew linear groups, M. SHIRVANI & B. WEHRFRITZ
119 Triangulated categories in the representation theory of finite-dimensional algebras, D. HAPPEL
120 Lectures on Fermat varieties, T. SHIODA
121 Proceedings of *Groups - St Andrews 1985*, E. ROBERTSON & C. CAMPBELL (eds)
122 Non-classical continuum mechanics, R.J. KNOPS & A.A. LACEY (eds)
123 Surveys in combinatorics 1987, C. WHITEHEAD (ed)
124 Lie groupoids and Lie algebroids in differential geometry, K. MACKENZIE
125 Commutator theory for congruence modular varieties, R. FREESE & R. MCKENZIE

London Mathematical Society Lecture Note Series. 117

Homotopy Theory

Proceedings of the Durham Symposium 1985

Edited by

E. REES
Department of Mathematics, University of Edinburgh

J.D.S. JONES
Mathematical Institute, University of Warwick

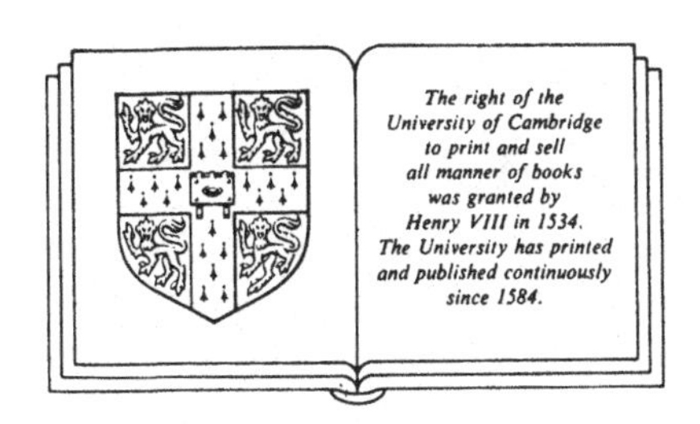

CAMBRIDGE UNIVERSITY PRESS
Cambridge
New York New Rochelle Melbourne Sydney

CAMBRIDGE UNIVERSITY PRESS
Cambridge, New York, Melbourne, Madrid, Cape Town,
Singapore, São Paulo, Delhi, Tokyo, Mexico City

Cambridge University Press
The Edinburgh Building, Cambridge CB2 8RU, UK

Published in the United States of America by Cambridge University Press, New York

www.cambridge.org
Information on this title: www.cambridge.org/9780521339469

First published 1987
Re-issued 2011

A catalogue record for this publication is available from the British Library

ISBN 978-0-521-33946-9 Paperback

Contents

Preface

An LMS Symposium, sponsored by the SERC, was held at Durham University from July 29 1985 until August 7 1985. There were many varied and interesting talks including reports on several spectacular recent advances in the subject. The timing of the Symposium was very fortunate in this respect. We therefore solicited expository papers on these new results from a number of the speakers and these papers are collected in this volume.

We would like to thank all the speakers and participants, as well as everyone who helped in the organization, for making the Symposium such a succesful one.

John Jones

Elmer Rees

ON THE COMPUTATIONAL COMPLEXITY OF RATIONAL HOMOTOPY

D.J. Anick
Department of Mathematics
Massachusetts Institute of Technology
Cambridge, MA 02139

> "[A]lthough the procedures developed for [computing homotopy groups] are finite, they are much too complicated to be considered practical."
>
> E. Brown, 1956[3]

The research summarized here may be viewed as a vindication of Brown's remarks. Given a finite simply-connected CW complex X, a common problem in algebraic topology is to evaluate $\pi_n(X)$. A much simpler problem is to determine the rank of $\pi_n(X)$. The latter problem is shown to belong to the class of $\#P$-hard problems, which are believed to require more than polynomial time to compute deterministically. Computing the Hilbert series of a graded algebra or the Poincaré series of a local Artinian ring is also $\#P$-hard.

Full details will be given in a forthcoming paper [2]. This note contains a brief overview of the results.

SUMMARY OF RESULTS

In computer science, a "problem" is a function f from a subset of $\mathbb{N}$, the non-negative integers, to $\mathbb{N}$. Sometimes the argument of this function, also called the "input", may be viewed naturally as a single integer, while at other times it may encode, through some fixed injection $\eta\colon \coprod_{j=1}^{\infty} \mathbb{Z}^j \to \mathbb{N}$, the finite description of some other object. (The map η is said to <u>$\mathbb{N}$-encode</u> the finite description.) Regardless of the proper interpretation of the input, a computer scientist may imagine that a machine or algorithm exists which can accept an arbitrary $N \in \mathcal{D}om(f)$ as input and after $\tau(N)$ steps deliver $f(N)$ as output. Given f, he or she may seek theoretical lower bounds on the function $\tau(N)$ or seek algorithms (machines) on which $\tau(N)$ exhibits a certain level of efficiency.

It is not at all obvious how one gives an efficient finite description of an arbitrary finite simply-connected CW complex X. Simplicial and semi-simplicial descriptions of X tend to be "too large."

If we restrict our attention to rational homotopy, however, Quillen's minimal Lie algebra model provides a complete description of the rational homotopy type of X, a description whose size is roughly comparable with the number of cells of X and with the complexity of the attaching maps. The problem "compute rank $(\pi_n(X))$" is therefore interpreted as follows: given an integer n and the Quillen model for a space X, both $\mathbb{N}$-encoded, determine $\dim_{\mathbb{Q}}(\pi_n(X)\otimes\mathbb{Q})$.

Related problems are to be viewed similarly. For example, "Compute the Hilbert series of a finitely presented connected graded algebra" means: given an $\mathbb{N}$-encoded list of generators and relations for a graded $\mathbb{Q}$-algebra $A = \bigoplus_{m=0}^{\infty} A_m$ and an integer n, determine $\dim_{\mathbb{Q}}(A_n)$.

Computer scientists have developed a scale or continuum along which various problems may be placed according to their complexity. Certain classes of problems, including the classes known as P, $\#P$-complete, and "computable in exponential time," serve as landmarks. In particular, there is a transitive, reflexive relation on the set of problems, called "Turing reducible in polynomial time" and denoted $\leq_T^P$, by which f_2 is at least as hard as f_1 if $f_1 \leq_T^P f_2$. The problems f_1 and f_2 are "Turing equivalent", denoted $f_1 \approx_T f_2$, if $f_1 \leq_T^P f_2$ and $f_2 \leq_T^P f_1$. The goal of this research is to locate the problem "compute rational homotopy" and some related problems upon this continuum.

Our results are summarized in Figure 1. More difficult problems are placed higher on the scale, and Turing equivalent problems have been bracketed. Problems at a specific difficulty level are marked by a bar crossing the vertical axis. Classes of problems which encompass a range along the continuum are labeled at the upper or lower end of their range.

The purpose of Figure 1 is to give a visual overview of our results, and some technical points were sacrificed for crispness. For instance, one might conclude from the picture that any problem unsolvable in exponential time is $\#P$-hard, but this has not been proved (and is probably false). Nor have we proved that the three Turing equivalence classes marked by bars crossing the axis must actually be distinct.

As to interpretation, bear in mind that the class $\#P$-hard starts near the bottom of Figure 1 even though this class is viewed by computer scientists as being "very difficult". In other words, except for

"$\mathcal{P}$", the section of the scale shown in Figure 1 actually starts far above such familiar computer mainstays as "solve a linear system of equations", "find the roots of a polynomial", or "factor the integer N."

There are no known algorithms which can evaluate a #$\mathcal{P}$-hard problem in less than exponential time. Algorithms which require exponential time are generally thought of as being beyond the scope of today's machinery to implement efficiently. Thus computing rational homotopy groups and the other problems listed in Figure 1 may truly be described as very complex problems.

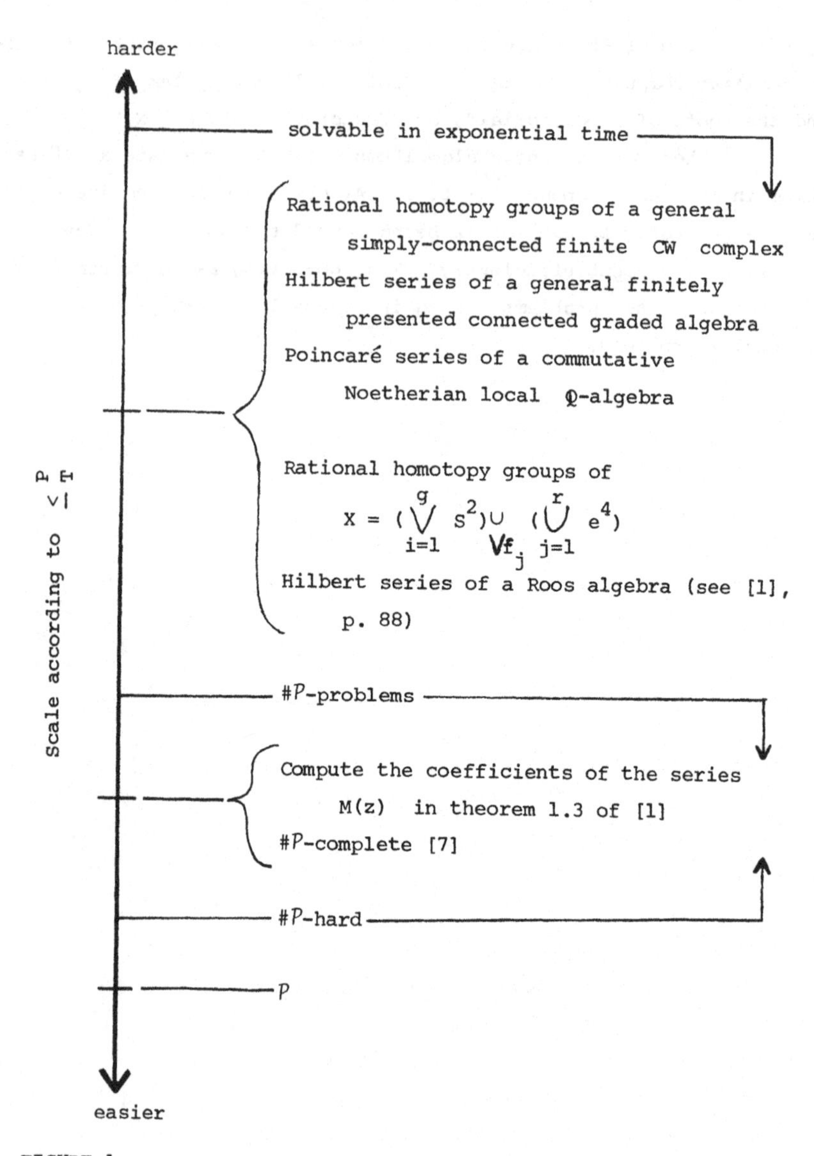
harder
solvable in exponential time
Rational homotopy groups of a general simply-connected finite CW complex
Hilbert series of a general finitely presented connected graded algebra
Poincaré series of a commutative Noetherian local ℚ-algebra
Rational homotopy groups of
$X = (\bigvee_{i=1}^{g} S^2) \cup_{\vee f_j} (\bigcup_{j=1}^{r} e^4)$
Hilbert series of a Roos algebra (see [1], p. 88)
#P-problems
Compute the coefficients of the series M(z) in theorem 1.3 of [1]
#P-complete [7]
#P-hard
P
easier
Scale according to $\leq_T^P$

FIGURE 1

REFERENCES

1. D. Anick, "Diophantine equations, Hilbert series, and undecidable spaces", Annals of Math. 122(1985), pp. 87-112.

2. D. Anick, "The computation of rational homotopy groups is #P-hard", to appear in the Proceedings of the Conference on Computers and Topology, Univ. of Illinois, 1986.

3. E. Brown, "Finite computability of Postnikov complexes", Annals of Math. 65(1957), pp. 1-20.

4. A.D. Gavrilov, "Effective computability of the rational homotopy type", Math. USSR Isv. 10(1976), no. 6, pp. 1239-1260.

5. J.E. Hopcroft and J.D. Ullman, Introduction to Automata Theory, Languages, and Computation, Addison-Wesley, 1979.

6. D.G. Quillen, "Rational homotopy theory", Annals of Math. 90(1969), pp. 205-295.

7. L.G. Valiant, "The complexity of computing the permanent", Theoretical Computer Science 8(1979), pp. 189-201.

SEGAL'S BURNSIDE RING CONJECTURE AND THE HOMOTOPY LIMIT PROBLEM

Gunnar Carlsson
Department of Mathematics
University of California, San Diego
La Jolla, California 92093

Supported in part by NSF Grant MCS-82-01125.
The author is an Alfred P. Sloan fellow.

This is a somewhat informal discussion of Segal's Burnside ring conjecture. We give an outline of how it is proved, and in later sections discuss how it can be applied to other problems of identification of a "homotopy fixed point set." We outline the paper. §1 is a discussion of Atiyah's theorem on the K-theory of classifying spaces of finite groups, which is the first theorem of this type. §2 discusses Dyer Lashof maps and the connection of $B\Sigma_n$ with $\Omega^\infty S^\infty$, and thereby motivates Segal's conjecture. In §3, we define and outline some of the properties of equivariant stable homotopy theory, which is the key piece of machinery in stating the conjecture in its correct generality, and is in fact the key ingredient in the reduction from the general p-group case to the elementary Abelian case. In §4, we outline the work of Lin (1980) and Adams et al. (1985), who prove the elementary Abelian case. In §5, we give an account of the results of my paper (Carlsson, 1984), in which the reduction to the elementary Abelian case is accomplished. In §6, we state the so-called "homotopy limit" problem, of which Segal's conjecture is a special case. Other cases include Sullivan's conjecture and the conjectured existence of a "descent spectral sequence" for algebraic K-theory. Finally, in §7, we show how Segal's conjecture may be applied to study these problems.

As stated above, the discussion is informal, and for complete statements and proofs of all theorems, one should refer to the original sources. Technical points are sometimes suppressed in order to continue the flow of the discussion; for instance, $\lim^1$-arguments are omitted.

A few words about references are in order. Atiyah's theorem can be found in Atiyah's original paper; a different (and perhaps more enlightening) proof of a more general result may be found in Atiyah and Segal (1969).

For the results of §2, the Dyer-Lashof maps and the connections with $B\Sigma_n$ are treated well in Madsen and Milgram (1979): more thorough treatments include the books of Adams (1978) and May (1972). The Burnside ring may be found in Tom Dieck (1979), as well as in the work of Dress (1969). Equivariant stable homotopy was defined by Segal in Segal (1970), and was presumably the motivation for the original conjecture. It is also discussed in Tom Dieck's book, and in Adams (1984). A more encyclopedic account due to Lewis, May, McClure, and Steinberger will appear in due course (Lewis et al., to appear). The references for the elementary Abelian case are Lin (1980), Gunawardena (1980), and Adams et al. (1985). The primary reference for §5 is Carlsson (1984). The Adams spectral sequence is discussed in Adams (1974) and Switzer (1975). The nerve construction for posets and categories is available in Quillen (1973), and results of the paper by Quillen (1978) shed much light on the reduction procedure. For the necessary results on group cohomology, we refer to Quillen and Venkov (1972). To my knowledge, the homotopy limit problem, as such, was formulated by Thomason in (1983). Sullivan's conjecture was posed in Sullivan (1970), and its proof in the case of trivial G-action by H. R. Miller appears in Miller (1974). For a discussion of the descent problem in algebraic K-theory, and its solution for "Bott periodic" K-theory, see Thomason's paper (to appear). More general discussions of algebraic K-theory appear in Quillen (1970,1973). The theorem of Suslin appears in Suslin (1983). Bousfield and Kan (1972) is an excellent source for cosimplicial spaces, and in particular for the cosimplicial space of a triple.

The author wishes to apologize for omissions, made in the interest of brevity. Segal and Stretch (1981) and Laitinen (1979) made earlier contributions to the study of the Segal conjecture for abelian groups. We have also omitted recent work on Sullivan's conjecture by Jackowski, as well as applications of Miller's result by McGibbon-Neisendorfer and Zabrodsky.

Finally, the author wishes to express his thanks to many people for valuable discussions on many of the subjects discussed in this paper. An incomplete list includes: J. F. Adams, E. Friedlander, M. J. Hopkins, W. C. Hsiang, M. Karoubi, I. Madsen, M. E. Mahowald, R. J. Milgram, H. R. Miller, V. P. Snaith, R. W. Thomason, and C. Weibel.

1 ATIYAH'S THEOREM

In 1960, Atiyah (1961) gave a complete description of the group $[BG.BU \times \mathbb{Z}]$ of homotopy classes of maps from BG to BU, where BG and BU denote the classifying spaces of a finite group G and the infinite unitary group, respectively. (The $\mathbb{Z}$ factor denotes $\mathbb{Z}$ viewed as a discrete space.) Since homotopy classification of maps in general is a very complicated problem. and classifying spaces of finite groups can be very complicated, the simplicity of the answer is striking. and we describe it.

We recall first that $BU \times \mathbb{Z}$ admits two pairings,
$\oplus : (BU \times \mathbb{Z}) \times (BU \times \mathbb{Z}) \to BU \times \mathbb{Z}$ and
$\otimes : (BU \times \mathbb{Z}) \times (BU \times \mathbb{Z}) \to BU \times \mathbb{Z}$, arising from Whitney sum and tensor product of vector bundles respectively, which satisfy the usual associativity and distributivity relations. Accordingly, $[BG, BU \times \mathbb{Z}]$ becomes a ring, and we will describe it in terms of a classical algebraic invariant of G . the complex representation ring. Recall how it is defined. Let Rep(G) denote the set of isomorphism classes of unitary representations of G ; equivalently, Rep(G) is just the set of complex characters of G. Rep(G) becomes an Abelian monoid under the direct sum operation, and a semiring under direct sum and tensor product. After adjoining formal additive inverses, we obtain the complex representation ring of G, R[G]. Classically, it was known as the character ring. It is a free, finitely generated Abelian group whose rank is equal to the number of conjugacy classes of elements of G. R[G] is contrafunctorial in G ; moreover, $R[\{e\}] \cong \mathbb{Z}$, so the inclusion $\{e\} \to G$ induces a ring homomorphism $\epsilon : R[G] \to \mathbb{Z}$, called the augmentation. The kernel of ϵ , I[G], is called the *augmentation ideal* in R[G].

Let ρ be a unitary representation of G ; thus, ρ is a homomorphism from G to U(n) for some n . The functoriality of the classifying space construction means that we have a map

$f_\rho : BG \xrightarrow{B\rho} BU(n) \xrightarrow{Bi} BU \times \mathbb{Z}$ where $i : U(n) \to U$ is the inclusion. It is elementary to check that f_ρ depends only on the isomorphism class of ρ , and that we therefore obtain a function $Rep[G] \to [BG, BU \times \mathbb{Z}]$. One checks that this induces a homomorphism of rings $R[G] \to [BG, BU \times \mathbb{Z}]$. It is tempting to guess that this map is an isomorphism; it isn't quite, but we may describe precisely what the necessary modifications to R[G] are. Let R be a commutative ring, and $I \subseteq R$ an ideal. Then by the I-adic completion of R, $\hat{R}_I$, we mean the inverse limit $\lim\limits_{\overleftarrow{n}} R/I^n$. Atiyah then proved that there is a factorization

$$\begin{array}{ccc} R[G] & \rightarrow & \hat{R}[G]_{I[G]} \\ & \searrow & \downarrow \\ & & [BG, BU] \end{array}$$

where the right hand vertical arrow is an isomorphism. This gives a completely algebraic description of the ring [BG, BU]. The reader may wonder, however, whether the ring $\hat{R}[G] = \hat{R}[G]_{I[G]}$ is a particularly tractable algebraic invariant. If G is a p-group, then $\hat{R}[G]$ may be described by the Cartesian square of rings.

$$\begin{array}{ccc} \hat{R}[G] & \rightarrow & R[G] \otimes_{\mathbb{Z}} \hat{\mathbb{Z}}_p \\ \downarrow & & \downarrow \epsilon \otimes \mathrm{id} \\ \mathbb{Z} & \rightarrow & \hat{\mathbb{Z}}_p \end{array} .$$

In the general (composite order) case, the map $R[G] \rightarrow \hat{R}[G]$ may not be injective, but $\hat{R}[G]$ may be described by induction theorems. In short, $\hat{R}[G]$ is quite a tractable invariant.

2 DYER LASHOF MAPS, THE BURNSIDE RING, AND SEGAL'S CONJECTURE

We let $Q(S^0) = \lim\limits_{\overrightarrow{n}} \Omega^n S^n$, where the maps in the directed system are induced by suspension. For a point $x \in \mathbb{R}^n$, and a number r, we let $f_{x,r}: B_r(x) \rightarrow \mathbb{R}^n \cup \{\infty\}$ be defined by $f_{x,r}(z) = \dfrac{z-x}{r-|z-x|}$. For any finite subset $S = \{x_1, ..., x_n\}$, and collection of radii $R = (r_1, ..., r_k)$, for which $B_{r_i}(x_i) \cap B_{r_j}(x_j) = 0$, we define

$$\begin{aligned} f_{S,R}(z) &= f_{x_i,r_i}(z) \quad \text{for } z \in B_{r_i}(x_i) \\ &= \infty \quad\quad\quad \text{if } z \notin \bigcup_i B_{r_i}(x_i) . \end{aligned}$$

Let $C_{n,k} = \{(x_1,...,x_k, r_1,...,r_k) | x_i \in \mathbb{R}^n, B_{r_i}(x_i) \cap B_{r_j}(x_j) = 0\}$. Then Σ_k act on $C_{n,k}$ by permutation of coordinates, and the action is clearly free. Moreover, the construction above gives a map $C_{n,k}/\Sigma_k \rightarrow (\Omega^n S^n)_k$, the component of $\Omega^n S^n$ of maps of degree k. Passing to the direct limit gives a map from $\lim_{\vec{n}} C_{n,k}/\Sigma_k \to (Q(S^o))_k \cdot \lim_{\vec{n}} C_{n,k}$ can be shown to be contractible, so the domain is in fact a model for $B\Sigma_k$. After translating by a map of degree $-k$, we get map $B\Sigma_k \xrightarrow{D_k} Q(S^o)$, which are homotopy-compatible with the inclusions $B\Sigma_k \to B\Sigma_{k+1}$. These are the Dyer-Lashof maps, which produce a map $B\Sigma_\infty \xrightarrow{D} Q(S^o)_0$.

Theorem (Barratt, Priddy, Quillen, see Priddy, 1971). *The map D induces an isomorphism on integral homology.*

Of course, the map is not a homotopy equivalence, since $Q(S^O)_0$ is not a $K(\Sigma_\infty, 1)$-space, but it suggests that the relationship between the symmetric groups and stable homotopy theory should be a strong one. In particular, we could ask a question quite analogous to the one studied by Atiyah, with $Q(S^O) \times \mathbb{Z}$ replacing $BU \times \mathbb{Z}$.

Question: *Can $[BG, Q(S^O) \times \mathbb{Z}]$ be described in purely algebraic terms, where G is a finite group?*

We first note that if $\rho : G \to \Sigma_k$ is any homomorphism, we may form the composite

$$f_\rho : BG \xrightarrow{B\rho} B\Sigma_k \xrightarrow{D_k} Q(S^o),$$

and obtain an element in $[BG, Q(S^O)]$. One is led to conjecture, by analogy with Atiyah's theorem, if in an appropriate sense all elements of $[BG, Q(S^O)]$ are obtained in this way. To understand what the "appropriate sense" is, we must first understand what the analogue of $R[G]$ is.

By a finite G set, we simply mean a finite set X together with an action of G; equivalently, we mean a homomorphism from G to $\Sigma(X)$, the group of all permutations of X. There is an evident notion of isomorphism of G-sets acting on a fixed set X, and the isomorphism classes correspond to conjugacy classes of homomorphisms $G \to \Sigma(X)$. If X_1 and X_2 are two finite G-set then the disjoint union and products of X_1 and X_2, $X_1 \amalg X_2$ and $X_1 \times X_2$, are also finite G-sets in the evident way. This fact means that the collection of isomorphism classes of finite G-sets form a commutative monoid under disjoint union, and a commutative semiring under disjoint union and products. After adjoining formal additive inverses, we actually obtain a commutative ring $A(G)$, called the *Burnside ring* of G.

It is not hard to see that $A(\{e\}) \cong \mathbb{Z}$, and that A is a contravariant functor in G, so we have an augmentation $\epsilon : A(G) \to \mathbb{Z}$ analogous to the augmentation for $R[G]$. We denote by $I(G)$ the kernel of ϵ. and call the ideal the augmentation ideal of $A(G)$.

$A(G)$ is a readily computable object. Every finite G-set X may be uniquely (up to isomorphism) decomposed as $X = \coprod_{i=1}^{s} G/K_i$, for some s, where each K is a subgroup, and G/K_i denotes the left G-set of right cosets of K_i. This decomposition shows that as an Abelian group, $A(G)$ is free of rank $c(G)$, where $c(G)$ is the number of conjugacy classes of subgroups of G, since $G/gKg^{-1} \cong G/K$ as G-sets. Moreover, the completion process at $I(G)$ is not too drastic. In particular, if G is a p-group, $\hat{A}(G)_{I(G)}$ is described by the following Cartesian square of rings:

$$\begin{array}{ccc} \hat{A}(G)_{I(G)} & \to & A(G) \otimes_{\mathbb{Z}} \hat{\mathbb{Z}}_p \\ \downarrow & & \downarrow \epsilon \otimes \mathrm{id} \\ \mathbb{Z} & \to & \hat{\mathbb{Z}}_p \end{array} .$$

$Q(S^0) \times \mathbb{Z}$ admits two pairings, one from addition of maps and one from smash product of maps. These structures induce a ring structure on $[X, Q(S^0) \times \mathbb{Z}]$ for any space X. The correspondence $\rho \to f_\rho$, which associates to a homomorphism $\rho : G \to \Sigma_k$ the map $D_k \circ B_\rho$, defines a function from the isomorphism classes of finite G-sets to $[BG, Q(S^0) \times \mathbb{Z}]$ which extends to a ring homomorphism $\phi : A(G) \to [BG, Q(S^0) \times \mathbb{Z}]$. The question becomes, as in the case of $R[G]$, how close is this homomorphism to an isomorphism? One may show that ϕ extends over $\hat{A}(G)_{I(G)}$, to produce a ring homomorphism $\hat{\phi} : \hat{A}(G)_{I(G)} \to [BG, Q(S^0) \times \mathbb{Z}]$. Segal made the following conjecture, based on the analogy with Atiyah's theorem.

Conjecture: *$\hat{\phi}$ is an isomorphism.*

This conjecture is difficult to study as it stands; one reason is that all inductive schemes one attempts to study involve, among other things, the groups $[\Sigma^k BG, Q(S^0) \times \mathbb{Z}]$, for positive values of k. In order to find a form of the conjecture which we can study, we will need, at least, to formulate the conjecture in such a way as to include all values of k. This brings us to the subject matter of the next section.

3 EQUIVARIANT STABLE HOMOTOPY THEORY

We recall that if X is a based space, we define the stable homotopy groups of X to be the homotopy groups of the space $Q(X) = \lim_{\vec{n}} \Omega^n \Sigma^n X$. Of course, if X is a based G-space, G a finite group, then Q(X) becomes a based G-space. It was Segal's observation that Q(X) equipped with this G-action is not the correct generalization of Q(X) for the category of G-spaces. The correct generalization involves a G-action on the suspension coordinates as well; we describe this generalization, and indicate in what sense it is the correct generalization of Q.

Let G be finite, and let X be a based G-complex. Let S^{kR} denote the one-point compactification of a k-fold direct sum of the regular real representation of G. Thus, S^{kR} is a $k|G|$-dimensional sphere, and we let $\{\infty\}$ be the distinguished base point. For any based G-complex X, $\Omega^{kR}X$ will denote the space of all continuous based maps $f : S^{kR} \to X$, equipped with G-action by $g \cdot f = gfg^{-1}$. Note that $S^{kR} \wedge S^{\ell R} \cong S^{(k+\ell)R}$. For a given k, we have a map of G-spaces $\Omega^{kR}(S^{kR} \wedge X) \to \Omega^{(k+1)R}(S^{(k+1)R} \wedge X)$ defined by $f \to id_{S^R} \wedge f$, where we identify $S^{(k+1)R} \wedge X$ with $S^R \wedge S^{kR} \wedge X$. The direct limit of these G-spaces is denoted $Q^G(X)$, and is the correct equivariant analogue of Q(X). We thus define the equivariant stable homotopy groups to be the homotopy groups of the fixed point set $(Q^G(X))^G$. One shows that for any fixed finite based G-complexes X and Y, there is a representation W so that the map $[X, \Omega^W(S^W \wedge Y)]^G \to [X, \Omega^{W+V}(S^{W\oplus V} \wedge Y)]^G$ is a bijection for all representations V. This means that the direct limit defining, say, the equivariant stable homotopy groups, is achieved at a finite stage for each value of n.

To see why this is a good generalization of the functor Q, we recall that by Thom transversality, $\pi_n^S(X^+) = \pi_n(Q(X^+))$ may be identified with the bordism group of n-dimensional framed manifolds over X. In the G-equivariant setting, we define a framed n-dimensional G-manifold to be a smooth G-manifold M with an isomorphism $\tau(M) \times V \cong M \times \mathbb{R}^n \times V$, where $\tau(M)$ is the tangent bundle of M (a G-bundle), G acts trivially on $\mathbb{R}^n$, and V is a representation of G. One can construct a bordism theory of these G-manifolds, and $\Omega_n^{fr,G}(X)$ will denote the n-dimensional bordism group. One can show that transversality behaves as well in these situation as in the non-equivariant one, and that therefore $\pi_n^G(X^+) \cong \Omega_n^{fr,G}(X)$. We take this to be good justification for the assertion that Q^G is the correct equivariant generalization of Q.

We now proceed with a brief discussion of the properties of this construction. Recall that Q is a linear functor, in the sense that if $X \subseteq Y$ is an inclusion of a based subcomplex, the sequence $Q(X) \to Q(Y) \to Q(Y/X)$ is a quasifibration sequence. Q^G shares this property, in the sense that if $X \subseteq Y$ is an inclusion of a based G-subcomplex, the sequence $(Q^G(X))^H \to (Q^G(Y))^H \to (Q^G(Y/X))^H$ is a quasifibration sequence for all subgroups $H \subseteq G$. We say that a based G-complex X is free if $G_x = \{e\}$ for all $x \neq *$. We now let X be a based G-complex, and describe $(Q^G(X))^G$.

Proposition. $(Q^G(X))^G \cong \prod_{K \subseteq G} Q(EW_G(K)^+ \wedge_{W_G(K)} X^K)$ *where*

(i) *The product is over all conjugacy classes of subgroups of* G.

(ii) *For a G-space* Z, Z^+ *denotes* Z *with a fixed disjoint base point added*

(iii) $W_G(K) = N_G(K)/K$, *where* N_G *denotes normalizer in* G

Note that $W_G(K)$ *acts on* X^K *in a natural way.*

Corollary $(Q^G(S^0))^G \cong \prod_{K \subseteq G} Q(BW(K)^+)$

This theorem was first proved by Tom Dieck (Tom Dieck, 1975). The most transparent way to understand the result is via equivariant framed bordism, as in Segal (1970), where this splitting corresponds to a canonical disjoint union decomposition of a framed G-manifold.

We may conclude from the above description that if X is a free G-complex, $(Q^G(X))^G \cong Q(X/G)$. If X and Y are G-complexes, we may define $\{X,Y\}^G$, the stable G-maps from X to Y, by $\{X,Y\}^G \cong [G, Q^G(Y)]^G$. If $Y = S^n$, we define $\{X; S^n\}^G = \pi_G^n(X)$. The definition is readily extended to negative values of n in a routine way, and we obtain a graded group $\pi_G^*(X)$. π_G^* and π_*^G are in fact now equivariant cohomology theories on the category of G-complexes. If X is free, there is a non-equivariant description of $\pi_G^*(X)$.

Proposition. *If* X *is free, we have* $\pi_G^*(X) \cong \pi_s^*(X/G)$.

Corollary $\pi_G^*(EG^+) \cong \pi_s^*(BG^+)$.

Finally, we wish to discuss questions involving change of groups. Let $H \subseteq G$, and let X be any based H-complex. Then we define the induced G-complex $e_H^G(X)$ to be $G^+ \wedge X/\simeq$ where $\simeq$ is the equivalence relation generated by the equivalences $gh \wedge x \simeq g \wedge hx$, $h \in H$. G acts on $e_H^G(X)$ by its action on the left factor. If X is a G-complex to start with, it is not hard to see that $e_H^G(X) \cong G/H^- \wedge X$. One can now show that the behavior of the theories π_*^G and π_G^* on $e_H^G(X)$ is predictable.

Proposition $\pi^{*}_{G}(e^{G}_{H}(X) \cong \pi^{*}_{H}(X)$ *and* $\pi^{G}_{*}(e^{G}_{H}(X)) \cong \pi^{H}_{*}(X)$.

These results permit one to compute equivariant stable homotopy and cohomotopy of any G-complex by induction on skeleta, for if X is a G-complex, $X^{(k+1)}/X^{(k)} \cong \bigvee_i (G/G_i{}^+ \wedge S^{k+1})$, where each $G_i \subseteq G$ is a subgroup, and $\pi^{*}_{G}(G/G_i{}^+ \wedge S^{k+1}) \cong \pi^{*}_{G_i}(S^{k+1})$ and $\pi^{G}_{*}(G/G_i{}^+ \wedge S^{k+1}) \cong \pi^{G_i}_{*}(S^{k+1})$.

We now show how to reformulate and generalize Segal's conjecture. We have seen above that $\pi^{0}_{G}(S^0) \cong \Pi_{K \subseteq G} \mathbb{Z}$, where K varies over all conjugacy classes of subgroups of G; just observe that each factor in the description of $\pi^{G}_{0}(S^0) = \pi^{0}_{G}(S^0)$ contributes a factor of $\mathbb{Z}$. By the description of the Burnside ring in §2, $A(G)$ and $\pi^{0}_{G}(S^0)$ are isomorphic as Abelian groups. Tom Dieck (1979), in fact observes that $\pi^{0}_{G}(S^0)$ is a ring in an evident way, and that there is a natural (in G) ring isomorphism $A(G) \xrightarrow{\approx} \pi^{0}_{G}(S^0)$. This means that $\pi^{*}_{G}(S^0)$, and, in fact, $\pi^{*}_{G}(X)$ for any based G-complex is an $A(G)$-module. Hence, we can talk about the $I(G)$-adic completion of $\pi^{*}_{G}(X)$. This suggests that Segal's conjecture should be the special case (n=0) of the following conjecture.

Segal Conjecture (Strong Form) *The map* $\pi^{n}_{G}(S^0) \to \pi^{n}_{G}(EG^+)$ *induced by the map* $EG^+ \to S^0$ *which sends* EG *to the non base point becomes an isomorphism after* $I(G)$*-adic completion.*

This is the form of the conjecture which turns out to be approachable by an inductive scheme. It gives a complete description of the stable cohomotopy groups of BG in terms of stable homotopy groups of classifying spaces of subquotients of G. In the next section, we will discuss the history of this conjecture.

4 THE WORK OF LIN, GUNAWARDENA, AND ADAMS-MILLER-GUNAWARDENA

The first group for which Segal's conjecture was studied (and resolved) was $G = \mathbb{Z}/2\mathbb{Z}$; the strong form of the conjecture was proved by Lin (1980) using Adams Spectral Sequence techniques. We give a brief outline.

If G is any finite group, and V is an orthogonal representation of G, then by BG^V we mean the quotient $EG \times_G D(V) / EG \times_G S(V)$, where $S(V)$ and $D(V)$ denote unit spheres and unit discs, respectively. For instance, if V is an n-dimensional representation of $\mathbb{Z}/2\mathbb{Z}$, with $Tv = -v \ \forall v \in V$, where T generates G, $BG^V \cong \mathbb{R}P^\infty/\mathbb{R}P^{n-1}$. It turns out that one may actually perform this construction for virtual representations of G, provided one accepts that the result can be a spectrum instead of a space. This construction has the following properties.

Proposition 1. *Let* α *be a virtual representation of a finite group* G, *so that* $\det g = 1 \ \forall \ g \in G$, *or let* $R = \mathbb{Z}/2\mathbb{Z}$ *below. Then*

(a) $H^*(BG^\alpha; R)$ *is in a natural way a graded* $H^*(BG; R)$*-module, where* R *is any commmutative ring.*

(b) As an $H^*(BG; R)$*-module,* $H^*(BG^\alpha, R)$ *is free on one generator* e_α *of dimension* $= \dim(\alpha)$.

(c) If α *and* β *are virtual representations, so that* α-β *is an "honest" representation* V, *then there is a map* $BG^\beta \to BG^\alpha$, *and the induced map on cohomology carries* e_α *to* $e(V) \cdot e_\beta$, *where* $e(V)$ *denotes the Euler class of* V.

For the case of $G = \mathbb{Z}/2\mathbb{Z}$, we will work with $H^*(\ ;\mathbb{Z}/2\mathbb{Z})$. Let $\mathbb{R}P_n^\infty$ denote BG^α, where $\alpha = n\xi$, where ξ is a one-dimensisonal nontrivial representation of G. Here, n may be positive or negative. For integers k and ℓ, $k \leqslant \ell$, we write $\mathbb{R}P_k^\ell$ for the spectrum $\Sigma^{-1}\mathbb{R}P_{\ell+1}^\infty/\mathbb{R}P_k^\infty$. One can show that the S-dual spectrum to $\mathbb{R}P_0^k$ is $\Sigma(\mathbb{R}P_{-k-1}^{-1})$; Lin's method was to study $\lim\limits_{\ell} \pi_*(\mathbb{R}P_{-\ell}^{-1}) \cong \lim\limits_{\ell} \pi^*(\mathbb{R}P_0^\ell)$.

We have cofibration sequences $\mathbb{R}P_{-\ell}^{-1} \to \mathbb{R}P_{-\ell}^\infty \to \mathbb{R}P_0^\infty$. Segal's conjecture predicts that $\lim\limits_{\ell} \pi_*(\mathbb{R}P_{-\ell}^{-1}) \cong \pi_*(S^{-1}) \oplus \pi_*(\Sigma^{-1}\mathbb{R}P_0^\infty)$. One checks that the map $\pi_*^G(S^0) \to \pi^*(\mathbb{R}P_0^\infty)$ is compatible with the boundary maps $\partial : \pi_*(\mathbb{R}P_0^\infty) \to \pi_*(\Sigma\mathbb{R}P_\ell^{-1})$ in the above cofibration sequences. It then remains to study $\lim\limits_{\ell} \pi_*(\mathbb{R}P_{-\ell}^\infty)$, and show that this provides the factor $\pi_*(S^{-1})$, and this is what Lin did. One can show that for inverse systems of spectra $\{X_i\}$, there is an Adams spectral sequence with E_2-term $\mathrm{Ext}_{A(p)}(\lim\limits_{i} H^*(X_i); \mathbb{Z}/p\mathbb{Z})$ converging to $\lim\limits_{i} \hat{\pi}_*(X_i)$, where $\hat{\pi}$ denotes homotopy groups P-adically completed. In our case, $X_i = \mathbb{R}P_{-i}^\infty$, and $H^*(X_i : \mathbb{Z}/2)$ is a free module over $H^*(\mathbb{R}P^\infty; \mathbb{Z}/2) = \mathbb{Z}/2[x]$ on a single generator e_{-i} in dimension $-i$, and the map $H^*(X_i; \mathbb{Z}/2) \to H^*(X_{i+1}; \mathbb{Z}/2)$ is given by $x^k \cdot e_{-1} \to x^{k+1} \cdot e_{-(i+1)} \cdot \lim\limits_{i} H^*(X_i : \mathbb{Z}/2)$ can now be identified with the ring of Laurent series $\mathbb{Z}/2[x, x^{-1}]$, with $A(2)$-action specified by $Sq^i(x) = x^2$ if $i = 1$, $Sq^i(x) = 0$ for $i \geqslant 2$, and the Cartan formula.

Lin now proved that

$$\mathrm{Ext}_{A(2)}^{s,t}(\mathbb{Z}/2[x,x^{-1}];\mathbb{Z}/2) \cong \mathrm{Ext}_{A(2)}^{s-1,t-1}(\mathbb{Z}/2,\mathbb{Z}/2),$$

and that there is a filtration-increasing map of Adams Spectral Sequences from the Adams Spectral Sequence for S^{-1} to that for $\varprojlim_{\ell} \pi_*(\mathbb{R}P^{\infty}_{-\ell})$, inducing the isomorphism. He concludes that $\varprojlim_{\ell} \pi_*(\mathbb{R}P^{\infty}_{-\ell}) \cong \pi_*(S^{-1})$, which proves Segal's conjecture for $G = \mathbb{Z}/2\mathbb{Z}$.

A very similar calculation can be carried out for $G = \mathbb{Z}/p\mathbb{Z}$; again, the conjecture predicts $\pi^*(B\mathbb{Z}/p^+) \cong \pi_*(S^0) \oplus (B\mathbb{Z}/p^+)$, and is reduced to studying the Ext-groups of a certain ring of Laurent series. This was carried out by J.H.C. Gunawardena in his J. T. Knight Prize Essay (Gunawardena, 1980).

D. Ravenel (1981), after constructing a modified Adams filtration well suited to the problem at hand, used these methods to establish the conjecture for all finite cyclic groups. This is the first situation where the conjecture is not true at the level of E_2-terms of Adams spectral sequences, but turns out to be true on the homotopy level.

Adams, Miller, and Gunawardena (in Adams et al., 1985) now combined forces to extend this computational approach to the case of elementary Abelian groups, $G \cong (\mathbb{Z}/p\mathbb{Z})^k$: The calculation becomes quite involved, but the key result is the calculation of the Tor-groups, over the Steenrod algebra, of $H^*(B\mathbb{Z}/p^k;\mathbb{Z}/p)(1/L)$, where L is the product of all 2-dimensional linear monomials if p is odd, and the product of all one-dimensional linear monomials if $p = 2$. More recently, Priddy and Wilkerson (1985) have shown how to simplify their proof considerably.

One can presumably proceed somewhat further using the computational method. assuming that one is willing to work with a modified Adams Spectral Sequence. It is, however, clear that one cannot expect to prove the full conjecture solely by these methods, since the input for the Adams Spectral Sequence involves $H^*(BG;\mathbb{Z}/p)$, which in general is quite an intractable object. Fortunately, there is an inductive procedure which reduces the full conjecture to that for elementary Abelian groups $(G = (\mathbb{Z}/p\mathbb{Z})^k)$. It is in this procedure that the equivariant stable homotopy theory is used in an essential way. We now describe this procedure.

5 THE REDUCTION TO THE ELEMENTARY ABELIAN CASE

In this section, we outline the contents of Carlsson (1984), in which the Segal conjecture is proved. We first describe some preliminary reductions, due to Lewis, May, and McClure.

Proposition 1. *The Segal conjecture holds for all finite groups* G *if and only if it holds for all finite p-groups.*

This is proved by observing that both $\pi_s^*(BG^+)$ and $\hat{\pi}_G^*(S^o)_{I(G)}$ are Mackey functors of a certain kind, which suffices to guarantee that the p-group case suffices.

Proposition 2. *The Segal conjecture is true for a p-group* G *if and only if the map* $\pi_G^*(S^o) \to \pi_G^*(EG^+)$ *is an isomorphism after p-adic completion.*

This comes from the Cartesian square diagram in §2, describing $\hat{A}(G)_{I(G)}$ in terms of $A(G) \otimes \hat{\mathbb{Z}}_p$.

Thus, we are free to work with the p-group case, where simplifications are possible due to the solvability of G , the existence of central elements, and a theorem of Serre on the cohomology of p-groups which we discuss below. Also, in the remainder of this section, all groups will be assumed to be p-adically completed.

We are now trying to prove that the map $\pi_G^*(S^o) \to \pi_G^*(EG^+)$ is an isomorphism. Since π_G^* is a cohomology theory, this is equivalent to showing that $\pi_G^*(\tilde{E}G) = 0$, where $\tilde{E}G$ denotes the mapping cone on the map $EG^+ \to S^o$. $\tilde{E}G$ turns out to be the (unreduced) suspension of EG , with, say, the upper cone point as base point. Of course, we may also study the cofibration sequence

$$EG^+ \to S^o \to \tilde{E}G$$

as the target of maps from $\tilde{E}G$; we obtain a long exact sequence

$$\cdots \xrightarrow{\partial} \{\tilde{E}G, EG^+\}_*^G \to \{\tilde{E}G, S^o\}_*^G \to \{\tilde{E}G, \tilde{E}G\}_*^G \xrightarrow{\partial} \cdots \qquad (A)$$

The middle group is the group which we are trying to show vanishes; unfortunately, neither of the other two groups are particularly well suited for computation. What, then, is suitable for computation in this situation? It turns out that the groups $\{S^V, EG^+\}_*^G$ are sufficiently computable for all purposes. That is, we can prove the following.

Proposition 3. $\{S^V.EG^-\}_*^G \cong \pi_*(BG^{-V})$, *where* BG^{-V} *was defined in* § 4 . *Moreover, if* $S^{\infty V}$ *denotes the union of all the one-point compactification* S^{kV}, *then* $\{S^{\infty V}.EG^+\}_*^G \cong \lim_k \pi_*(BG^{-kV})$.

Corollary 4. *If the Euler class of* V *in* $H^*(BG;\mathbb{Z}/p\mathbb{Z})$ *is nilpotent, then* $\{S^{\infty V}.EG^+\}_*^G = 0$.

For, as in §4, we have an Adams spectral sequence with E_2-term $\mathrm{Ext}_{A(p)}\left(H^*(BG;\mathbb{Z}/p)\left[\frac{1}{e(V)}\right];\mathbb{Z}/p\right)$ converging to $\lim_k \pi_*(BG^{-kV})$, and the nilpotence of $e(V)$ shows that the localized cohomology ring vanishes.

Thus, if in the long exact sequence (A), we could replace $\{\hat{E}G, EG^+\}_*^G$ by $\{S^{\infty V}, EG^+\}_*^G$, for some representation V with nilpotent Euler class, we would guarantee the vanishing of one of the relevant groups. Fortunately, we may make this replacement, due to the following.

Proposition 5.

(a) *If the Segal conjecture holds for all p-groups* H , *with* $|H| < |G|$, *and if* $\pi_G^*(S^{\infty V}) = 0$ *for some representation* V *with no trivial summands, then it holds for* G .

(b) *If the Segal conjecture holds for all p-groups* H , *with* $|H| < |G|$, *and it holds for* G , *then* $\pi_G^*(S^{\infty V}) = 0$ *for all representations* $V \neq \{0\}$.

This is Theorem A of Carlsson (1984); its proof is a routine, using the skeletal filtrations of EG and $S^{\infty V}$.

Thus, if we assume Segal's conjecture to be true for all p-groups of order smaller than $|G|$, it will suffice to show that $\pi_G^*(S^{\infty V}) = 0$ for some representation with no trivial summands. We now have the long exact sequence

$$\cdots \xrightarrow{\partial} \{S^{\infty V}, EG^+\}_* \to \pi_G^*(S^{\infty V}) \to \{S^{\infty V}, \tilde{E}G\}_*^G \xrightarrow{\partial} \cdots$$

If we can select such a representation V for G , and its Euler class e(V) is nilpotent, we can guarantee that $\{S^{\infty V}, EG^+\}_*^G = 0$. This will leave us only the group $\{S^{\infty V}, \hat{E}G\}_*^G$ to analyze. We are free to do this, by virtue of the following proposition, implicit in the work of Serre (1964).

Proposition 6. *Let* G *be a p-group, and suppose* G *is not elementary Abelian. Then there is a representation* V *of* G , *containing no trivial summands, for which* e(V) *is a nilpotent element of* $H^*(G,\mathbb{Z}/p\mathbb{Z})$.

One choice of V may be obtained as follows: Let $\overline{G}$ denote the "Frattini quotient" of G . i.e.. the maximal elementary Abelian image of G . Let $\hat{V}$ denote the sum of all irreducible representations of $\overline{G}$, and let V be the $\hat{V}$ pulled back along the projection $G \to \overline{G}$. One shows that after restricting V to any proper subgroup. V contains a trivial summand. Hence $i^*(e(V)) = 0$ for any inclusion $i : A \to G$, where A is elementary Abelian. The theorem of Quillen and Venkov (1972) now asserts that $e(V)$ is nilpotent.

We are now reduced to showing that the group $\{S^{\infty V}, \tilde{E}G\}_*^G$ vanishes. To accomplish this, we first need a lemma from equivariant homotopy theory.

Lemma 7. *Let* X *and* Y *based* G-*complexes, and suppose that* Y *is contractible as a space. (We are not requiring that* Y *be* G-*contractible, simply that it is contractible non-equivariantly.) Let* $f: G^+ \wedge S^{n-1} \to X$ *be a based* G-*map. Then the function*

$$\theta : \left[X \cup_f (G^+ \wedge e^n), Y \right]^G \to [X, Y]^G$$

is bijection, where $X \cup_f (G^+ \wedge e^n)$ *denotes the mapping cone of* f *.*
$X \cup_f (G^+ \wedge e^n)$ *is a* G-*complex in a natural way.*

One proves this by observing that the obstruction to extend a G-map $f : X \to Y$ over $X \cup_f (G^+ \wedge e^{n+1})$ is an element in $[G^+ \wedge S^n, Y]^G$. $[G^+ \wedge S^n, Y]^G$ is easily seen to be in bijective correspondence with $[S^n, Y] \cong \pi_n(Y)$, and $\pi_n(Y) = 0$, since Y is contractible. This guarantees the surjectivity of θ . The injectivity is obtained by a similar extension of a homotopy over $(X \cup_f G^+ \wedge e^n) \times I$.

Corollary 8. *Let* X *and* Y *be* G-*complexes, and suppose* Y *is non-equivariantly contractible. Then the map*

$$\theta : [X, Y]^G \to [\Sigma(X), Y]^G$$

is bijective, where

$$\Sigma(X) = \bigcup_{\substack{H \subseteq G \\ H \neq \{e\}}} X^H \subseteq X.$$

For, every G complex is obtained from its singular locus $\Sigma(X)$ by attaching G-cells of the form $G^+ \wedge e^n$.

Now, we are attempting to study the stable mapping groups $\{S^{\infty V}, \tilde{E}G\}_*^G = \lim_k \{S^{kV}, \tilde{E}G\}_*^G$. Since S^{kV} is a finite G-complex, we have seen in §3 that there is a representation W so that $\{S^{kV}, \hat{E}G\}_0^G = [S^{kV \oplus W}, S^W \wedge \tilde{E}G]^G$. Similarly each group $\{S^{kV}, \tilde{E}G\}_n^G$ is given by $[S^\ell \wedge S^{kV} \wedge S^W, S^m \wedge S^W \wedge \tilde{E}G]^G$ for some ℓ, m, and W, where S^ℓ and S^m denote the corresponding spheres with trivial G-action. The preceding corollary shows that, since $\tilde{E}G$ is non-equivariantly contractible,
$[S^\ell \wedge S^{kV} \wedge S^W, S^m \wedge S^W \wedge \hat{E}G]^G \cong \left[\Sigma(S^\ell \wedge S^{kV} \wedge S^W), S^m \wedge S^W \wedge \tilde{E}G\right]^G$.
On the other hand, $\Sigma(S^m \wedge S^W \wedge \hat{E}G) \cong S^m \wedge \Sigma(S^W)$, since $(EG)^H = S^0 \ \forall H \neq \{e\}$, and any equivariant map from $\Sigma(S^\ell \wedge S^{kV} \wedge S^W)$ must automatically have its image in $\Sigma(S^m \wedge S^W \wedge \tilde{E}G)$, since $f(X^H) \subseteq Y^H$ for any equivariant map $f: X \to Y$. The conclusion is that $\{S^{kV}, \tilde{E}G\}_n^G$ may be identified with $\left[\Sigma(S^\ell \wedge S^{kV} \wedge S^W), S^m \wedge S^W\right]^G$ for appropriately chosen ℓ, m, and W. In order to understand this group, i.e., to reduce its calculation to other groups whose calculation may be obtained from the Segal conjecture for smaller groups, we must filter the space $\Sigma(S^\ell \wedge S^{kV} \wedge S^W)$ in a useful way. This requires first "blowing up" the singular locus.

To motivate this procedure, we recall how one can construct the Mayer-Vietoris sequence on the space level, using a suitable "blowing up" of a space $Z = X \cup Y$. Let X and Y be two subcomplexes of a complex Z, with intersection $X \cap Y$. Then we construct a space $\hat{Z} \subseteq Z \times I$ by setting $\hat{Z} = X \times \{0\} \cup Y \times \{1\} \cup X \cap Y \times I$. There is a natural map from $\hat{Z}$ to Z, which "forgets" the I-coordinate, and using standard properties of CW-complexes, this map is an equivalence. Let $* \in X \cap Y$ be a basepoint. We now have a cofibration sequence $X \vee Y \to \hat{Z} \to \Sigma X \cap Y$, where $X \vee Y$ is homotopy equivalent to $X \times \{0\} \cup Y \times \{1\} \cup \{*\} \times I$. The Mayer-Victoris sequence is simply the long exact sequence associated to this cofibration. Let $X_1, \ldots, X_n \subseteq Z$ be a family of based subcomplexes of a based complex Z; what is the correct generalization of the above construction? Let Δ denote the (n-1)-simplex, with vertices $\{1, \ldots, n\}$. We now define $\hat{Z} \subseteq Z \times \Delta$ by $\hat{Z} = \bigcup_S X(S) \times \Delta[S]$, where S runs over all subsets of $\{1, \ldots, n\}$, $X(\{i_1, \ldots, i_s\}) = X_{i_1} \cap \cdots \cap X_{i_s}$, and $\Delta[S]$ denotes the corresponding face of Δ. Again, there is a natural map $\hat{Z} \to Z$, and it can again be shown to be an equivalence by standard techniques. Note that if we filter $\hat{Z}$ by the skeleta of Δ. $\Delta^{(k)}$, we find that

$$\frac{\hat{Z} \cap (Z \times \Delta^{(k)})}{\hat{Z} \cap (Z \times \Delta^{(k-1)})} \cong \bigvee_{|S|=k} S^{k-1} \wedge X(S).$$

This gives rise to a Mayer-Vietoris spectral sequence converging to $H_*(Z)$.

We wish to study $\Sigma(X)$, where X is a G-complex. Recall that $\Sigma(X) = \bigcup_{\substack{H \subseteq G \\ H \neq G}} X^H$; the construction that now suggests itself is the following. Let Δ be a simplex with a vertex for each $H \subseteq G$, $H \neq \{e\}$. Define $\hat{\Sigma}(X) \subseteq \Sigma(X) \times \Delta$ by

$$\hat{\Sigma}(X) = \bigcup_p \bigcup_{\{e\} \neq H_1 \subseteq \cdots \subseteq H_p} X^{H_1} \cap \cdots \cap X^{H_p} \times \Delta[H_1, \ldots, H_p],$$

where $\Delta[H_1, \ldots, H_p]$ is the face corresponding to the subset $\{H_1, \ldots, H_p\}$ of the family of all non-trivial subgroups. We must impose a G-action on $\hat{\Sigma}(X)$, which requires that we allow G to act on the vertex set of all non-trivial subgroups of G by conjugation. This is the natural action because of the equality $gX^k = X^{gKg^{-1}}$, for any $K \subseteq G$. Once we adopt this action, the projection $\hat{\Sigma}(X) \to \Sigma(X)$ becomes a G-equivalence. Note also that $X^{H_1} \cap \cdots \cap X^{H_p} = X^{H_p}$, so we may actually write $\hat{\Sigma}(X) = \bigcup_p \bigcup_{\{e\} \neq H_1 \subseteq \cdots \subseteq H_p} X^{H_p} \times \Delta[H_1, \ldots, H_p]$. Although this construction is extremely natural and does give a "blowing up" of $\Sigma(X)$, it must be further modified to be of use. $\hat{\Sigma}$ is based on the use of the partially ordered set of all non-trivial subgroups of G; in order for us to obtain a result based on the Segal conjecture for smaller groups, we must base our construction on only the <u>proper</u> non-trivial subgroups.

Thus, let Δ^* denote a simplex with a vertex for every proper non-trivial subgroup $H \subseteq K$, and let $\hat{\Sigma}^*(X) \subseteq \Sigma(X) \times \Delta^*$ be defined by $\Sigma^*(X) = \bigcup_p \bigcup_{\{e\} \neq H_1 \subseteq \cdots \subseteq H_p \subsetneq G} X^{H_p} \times \Delta^*[H_1, \ldots, H_p]$, where $\Delta^*[H_1, \ldots, H_p]$ is the face corresponding to the subset $\{H_1, \ldots, H_p\}$. Again, G acts on Δ^* by conjugation of subgroups, and we have a natural projection map $\Sigma^*(X) \to \Sigma(X)$. For general p-groups G, this map is <u>not</u> a G-homotopy equivalence. However, one can show the following.

Proposition 9. *If* G *is a p-group, and* G *is not elementary Abelian (i.e.,* $G \cong (\mathbb{Z}/p\mathbb{Z})^k$*), then the map* $\Sigma^*(X) \to \Sigma(X)$ *is a G-homotopy equivalence.*

This is Proposition V.5(b) of Carlsson (1984). Notice that the hypothesis $G \not\cong (\mathbb{Z}/p\mathbb{Z})^k$ is precisely the one needed to guarantee the existence of the representation V in Proposition 6. We will not describe the whole proof of this; however, it turns out that the case $X = \{*\}$, with trivial G-action, is the central case. For this case, we sketch the proof. Of course, $\Sigma^*(*) \cong \Delta^*$ as G-spaces, so it suffices to show that Δ^* is G-contractible. We have already stated that $\hat{\Sigma}(X) \cong \Sigma(X)$, so that Δ is equivariantly contractible. There is a natural inclusion $\Delta^* \xrightarrow{i} \Delta$, for which we wish to produce a G-homotopy inverse. In order to do this, we observe that from standard theory of p-groups, and the fact that G is not elementary Abelian, there is a central element of G, say T, which projects trivially to the Frattini quotient $\overline{G}$ of G. Let $< T >$ denote the subgroup generated by T. Then, again by standard group theory, if $H \subseteq G$ is a subgroup, then H is proper if and only if $H \cdot < T >$ is. These algebraic facts are proved in §5 of Carlsson (1984). The desired homotopy inverse to i, j, is defined on simplices by
$(\{H_0, H_1, ..., H_s\}) = \{H_0 < T > , H_1 < T > , ..., H_s < T > \}$, for
$H_0 \subseteq H_1 \subseteq ... \subseteq H_s$, $H_0 \neq \{e\}$. Thus, $j \circ i$ is defined by the same formula, and the fact that $j \circ i$ is homotopic to the identity arises from the fact that $H \subseteq H \cdot < T >$, in the following way. Δ^* can be viewed as the nerve of the poset of all proper nontrivial subgroups of G (see Quillen, 1973, for definitions). The map $j \circ i$ is induced by the endomorphism $J : H \to H \cdot < T >$ of this poset. The inequality $H \subseteq H < T >$ gives a natural transformation from the identify functor to the functor J, where we view the poset as a category by assigning a unique morphism to every inequality $H_1 \subseteq H_2$, and such natural transformations construct homotopies. Again, see Carlsson (1984) for a thorough discussion of these matters.

Why is $\Sigma^*(X)$ so useful? The reason becomes clear after we factor $\Sigma^*(X)$ by the subspace $\Sigma_k^*(X) = \Sigma^*(X) \cap X \times \Delta^{*(k)}$, where $\Delta^{*(k)}$ denotes the k-skeleton of Δ^*. Fixing an integer p, we describe $\Sigma_p^*(X)/\Sigma_{(p-1)}^*(X)$. First, let $\{\delta_1, ..., \delta_s\}$ denote a system of orbit representatives for the G-action on the set of all p-simplices of Δ^*, exactly one for each orbit, and let $K_1, ..., K_s$ denote the stabilizers of $\delta_1, ..., \delta_s$ respectively. Suppse δ_i is the face corresponding to the chain $H_0 \subseteq H_1 \subseteq \cdots \subseteq H_p$, and let $\aleph_i = H_p$. Then, after examining the situation, we see that $\Sigma_p^*(X)/\Sigma_{(p-1)}^*(X) \cong \bigvee_{i=1}^{s} e_{K_i}^G(S^p \wedge X^{\mathcal{H}_i})$. K_i must normalize $\mathcal{H}_i$, since it stabilizes the simplex corresponding to $H_0 \subsetneq H_1 \subsetneq \dots \subsetneq H_p$, and the action is defined by conjugation, so $S^p \wedge X^{\mathcal{H}_i}$ is a K_i-complex.

We are attempting to study the group $[\Sigma(S^\ell \wedge S^{kV} \wedge S^W), S^m \wedge S^W]^G$. We may now replace $\Sigma(S^\ell \wedge S^{kV} \wedge S^W)$ by $\Sigma^*(S^\ell \wedge S^{kV} \wedge S^W)$, if G is not elementary Abelian. We now filter Σ^* by its skeleta, and examine $\left[\Sigma_p^*(S^\ell \wedge S^{kV} \wedge S^W)/\Sigma_{(p-1)}^*(S^\ell \wedge S^{kV} S^W), S^m \wedge S^W\right]$. By the above analysis, we see that, if we let $V_i = V^{\mathcal{H}_i}$, $W_i = W^{\mathcal{H}_i}$, then

$$\left[\Sigma_p^*(S^\ell \wedge S^{kV} \wedge S^W)/\Sigma_{(p-1)}^*(S^\ell \wedge S^{kV} \wedge S^W), S^m \wedge S^W\right]^G$$

$$\cong \prod_{i=1}^{s} \left[e_{K_i}^G(S^{p+\ell} \wedge (S^{kV} \wedge S^W)^{\mathcal{H}_i}), S^m \wedge S^W\right]^G$$

$$\cong \prod_{i=1}^{s} \left[S^{p+\ell} \wedge (S^{kV} \wedge S^W)^{\mathcal{H}_i}, S^m \wedge S^W\right]^{K_i}$$

$$\cong \prod_{i=1}^{s} \left[S^{p+\ell} \wedge S^{kV_i} \wedge S^{W_i}, S^m \wedge S^{W_i}\right]^{K_i/\mathcal{H}_i}$$

We now permit k to increase, and we find that there is a spectral sequence arising from the filtration of Σ^* by the subspaces Σ_p^* , converging to $\{S^{\infty V}, \tilde{E}G\}_*^G$, where E_1^p term is $\bigoplus_{i=1}^{s} \pi_{K_i/\mathcal{H}_i}^{*-p}(S^{\infty V_i})$. We remarked in the proof of Proposition 6 that V_i is non-empty, since $\mathcal{H}_i$ is a <u>proper</u> subgroup of G (this is where we see that it is essential to replace $\hat{\Sigma}$ by Σ^*), and $|K_i/\mathcal{H}_i| < |G|$ since $\mathcal{H}_i$ is non-trivial. Therefore, if we know that the Segal conjecture is true for all groups of order smaller than $|G|$, we conclude that $\pi_{K_i/\mathcal{H}_i}^*(S^{\infty V_i}) = 0$ by part (b) of Proposition 5. The spectral sequence now shows that $\{S^{\infty V}, \tilde{E}G\}_*^G = 0$, and we conclude the theorem from part (a) of Proposition 5.

6 THE HOMOTOPY LIMIT PROBLEM

In this section, we discuss a family of homotopy theoretic problems, of which Segal's conjecture is one, which are playing an increasingly central role in algebraic topology. The observation that this family of problems unified several open questions in topology is due to Bob Thomason (1983). He referred to all these problems as *homotopy limit problems*. We now define what is meant by this. Let G be a group, and let EG denote a contractible space on which G acts freely. Let $*$ denote the one point space with trivial G-action, and let X be a G-space. If Y_1 and Y_2 are G-spaces, then the mapping space $F(Y_1, Y_2)$ becomes a G-space under the G-action $(g \cdot f)(xy) = gf(g^{-1}y)$. The space F(EG, X) thus becomes a G-space, and its fixed point set, $F(EG, X)^G$, is called the homotopy fixed point set of X. Its points consist of equivariant maps from EG to X. On the other hand, $F(*,X)$ also becomes a G-space, canonically isomorphic to X. The equivariant map $EG \to *$ induces an equivariant map $\epsilon : X \cong F(*,X) \to F(EG,X)$ and hence $\epsilon^G : X^G \to F(EG,X)^G$. We now state the problem.

Homotopy Limit Problem: *Describe the homotopy fixed point set* $F(EG; X)^G$. *In particular, analyze the map* ϵ^G *and describe to what extent it differs from a homotopy equivalence.*

We now describe how the Segal conjecture is a homotopy limit problem, and then discuss some other homotopy limit problems of importance in topology. Recall that for a p-group, the Segal conjecture asserts that the map $\pi_G^*(S^0) \to \pi_G^*(EG^+)$ is isomorphism after p-adic completion. On the space level, this amounts to the assertion that the map $(Q^G(S^0))^G \to (F_0(EG^+, Q^G(S^0)))^G$ is an equivalence after p-adic completion, where F_0 denotes the space of based maps. It is easy to see that $F_0(EG^+, Q^G(S^0)) \cong F(EG, Q^G(S^0))$, and we wish to show that $[Q^G(S^0)]^G \to [F(EG, Q^G(S^0))]^G$ is a p-adic equivalence. This is precisely the homotopy limit problem for G a finite p-group, and with $X = Q^G(S^0)$.

As another example of the homotopy limit problem, we consider the Sullivan conjecture. Recall that one form of this conjecture is that the space of based maps from the classifying space of a finite group to a finite complex is contractible. In this form, it was proved in 1983 by Haynes Miller (1984). Let us now try to interpret this theorem as a homotopy limit problem, and simultaneously generalize the question to the one originally posed by Sullivan in 1970. Let the finite complex X be viewed as a G-space with trivial action. Then an equivariant map from EG to X is precisely the same thing as a map from $BG = EG/G$ to X. In other words, the mapping spaces $F(EG, X)^G$ and F(BG, X) are homeomorphic. We now have the diagram

$$\begin{array}{ccccc} & & & & F_0(BG,X) \\ & & & & \downarrow \\ X = X^G & \xrightarrow{\epsilon^G} & F(EG,X)^G & \xrightarrow{\approx} & F(BG,X) \\ & & & & \downarrow \eta \\ & & & & X \end{array}$$

where η is evaluation at a basepoint, and $F_o(BG, X)$ is the space of based maps from BG to X. If we can show that ϵ^G is an equivalence, then $F_o(BG, X)$ will clearly be contractible, and conversely. Thus, the Sullivan conjecture, as proved by Miller, is the homotopy limit problem for G a finite group, and X a finite complex with trivial G-action. Immediately, one asks what happens for finite G-complexes with non-trivial action, and it was in fact in this form that Sullivan originally posed the conjecture. He was interested in the case of a real algebraic variety V, with $G = \mathbb{Z}/2$ acting by complex conjugation on the set of complex points of V, $V_{\mathbb{C}}$, with fixed point set $V_{\mathbb{R}}$. An affirmative answer to this question has implications for the étale homotopy type (Friedlander, 1982) of a real algebraic variety.

In the form proved by Miller in 1983, this conjecture has already had several important applications, notably the resolution by McGibbon and Neisendorfer of an old conjecture of Serre's. An interesting application of the generalized version, pointed out to the author by M. Hopkins and M. Mahowald, is the existence of a "destabilization" spectral sequence. To see this, consider the n-sphere S^n as $S^o * S^{n-1}$, and allow $G = \mathbb{Z}/2Z$ to act by permuting the points of S^o, so $(S^n)^G = S^{n-1}$. This is a finite G-complex, and if Sullivan's conjecture holds, say after 2-adic completion, then $F(EG; S^n)^G$ is 2-adically equivalent to S^{n-1}. Filtering EG by its skeleta produces a tower of fibrations converging to $F(EG; S^n)^G$, which produces a spectral sequence with E_2-term $H^*(G; \pi_{-*}(S^n))$ converging to $\pi_*(S^{n-1})$ at 2. This is the destabilization spectral sequence, which seems to be a very interesting construction.

We now look at a final example of problems of this type. In the first two problems, the homotopy fixed point set, $F(BG, Q(S^o))$ in the case of the Segal conjecture and $F(BG, X)$ in the case of the Sullivan conjecture, have been viewed as the *a priori* uncomputable objects, which become computable due to the theorem. In the third case, the opposite will be true; that is, the fixed point set is the uncomputable object, and becomes computable due to a theorem equating a fixed point set with a homotopy fixed point set.

Recall that Quillen (1973) defined the algebraic K-theory of a ring R to be the homotopy groups of $(BGL(R))^+$, where $GL(R)$ denotes the infinite general linear group over R, $BGL(R)$ its classifying space, and $^+$ Quillen's plus construction. Quillen's plus construction Abelianizes the fundamental group of $BGL(R)$ while not changing the homology. K_*R has turned out to be a very difficult invariant to compute, although the functor K_* has very good formal properties. One of these properties is the existence of a localization long exact sequence. In the case of R = $\mathbb{Z}$, for instance, this sequence has the form

$$\to \coprod_p K_*(\mathbb{F}_p) \xrightarrow{\partial} K_*(\mathbb{Z}) \to K_*(\mathbb{Q}) \xrightarrow{\partial} .$$

and so, modulo extension problems, the computation of $K_*(\mathbb{Z})$ is reduced to a question about the K-theory of fields. Similarly, for a large class of commutative rings, there is a spectral sequence involving only the K-theory of residue clas fields of localizations of the ring, converging to the K-theory of the ring. These facts show that in computational questions in algebraic K-theory, the K-theory of fields plays a central role. However, even for fields, the problem is difficult.

Recently, A. A. Suslin (1983) showed that the K-theory with finite coefficients of algebraically closed fields was computable, and that it depends only on the characteristic of the field. Thus, K-theory with finite coefficients of an algebraically closed field of characteristic zero is isomorphic to the homotopy groups of BU with finite coefficients, and the K-theory with finite coefficients of an algebraically closed field of characteristic p is isomorphic to the K-theory with finite coefficients of the field $\overline{\mathbb{F}}_p$, the algebraic closure of the prime field $\mathbb{F}_p$, which was computed by Quillen.

Let L be a field, $\bar{L}$ its algebraic closure, and let G denote the Galois group of $\bar{L}$ over L . This is a profinite group, and G acts on the representing space for $K_*(\bar{L};\mathbb{Z}/n\mathbb{Z})$, and the fixed points of this action are the representing space for $K_*(L;\mathbb{Z}/n\mathbb{Z})$. If, in this setting, we could guarantee that the fixed point set is equal to the homotopy fixed point set, then we would have a "descent spectral sequence" with $E_2 \cong H^*(G;K_*(\bar{L};\mathbb{Z}/n\mathbb{Z}))$ $\Rightarrow K_*(L;\mathbb{Z}/n\mathbb{Z})$. Since, for many fields, the Galois group has finite cohomological dimension, and E_2 is computable using Suslin's result, this would be quite an effective computational device.

Important work has already been done in this direction. If one takes algebraic K-theory "with Bott element inverted", then the spectral sequence is known to exist by work of Thomason (to appear), who has used it to prove the Lichtenbaum conjectures for this variant of K-theory. It remains to show that it holds for the actual Quillen K-theory.

As a first step, one might attempt to study the situation of a finite Galois extension of degree a power of p , between number fields or function fields over finite fields, and in the next section we will indicate how to use the Segal conjecture to attack this problem.

7 APPROXIMATION SCHEMES: COSIMPLICIAL SPACES

In the last section, we discussed several homotopy limit problems. Since we have actually solved one of them, namely the case of $Q^G(X)$, where X is a finite based G-complex, we should attempt to understand how much this says about other such problems. One common way to try to solve such a problem is, for instance, by approximating a space by its skeleta, or by some direct limit of spaces for which one understands the problem. In this case, however, we will be mapping an infinite G-complex, namely EG, into our space. This means that there is no guarantee that $[EG, X]^G = \varinjlim_k [EG, X^{(k)}]^G$, where $X^{(k)}$ is a direct system of spaces approximating X in some sense, and makes the whole procedure impossible to carry out. A drastic counter-example occurs when one studies $[EG, Q^G(EG^+)]^G$ where G is a p-group. One finds that, using the Segal conjecture, $[EG, Q^G(EG^+)]^G \cong [EG, Q^{G}(S^0)]^G \cong \hat{A}[G]$, whereas $\varinjlim_k [EG, Q^G(EG^{(k)+})]^G \cong \varinjlim \Pi_0(Q(BG^{(k)+})) = \hat{\mathbb{Z}}_p$. Thus, approximating by direct limits is not a viable procedure. However, approximating by inverse limits is. The objects which produce inverse systems of spaces are called cosimplicial spaces.

Recall that a simplicial object in a category C can be defined as a contravariant functor from Δ to C, where Δ is the category whose objects are the sets $\{0, 1, ..., n\}$, for $0 \leqslant n < \infty$, and whose morphisms are the order preserving maps. A cosimplicial object in C is simply a covariant functor from Δ to C. It is specified by a choice of object $C_n \in$ ob C for every n, $0 \leqslant n < \infty$, and coface maps $\delta_0, ..., \delta_n : C_n \to C_{n+1}$ and codegeneracy maps $\sigma_0, ..., \sigma_n : C_{n+1} \to C_n$, which satisfy certain commutation relations. Of course, these relations are simply the opposite of those satisfied for simplicial objects. If C is the category of spaces, then we know that there is a <u>geometric realization</u> functor $|\ |$ from the category of simplicial spaces to spaces. This functor comes naturally filtered by the skeleta $|X.|_k$ of the simplicial space, and so $|X.|$ is obtained as a direct limit of subspaces $|X.|_k$, in other words, as a direct limit of cofibrations. The analogous construction for cosimplicial spaces is called $\mathrm{Tot}(X^\cdot)$. $\mathrm{Tot}(X^\cdot)$ is obtained as an inverse limit of a tower of fibrations

$$\cdots \to \mathrm{Tot}_s(X^\cdot) \to \mathrm{Tot}_{s-1}(X^\cdot) \to \cdots$$

and should be viewed as dual to geometric realization of simplicial spaces. We give some examples of cosimplicial spaces.

(A) Let $X.$ be any simplicial set, and let Z be a fixed space. Then, by applying the contravariant functor $X \to F(\ ;Z)$, where F denotes function space. we obtain a cosimplicial space. $X.$ is a simplicial set, so each X_n is a discrete space, hence $F(X_n;Z) = \prod_{X_n} Z$. Where $X.$ is the simplicial circle, with a fixed basepoint, we obtain a cosimplicial space $X^{\cdot}$ for which $X^n = Z^n$. This cosimplicial space has $|\mathrm{Tot}\, X^{\cdot}| \cong |\Omega X|$, and is the basis for the geometric version of the Eilenberg-Moore spectral sequence (Rector, 1970).

(B) Let $X.$ be any simplicial space. Then the same construction yields a cosimplicial space $Y^{\cdot}$, and under favorable circumstances

$$|\mathrm{Tot}\, Y^{\cdot}| \cong F(|X.|, Z).$$

(C) Let T be a triple on the category of topological spaces (Bousfield & Kan, 1972). Recall that a triple on a category C is a functor $T : C \to C$ together with natural transformations $T^2 \to T$ and $\mathrm{Id} \to T$, satisfying certain relations; for instance, the diagram

$$\begin{array}{ccc} T & \to & T^2 \\ & \searrow^{\wr} & \downarrow \\ & & T \end{array}$$

should commute. What these relations should be becomes clear from the standard examples one has in mind; one is S^∞, the infinite symmetric product functor which assigns to every space its infinite symmetric product. Another is $Q = \Omega^\infty \Sigma^\infty$, and a third we will call $\mathbb{F}_p[\ ]$. $\mathbb{F}_p[\ ]$ is best defined on the category of simplicial sets, where it is the functor which assigns to every set the $\mathbb{F}_p$–vector space with that set as basis. Given any triple T, and a space X, one may define a cosimplicial space $T^{\cdot}X$ by setting $T^n X = \underbrace{T \circ T \circ \cdots \circ T}_{n+1 \text{ factors}}(X)$. The coface maps arise from the transformation $\mathrm{Id} \to T$, and the codegeneracies from the transformation $T^2 \to T$.

It is example (C) which will give us our approximation scheme. The point is that by iterating the transformation $\mathrm{Id} \to T \to T^2 \to \ldots \to T^n$, we obtain a map from the constant cosimplicial space with value X, $C^{\cdot}X$, to $T^{\cdot}X$, and that under suitable conditions, this becomes an equivalence. In particular, one may show that if $T = S^\infty$, $\pi_1(X)$ is nilpotent, and $\pi_1(X)$ acts nilpotently on the higher homotopy groups of X, then the map $X \to \mathrm{Tot}\, S^{\infty \cdot} X$ is a weak equivalence. In general, $\mathrm{Tot}\, S^{\infty \cdot} X$ is a functor of X, and is called the $\mathbb{Z}$–completion of X. It can be shown, similarly, that $\mathrm{Tot}\, Q^{\cdot}X$ is also a $\mathbb{Z}$–completion of X, so that $\mathrm{Tot}\, Q^{\cdot}X \to \mathrm{Tot}\, S^{\infty\cdot}X$ is a weak equivalence. An important fact about cosimplicial spaces is the following: Let X be any space, and let $Y^{\cdot}$ denote a cosimplicial space. Then, if F denotes function space, we define a new cosimplicial space $F(X, Y^{\cdot})$, which in codimension n is $F(X,Y^n)$. It is immediate from the definitions that $\mathrm{Tot}\, F(X, Y^{\cdot}) \cong F(X, \mathrm{Tot}\, Y^{\cdot})$, so in a sense, the study of maps into the total space is reduced to the study of maps into each term in the cosimplicial space.

A strategy for applying Segal's conjecture now suggests itself. Given a homotopy limit problem for a finite group G and G-complex X, we can construct the cosimplicial space of the triple Q^G on the category of G-complexes. One might then hope that the maps of EG into X might be analyzed using the maps of EG into $(Q^G)^n(X)$, for all n. To carry this program through, we must do two things. The first is to carry out the analysis of $F(EG, (Q^G)^n(X))$; this requires the equivariant Snaith decomposition, which we discuss briefly below. The second is to analyze the total space of the cosimplicial space of the triple Q^G, and to show that for "most" G-spaces X, this total space is weakly G-equivalent to X.

Let G be a group, and $\Gamma \subseteq G$ a normal subgroup. Then we say a G-space X is an $E_G\Gamma$ is space if: (a) $(E_G\Gamma)^H = \phi$ for all $H \subseteq G$ s.t. $H \cap \Gamma \neq \{e\}$, and (b) $(E_G\Gamma)^H$ is contractible for all $H \subseteq G$ such that $H \cap \Gamma = \{e\}$. Tom Dieck (1979) shows that such spaces exist and are unique up to weak G-homotopy type. We now have the equivariant analogue of the Snaith decomposition, proved by Lewis et al. (to appear), which asserts that there is a G-equivalence $Q^G(Q^G(X)) \cong \prod_{n=1}^{\infty} Q^G\left(E_{G\times\Sigma_n}\Sigma_n^+ \wedge_{\Sigma_n} X^{\wedge n}\right)$,
where $X^{\wedge n}$ denotes the n-fold smash product of X with itself.
(Note that $X^{\wedge n}$ is a $G \times \Sigma_n$-space in a natural way, if G acts on X by the diagonal action, so $E_{G\times\Sigma_n}\Sigma_n^+ \wedge X^{\wedge n}$ is a $G \times \Sigma_n$ space. If Z is any $G \times \Sigma_n$ space, Z/Σ_n becomes a G-space.) Moreover, it is possible to find prove a generalization of the Segal conjecture, which precisely applies in this situation.

Theorem 1. *Let* G *be any finite group, and* Γ *a subgroup of prime power index. Let* X *be a* G*-complex. Then, the map* $Q^{G/\Gamma}(E_G\Gamma^+ \underset{\Gamma}{\wedge} X) \to F\left(EG/\Gamma^+, Q^{G/\Gamma}(E_G\Gamma^+ \underset{\Gamma}{\wedge} X)\right)$ *is a* p*-adic weak* G*-equivalence.*

To see this, one proves a generalization of the splitting theorem for $(Q^G(X))^G$ and finds that the space mentioned in the theorem appears as a factor in a G/Γ-equivariant splitting of the G/Γ-space $(Q^G(X))^\Gamma$. This fact, together with the Segal conjecture for a p-Sylow subgroup of G and X allows one to conclude the result. In particular, this tells us that the map

$$Q^G\left(E_{G\times\Sigma_n}\Sigma_n^+ \underset{\Sigma_n}{\wedge} X^{\wedge n}\right) \to F\left(EG^+, Q^G(E_{G\times\Sigma_n}\Sigma_n^+ \underset{\Sigma_n}{\wedge} X^{\wedge n})\right)$$

is a weak G-equivalence after p-adic completion, if G is a p-group. If we now study the equivariant stable Snaith splitting, we see that the map

$$Q^G(Q^G(X)) \to F(EG^+, Q^G(Q^G(X)))$$

is a p-adic weak G-equivalence if G is a p-group, and X is a finite G-complex. It is now a simple matter to generalize this to $(Q^G)^n(X)$.

The second ingredient is the proof that $\mathrm{Tot}\, Q^{G^\cdot}(X)$ is weakly G-equivalent to X for a "large" class of spaces. This is achieved in two steps. We first study the functor $\overline{Q}^G = \lim_{\to} \Omega^n \Sigma^n$, which is a functor from G-spaces to G-spaces. It is simply the usual Q, with no group action on the suspension coordinate. One can now prove that for each subgroup H of G, $(\mathrm{Tot}\,\overline{Q}^{G^\cdot} X)$ is a $\mathbb{Z}$-completion of X^H. This is accomplished by observing that $\overline{Q}^G(X)^H = Q(X^H)$, and proving that for any space X, $\mathrm{Tot}\, Q X$ is a $\mathbb{Z}$-completion of X. Proving the analogous fact for Q^G involves the comparison of $Q^{G^\cdot}$ with $\overline{Q}^{G^\cdot}$; the usual way of doing this is via bicosimplicial spaces, and it works in this case, too. The bicosimplicial space we study has $X^{m,n} = \overline{Q}^{G^m}(Q^{G^n}(X))$, and we are able to prove the following.

Theorem 2. *For every finite* G, *and subgroup* $H \subseteq G$, *the space* $(\mathrm{Tot}(Q^{G^\cdot}X))^H$ *is a* $\mathbb{Z}$-completion *of* X^H.

We now see that the Sullivan conjecture for G-spaces with simply connected fixed point sets is an immediate corollary. For in that case, the mapping space $(F(\tilde{E}G, X)^G)_p^\wedge \cong \mathrm{Tot}(F(\tilde{E}G, Q^{G^\cdot}X)^G)_p^\wedge$, and the Segal conjecture together with the earlier discussion of the Snaith decomposition asserts that $(F(\tilde{E}G; Q^{G^*}(X))^G)_p^\wedge$ is weakly contractible, hence that $F(\tilde{E}G; X)^G$ is weakly p-adically contractible. The scheme also yields information when X and X^H are not necessarily simply connected; see Carlsson (1986a) for details. The main point here is that we have approximated X by a cosimplicial space whose spaces in individual codimensions are amenable to analysis by the Segal conjecture and its elaboration (Theorem 1 above).

We say a few words about what additional information is necessary to study the algebraic K-theory example mentioned above. Theorem 1, applied in the case where $X = S^0$, shows that the map $Q^{G/\Gamma}(E_G\Gamma^+/\Gamma) \to F(EG/\Gamma, Q^{G/\Gamma}(E_G\Gamma^+/\Gamma))$ is a weak G/Γ-equivalence after p-adic completion if G is a finite group, and Γ is a normal subgroup of index p^k for some k. From this point on, we let $B_G\Gamma$ denote $E_G\Gamma/\Gamma$. We will want to extend the above mentioned result to permit the study of certain infinite groups G, with normal subgroups Γ of finite, p-power index. We prove the following result.

Theorem 3. *Let G be a discrete group, not necessarily finite, and suppose there is a normal subgroup Γ of index p^α, where p is a prime. Suppose further that there is a finite G-complex X, having the following properties.*

(a) *For any $x \in X$, G_x is a finite subgroup of G.*

(b) *For any $H \subseteq G$, such that $H \cap \Gamma = \{e\}$, X^H is contractible.*

Then the map $Q^{G/\Gamma}(B_G\Gamma^+) \to F(EG/\Gamma, Q^{G/\Gamma}(B_G\Gamma^+))$ is a weak p-adic G/Γ-equivalence.

To see this, we note first that $E_G\Gamma \underset{\Gamma}{\times} X = B_G\Gamma$. The point is that as G-space $E_G\Gamma \times X$ is also an $E_G\Gamma$-space, since if $H \cap \Gamma = \{e\}$, $(E_G\Gamma \times X)^H = E_G\Gamma^H \times X^H = \{*\}$, and if

$H \cap \Gamma \neq \{e\}$, $(E_G\Gamma \times X)^H = \Phi \times X^H = \Phi$. If we filter X by its skeleta, we find that the subquotients are wedge sums of spaces of the form $S^n \wedge (G/\overline{G}^+)$, where $\overline{G}$ is a finite subgroup of G. Accordingly, it will suffice to prove the result for $S^n \wedge (E_G\Gamma^+ \underset{\Gamma}{\wedge} (G/\overline{G}^+))$. To do this, we observe that

$$E_G\Gamma \underset{\Gamma}{\times} G/\overline{G} \cong G/\Gamma \underset{\overline{G}\Gamma/\Gamma}{\times} B_{\overline{G}}(\overline{G} \cap \Gamma).$$

This is a straightforward consequence of the definitions, it will appear in Carlsson (1986b). Now $\overline{G}$ is finite and $\overline{G}\Gamma/\Gamma$ is finite, so we are reduced to the case where G is finite, which gives the result.

The relevant group in the case of algebraic K-theory is $G \tilde{\times} \Gamma$, where G is the Galois group in a finite Galois extension L over K of prime power degree p^α, and Γ is $SL_k(\mathcal{O}_L\frac{1}{N})$, where $\mathcal{O}_L$ denotes the ring of integers in L, and N is an integer so that all the ramified primes and p divide N. The normal subgroup will be Γ. We need only to produce the space X; it is provided for us by Borel & Serre (1974.1976) and Bruhat & Tits (1972). See Carlsson (1986b) for details of how these constructions are used. The conclusion is that in this context, the map $Q^G(B_{G \tilde{\times} \Gamma}\Gamma) \to F(EG, Q^G(B_{G \tilde{\times} \Gamma}\Gamma))$ is a p-adic weak G-equivalence.

One next shows that as a G space, $E_{G \tilde{\times} \Gamma}\Gamma$ is homotopy equivalent to $B\Gamma$, equipped with G action via the Galois action of G on $\mathcal{O}_L \frac{1}{N}$. This is a consequence of facts from the theory of non-Abelian cohomology, again, see Carlsson (1986b). Using the cosimplicial space of a triple, and Theorem 3, one may show that the map

$$BSL_k^+(\mathcal{O}_K \frac{1}{N}) \to F(EG, BSL_k^+(\mathcal{O}_L \frac{1}{N}))$$

is a p-adic weak equivalence (in this case, + refers to Quillen's plus construction). More can be said in this area; see Carlsson (1986b) for details.

8 SUMMARY

We have discussed how Segal's Burnside ring conjecture was motivated, and how it was proved. Further, we have seen that it fits naturally as a special case of the general homotopy limit problem, and that it is a particularly central special case which may be applied to other homotopy limit questions for finite groups. I believe that homotopy limit problems will arise in many different contexts, and therefore that this will be an extremely fruitful direction of research. Even now, the study of the homotopy fixed point set of the S^1-action on $Q(X^{S^1})$, which can be studied using the Segal conjecture, is playing a part in the analysis of Waldhausen's $A(X)$ for simply connected spaces.

In conclusion, I would like to suggest two important directions for further work. The first is the case of a compact Lie group, where the analogue of the Segal conjecture is not well understood. Feshbach (to appear) has proved that in an appropriate sense, the finite subgroups of a compact Lie group "tell all," but this does not give explicit answers. Secondly, when one has a profinite group G, such as the absolute Galois group of Q, to deal with, it is of interest to understand the precise nature of the failure of the conjecture, or rather, how the conjecture should be modified to accommodate this case. This leads rather naturally to the homotopy limit problem for general discrete groups. Presumably a study of individual cases is the most logical first step.

REFERENCES

Adams, J.F. (1974). *Stable Homotopy and Generalized Homology.* University of Chicago Press. Chicago. Illinois.

Adams, J.F. (1978). *Infinite Loop Spaces.* Princeton University Press, Princeton, NJ.

Adams, J.F. (1982). Graeme Segal's Burnside ring conjecture. Bull. A.M.S. 6, no.2, 201-210.

Adams, J.F. (1984). Prerequisites for Carlsson's lecture. Aarhus Algebraic Topology Symposium, Lecture Notes in Math. 1051, Springer-Verlag.

Adams, J.F., Miller, H.R. & Gunawardena, J.H.C. (1985). The Segal conjecture for elementary Abelian p-groups. Topology 24, 435-460.

Atiyah, M.F. (1961). Characters and cohomology of finite groups. Inst. Hautes Etudes Sci. Publ. Math. 9, 23-64.

Atiyah, M.F. & Segal, G.B. (1969). Equivariant K-theory and completion. J. Diff. Geom 3, 1-18.

Borel, A. & Serre, J.P. (1974). Corners and arithmetic groups. Comm. Math. Helv. 48, 244-297.

Borel, A. & Serre, J.P. (1976). Cohomologie d'immeubles et de groupes S-arithmétiques. Topology, 15, 211-232.

Bousfield, A.K. & Kan, D.M. (1972). *Homotopy Limits Completions and Localizations.* Lecture Notes in Mathematics 304, Springer-Verlag.

Bruhat, F. & Tits, J. (1972). Groups réductifs sur un corps local. Chap. I, Publ. Math. I.H.E.S. 41, 1-251.

Carlsson, G. (1984). Equivariant stable homotopy theory and Segal's Burnside ring conjecture. Annals of Math. 120, 189-224.

Carlsson, G. (1986a). Equivariant stable homotopy and Sullivan's conjecture. (to appear)

Carlsson, G. (1986b). Equivariant stable homotopy and the descent problem for algebraic K-theory. (to appear)

Dress, A. (1969). A characterization of soluable groups. Math. Z. 110, 213-217.

Feshbach, M. (1986). The Segal conjecture for compact Lie groups. (to appear)

Friedlander, E. (1982). *Etale Homotopy Theory of Simplicial Schemes.* Annals of Mathematics Studies, no.104, Princeton University Press.

Gunawardena, J.H.C. (1980). Segal's conjecture for cyclic groups of (odd) prime order. J.T. Knight Prize Essay, Cambridge University.

Laitinen, E. (1979). On the Burnside ring and stable cohomotopy of a finite group. Math. Scand. 44, 37-72.

Lewis, L.G., May, J.P. & McClure, J.E. (1981). Classifying G-spaces and the Segal conjecture. Current Trends in Algebraic Topology, C.M.S. Conference Proceedings, Part II, 165-180.

Lin, W.H. (1980). On conjectures of Mahowald, Segal, and Sullivan. Math. Proc. Comb. Phil. Soc. 87, 449-458.

Madsen, I. & Milgram, R.J. (1979). *The Classifying Spaces for Surgery and Cobordism of Manifolds.* Annals of Math. Studies #92, Princeton University Press.

May, J.P. (1972). *The Geometry of Iterated Loop Spaces.* Lecture Notes in Mathematics, #271.
May, J.P. & McClure, J.E. (1982). A reduction of the Segal conjecture. Canadian Math. Soc. Conf. Proc., Vol. 2, Part 2, 209-222.
Miller, H.R.(1984). The Sullivan conjecture on maps from classifying spaces. Annals of Mathematics 120, 39-87.
Priddy, S.B. (1971). On $\Omega^{\infty} S^{\infty}$ and the infinite symmetric group. Proceedings of Symposia in Pure Mathematics. Vol.XXII, Am. Math. Soc., 217-220.
Priddy, S.B. & Wilkerson, C. (1985). Hilbert's theorem 90 and the Segal conjecture for elementary Abelian p-groups. Am. J. of Math. 107, 775-786.
Quillen, D.G. (1970). Cohomology of groups. Proc. of the Int'l Cong. of Mathematicians at Nice.
Quillen, D.G. (1973). Higher algebraic K-theory I. Springer Lecture Notes, no.341, Springer-Verlag.
Quillen, D.G. (1978). Homotopy properties of the poset of non-trivial p-subgroups of a group. Adv. in Math. 28, 101-128.
Quillen, D.G. & Venkov, B. (1972). Cohomology of finite groups and elementary Abelian subgroups. Topology 11, 552-568.
Ravenel, D. (1981). The Segal conjecture for cyclic groups. Bull. Lond. Math. Soc. 13, 42-44.
Rector, D. (1970). Steenrod operations in the Eilenberg-Moore spectral sequence. Comm. Math. Helv. 45, 540-552.
Segal, G.B. (1970). Equivariant stable homotopy theory. Proc. of Int'l Congress of Mathematicians at Nice.
Segal, G.B. & Stretch, C.T. (1981). Characteristic classes for permutation representations. Math. Proc. Comb. Phil. Soc. 90, 265-272.
Serre, J.P. (1964). Sur la dimension cohomologique des groupes profinis. Topology 3, 413-420.
Sullivan, D. (1970). Geometric topology, Part I: localization, periodicity, and Galois symmetry. M.I.T. Press.
Suslin, A.A. (1983). On the K-theory of algebraically closed fields. Inventiones Math. 73, 241,245.
Switzer, R. (1975). *Algebraic Topology -- Homotopy and Homology.* Springer-Verlag.
Thomason, R. (1983). The homotopy limit problem. Proc. of the Northwestern Homotopy Theory Conf., Contemporary Mathematics, 19, 407-420.
Thomason, R. (1986). Algebraic K-theory and etale cohomology. (to appear)
Tom Dieck, T. (1975). Orbittypen und äquivariante homologie II. Arch. Math. 26, 650-662.
Tom Dieck, T. (1979). *Transformation Groups and Representation Theory.* Lecture Notes in Mathematics, 766, Springer-Verlag.

Addendum: J. Lannes has also independently proved a form of Sullivan's conjecture. Also, H. Miller and C. Wilkerson carried out the computational approach to Segal's conjecture, mentioned in the introduction, for groups with periodic cohomology.

PSEUDO - ISOTOPIES, K - THEORY, and HOMOTOPY THEORY

Ralph L. Cohen

The problem we address in this paper is that of studying the space of *pseudo - isotopies* of a manifold. This group, which we denote by $P(M)$, is defined to be the group of diffeomorphisms

$$P(M) = \mathrm{Diff}\,(M \times I,\ \partial M \times I \cup M \times \{0\})$$

where I denotes the unit interval, ∂M denotes the boundary of M, and where the above notation refers to diffeomorphisms of $M \times I$ that equal the identity in a neighborhood of $\partial M \times I \cup M \times \{0\}$. In this paper we will give an expository account of certain advances in the past fifteen years in understanding the homotopy type of the pseudo - isotopy space, $P(M)$, using K - theoretic and homotopy theoretic methods.

The story begins with the seminal paper of Cerf [10] in 1970. In that paper he addressed what he called the "pseudo - isotopy question". This question can be described as follows: The group $P(M)$ has an obvious action on the group of diffeomorphisms, $\mathrm{Diff}(M)$. Namely, if $H \in P(M)$, let H_1 denote the diffeomorphism of M given by the restriction of H to $M \times \{1\}$. The action of H on $\mathrm{Diff}(M)$ is given by $Hf = f \cdot H_1$. Two diffeomorphisms f_1 and f_2 in $\mathrm{Diff}(M)$ are said to be <u>pseudo - isotopic</u> if they lie in the same orbit under this group action. Recall also that the diffeomorphisms f_1 and f_2 are said to be <u>isotopic</u> if they lie in the same path component of $\mathrm{Diff}(M)$. The "pseudo - isotopy" question is the following:

<u>Question.</u> Suppose f_1 and f_2 are pseudo - isotopic diffeomorphisms of a manifod M. Are they isotopic?

Since the isotopy equivalence relation is determined by path components, then it is clear that to answer this question it is necessary to compute the set of path components, $\pi_0 P(M)$. In [10] Cerf proved the following:

<u>Theorem 1.</u> Let M be a simply connected, C^∞, closed, n - dimensional manifold with $n \geqslant 6$. Then $\pi_0 P(M) = 0$. Thus the answer to the pseudo -

---

During the preparation of this work the author was supported by an N.S.F. grant and an N.S.F. - P.Y.I. award.

isotopy question in this case is yes.

Soon thereafter, Hatcher and Wagoner computed $\pi_0 P(M)$ for M a manifold of dimension ≥6, that is not necessarily simply connected [22]. We will not state their results precisely here, but suffice it say that this group is not, in general zero, and it is very much related to the algebraic K - groups of the group ring of the fundamental group π of the manifold, $K_*(\mathbb{Z}[\pi])$. The relationship between pseudo - isotopy theory and algebraic K - theory has been studied and developed fully in the foundational work of F. Waldhausen [35 -39]. Waldhausen introduced the notion of the "Algebraic K - theory of a Space" and proved theorems showing how it is related to the pseudo - isotopy space, P(M). This notion of the algebraic K - theory of a space can be viewed as the algebraic K - theory (ala Quillen) of a certain homotopy theoretic version of the group ring of the fundamental group. This functor is not an algebraic functor of the fundamental group $\pi_1(M) = \pi_0(\Omega M)$, but rather is a functor of the stable homotopy type of the based loop space, $\Omega(M)$. He then showed how this functor completely describes the homotopy type of a "stable" version of P(M), which agrees with the homotopy type of P(M) in a range of dimensions depending on the dimension of M. We will now be more explicit.

Consider the "suspension" map

$$\sigma\colon P(M) \longrightarrow P(M \times I)$$

defined essentially by letting $\sigma(H) = H \times 1$. We say "essentially" because $H \times 1$ will not satisfy the requisite boundary conditions to be a pseudo - isotopy of $M \times I$. However there is a well defined "smoothing" process to deform $h \times 1$ so that it will satisfy the requisite boundary conditions. (See [21] for details.) In any case, Igusa has recently completed a program begun by Hatcher, the outcome of which is that the suspension map σ is roughly n/3 - connected ($n = \dim(M)$) [21,27]. Thus in a range of dimensions depending on the dimension of the manifold, to study the homotopy type of P(M) it is sufficient to study the homotopy type of the stable pseudo - isotopy space, $\mathbb{P}(M)$ defined by

$$\mathbb{P}(M) = \varinjlim_k P(M \times I^k).$$

One of the reasons it is of interest to study the homotopy type of $\mathbb{P}(M)$

is the following weak (and vaguely stated) version of a theorem of Waldhausen [35]:

<u>Theorem 2.</u> $\mathbb{P}(M)$ is an infinite loop space that depends only on the stable homotopy type of the based loop space ΩM. Moreover, the homotopy type of $\mathbb{P}(M)$ can be studied K - theoretically.

We will now make the K - theory statement in the above theorem more precise. The idea is the following. Given a space X, one wants to define and study the algebraic K - theory of the "ring up to homotopy" $Q((\Omega X)_+) = \varinjlim_k \Omega^k \Sigma^k ((\Omega X)_+)$, where the subscript "+" denotes the addition of a disjoint basepoint. The addition in this "ring" is given by the loop addition coming from the functor $Q(\ )$, and the multiplication is induced by the H - group pairing in ΩX. Notice that $Q((\Omega X)_+)$ can be viewed as in some sense, a homotopy theoretic version of a group ring. In any case notice that the set of path components is given by

$$\pi_0 Q((\Omega X)_+) = \mathbb{Z}[\pi]$$

where $\pi = \pi_0 \Omega X = \pi_1 X$.

In order to define what is meant by the algebraic K - theory of $Q((\Omega X)_+)$, one needs to make precise what is meant by the "general linear group" $GL_n(Q((\Omega X)_+)$. Waldhausen does this as follows: (See [35, 36] for details.) Let G denote an associative group of the same homotopy type (as H - spaces) as the loop space ΩX. This can be constructed by realizing Kan's simplicial loop functor $G(X)$ [29]. Let $H_{n,k}$ be the G - equivariant mapping space

$$H_{n,k} = \mathrm{Aut}^G(\bigvee_n S^k \wedge G_+)$$

where the Aut^G notation denotes those G - equivariant self maps of $\bigvee_n S^k \wedge G_+)$ that are weak homotopy equivalences. (Notice that one does not require that these maps be equivariant homotopy equivalences; that is that they have equivariant homotopy inverses. We only require that these equivariant maps be weak equivalences in the nonequivariant sense.) We then define $GL_n(Q((\Omega X)_+))$ to be the limit

$$GL_n(Q((\Omega X)_+)) = \varinjlim_k H_{n,k}.$$

If we then let $GL(Q((\Omega X)_+)) = \varinjlim_n GL_n(Q((\Omega X_+))$, then Waldhausen defines the space $A(X)$ to be the Quillen's plus construction applied to the classifying space of the monoid $GL(Q((\Omega X)_+))$. ($GL(Q((\Omega X)_+))$ is a monoid under composition of equivariant maps.) That is,

$$A(X) = BGL(Q((\Omega X)_+))^+.$$

Some of the notation above is a bit different than that of Waldhausen. However the definition of A(X) is the same as that given by Waldhausen in [35, 36]. The reader is referred to [35, 36] for details.

The connection between the space $A(X)$ and pseudo - isotopies of a manifold is given by the following theorem of Waldhausen [35,39]:

Theorem 3. There is a splitting of infinite loop spaces,

$$A(X) \simeq Wh(X) \times Q(X_+)$$

where Wh(_) is a homotopy functor from spaces to infinite loop spaces with the property that if X is homotopy equivalent to a compact manifold M (with or without boundary), then

$$\Omega^2 Wh(X) = \mathbb{P}(M).$$

Thus the study of the stable pseudo - isotopy space $\mathbb{P}(M)$ or equivalently the Whitehead space $Wh(M)$ can be approached using K - theoretic techniques, by means of understanding the homotopy type of the spaces A(X). Now there have been several successful calculations made of the rational homotopy type of A(X). (See for example [3,25,26,19].) Part of the reason for this success is that the homotopy ring in question $Q((\Omega X)_+)$ has the same rational homotopy type (as homotopy rings) as the rational topological group ring $\mathbb{Q}[G]$, where G is a topological group of the same homotopy type (as h - spaces) as ΩX. Said in another, but equivalent way, $\pi_*(A(X)) \otimes \mathbb{Q}$ is the algebraic K - theory of the rational simplicial group ring of the Kan simplicial loop group. Since the K - theory of simplicial rings over a field of characteristic zero is essentially an algebraic entity (as opposed to a homotopy theoretic object) these groups have been more tractible computationally than the torsion subgroups of $\pi_* A(X)$. Indeed until very recently very little has been known about the global homotopy theoretic structure of $A(X)$ and even less known about the global structure of the (unstable) pseudo - isotopy space of a manifold $P(M)$.

In this paper we will describe some recent advances in understanding the global homotopy type of the spaces $A(X)$, $Wh(X)$, and therefore the stable pseudo - isotopy space $\mathbb{P}(M)$. These results are contained in several papers stemming from a two year collaboration between the author, G. Carlsson, and W. c. Hsiang [7,8,9]. Included in these is also a collaboration with T. Goodwillie whose innovative work in this area has been one of the main influences on this entire project.

Our results so far have concerned only the "reduced" versions of the functors $A(X)$ and $Wh(X)$. More precisely, let $\tilde{A}(X)$ and $\tilde{W}h(X)$ respectively denote the homotopy fibers of the maps $A(X) \longrightarrow A(\text{point})$ and $Wh(X) \longrightarrow Wh(\text{point})$ induced by the projection onto the basepoint $X \longrightarrow \text{point}$. Notice that since these maps are infinite loop maps with obvious infinite loop sections, we have that

$$A(X) \simeq \tilde{A}(X) \times A(\text{point}) \quad \text{and} \quad Wh(X) \simeq \tilde{W}h(X) \times Wh(\text{point}).$$

Our main theorem describes the homotopy type of the reduced Whitehead space $\tilde{W}h(X)$ in terms of the unbased loop space (or "free loop space")

$$\Lambda(X) = \text{Maps}(S^1, X).$$

More specifically, let $B(X)$ denote the homotopy functor of X defined by

$$B(X) = Q\Sigma(ES^1 \times_{S^1} \Lambda(X))$$

where ES^1 is a contractible space which has a free S^1 action, and where the S^1 action on $\Lambda(X)$ is given by rotation of loops. As before let $\tilde{B}(X)$ denote the reduced version of this functor. Namely, $\tilde{B}(X)$ is the homotopy fiber of the map $B(X) \longrightarrow B(\text{point})$. Notice that since $B(\text{point}) = Q\Sigma(BS^1)$, it follows that $\tilde{B}(X) \simeq Q\Sigma(ES^1{}_+ \wedge_{S^1} \Lambda(X))$ where "$\wedge$" denotes smash product. The following theorem was proved in the case when the space X is the homotopy type of a suspension space in [9], and in the case when X is an arbitrary simply connected space in [8].

Theorem 4. For X simply connected, there is a homotopy fibration sequence

$$\widetilde{Wh}(X) \longrightarrow \tilde{B}(X) \longrightarrow Q(X).$$

Moreover, if X is the homotopy type of a suspension space, say $X \simeq \Sigma Y$, then this fibration is trivial. That is, $\tilde{B}(X) \simeq \widetilde{Wh}(X) \times Q(X)$.

It is interesting to compare this theorem to the above theorem of Waldhausen which yields a splitting $\tilde{A}(X) \simeq \widetilde{Wh}(X) \times Q(X)$. Thus $\tilde{A}(X)$ and $\tilde{B}(X)$ are equivalent functors when X is a suspension, and in general, $\tilde{B}(X)$ is a "twisted" version of $\tilde{A}(X)$. However, on the face of it, $B(X)$ has a much more computible homotopy type than $A(X)$. There are several techniques for computing its homology groups and its homotopy groups, some of which we will discuss in the body of this paper. These techniques, together with this theorem, should then lead to relatively straightforward calculations of the homotopy type of $\widetilde{Wh}(X)$ and $\tilde{A}(X)$.

The organization of this paper is as follows: In section one we will recall in more detail the relationship between the Whitehead space and Waldhausen's K - theory. We will then describe some of the results of Dwyer, Hsiang, and Staffeldt concerning how rationally, $\pi_* A(X)$ can be computed in terms of Lie algebra homology and invariant theory. In section 2 we recall results of Loday and Quillen about cyclic homology theory and how it is related to the Lie algebra homology of matrices. We then observe how when this is combined with the work of Dwyer, Hsiang, and Staffeldt, together with results of Goodwillie and Burghelea concerning how cyclic homology theory is related is related to the free loop space, one sees a strong relationship between the rational homotopy types of $\tilde{A}(X)$ and $\tilde{B}(X)$. The fact that they are rationally equivalent follows easily from the work of Burghelea [4,5]. In section 3 we describe some of the author and Carlsson's work on the stable homotopy of $B(X)$ using configuration space models. We also discuss other important combinatorial models for the free loop space. In section 4 we describe the motivation for suspecting that the relationship between $A(X)$ and $B(X)$ is more than just a rational equivalence of their reduced functors. Namely, we describe some of Goodwillie's theory of the "Calculus of functors" and study how it applies to the functors $A(X)$ and $B(X)$. In section 5 we discuss the map giving the equivalence between $\widetilde{Wh}(X)$ and the homotopy fiber of the map

$s : \tilde{B}(X) \longrightarrow Q(X)$. This is, in a sense that can be made precise, a realization of the Dennis trace map from algebraic K - theory to Hochschild homology, which was constructed by the author and Jones [12].

The aim of this paper is to describe in a cohesive manner recent results of many people in this exciting area of research. It is purely expository. The details of the proofs all appear in other papers. I'd like to thank(?) John D.S. Jones for suggesting that I take on this project.

Section 1. The Whitehead space, Algebraic K - theory, and Lie Algebra Homology

We begin this section by recalling a geometric version of the definition of the Whitehead space do to Waldhausen [35]. Having done this we will then recall the results of Dwyer, Hsiang, and Staffeldt [17] which gives Lie algebra techniques for computing the rationalization of Waldhausen's K - theory.

The motivation for this description of the Whitehead space was the work of Cerf, and of Hatcher and Wagoner on pseudo - isotopies [10, 22]. Given a pseudo - isotopy of a manifold M^n they showed how to produce a one parameter family of generalized Morse functions (functions whose critical points are all either nondegenerate or "birth - death" see [22] for strict definitions) on M. These gave rise to a one parameter family of handlebodies. Hatcher and Wagoner then showed how to associate to a one parameter family of handlebodies an element of $Wh_2(\pi_1 M)$, a certain quotient of $K_2(\mathbb{Z}[\pi_1 M])$. They were then able to compute $\pi_0(P(M)$ for dim $M > 5$.

In the following definition of the Whitehead space of a manifold M, these notions are generalized and expressed in terms of the simplicial category of k - parameter families of "rigid handlebodies" on M. We now describe this category in some detail. Our exposition follows that of [23, 24].

Let Y be an $n + k$ dimensional manifold, and $p : Y \longrightarrow \Delta^k$ be a fiber bundle, where Δ^k is the k - dimensional simplex. Let $\partial_0 \subset \partial Y$ be a codimension zero submanifold of the boundary of Y with the property that the restriction of the fibration p to ∂_0 has $n - 1$ dimensional manifolds as its fibers. We say that Y is a *k - parameter family of rigid handlebodies on* ∂_0 if there is a filtration of C.W. complexes

$$\partial_0 = Y_0 \subset Y_1 \subset Y_2 \subset \dots \subset Y_r = Y$$

satisfying the following properties:

a. For each $i > 0$, there is an embedding

$$e_i : S^{k_i - 1} \times D^{n-j_i} \times \Delta^k \longrightarrow Y^{i-1}$$

and a homeomorphism

$$Y^i \xrightarrow[\cong]{h_i} Y^{i-1} \cup_{e_i} D^{k_i} \times D^{n-j_i} \times \Delta^k$$

that fixes Y^{i-1} such that e_i and h_i preserve the projection onto Δ^k.

b. The space $Y^i \cup_{\partial_0 \times \{1\}} \partial_0 \times I$ is a manifold, which we call M^i.

c. Let $\partial_\Delta M^i$ be the subspace of ∂M^i that lies over $\partial\Delta^k$, and let
$\partial_+ Y^i = cl(\partial M^i - (\partial_0 \times \{0\} \cup \partial_\Delta M^i))$.

We then require that e_i: $S^{k_i - 1} \times D^{n-k_i} \times \Delta^k \longrightarrow \partial_+ Y^i$ be a differentiable embedding. Moreover the obvious extension of h_i to $M^i = M^{i-1} \cup D^{k_i} \times D^{n-k_i} \times \Delta^k$ is required to be a diffeomorphism.

One may then define a category $\mathcal{E}_{n,k}$ whose objects are k - parameter families of rigid handlebodies and whose morphisms consist of compositions of isomorphisms and certain handle cancelling operations. See [24] for details. Moreover there exist fairly obvious candidates for face and degeneracy maps that then makes $\mathcal{E}_{n,*}$ into a simplicial category.

Given a space X and a virtual vector bundle ξ over X one then defines a the category $\mathcal{E}_k(X, \xi)^n$ as follows. The objects are maps

$$f : (Y, \partial_0) \longrightarrow X$$

where (Y, ∂_0) is an object of $\mathcal{E}_{n,k}$ such that $\partial_0 \subset Y$ is a homotopy equivalence, f is a continuous map, and one is equipped with a bundle isomorphism ψ from the tangent bundle τY to the pullback bundle $f^*(\xi)$. The morphisms are induced by the morphisms of $\mathcal{E}_{n,k}$ in the natural way. By crossing with the unit interval one can form a "stable" category $\mathcal{E}_*(X, \xi)$ upon which one can perform the " S - construction" which is an offshoot of Quillen's Q -construction. The Whitehead space, Wh(X) is the realization of the resulting category. The reader is referred to [35, 24, 23] for details.

As mentioned in the introduction the following remarkable theorem of Waldhausen describes the relationship between the space Wh(X) and the K -

theory space $A(X)$.

Theorem 1.1. There is a fibration of infinite loop spaces

$$s: A(X) \longrightarrow Q(X_+)$$

whose fiber is homotopy equivalent to $Wh(X)$. Moreover this fibration has a section given by a map of infinite loop spaces, $j: Q(X_+) \longrightarrow A(X)$.

Thus the homotopy invariants of $Wh(X)$ (and therefore of the stable pseudo - isotopy space $\mathbb{P}(X)$) are completely determined by the homotopy invariants of $A(X)$ and the stable homotopy invariants of the space X. We now turn to some of the recent calculations of these invariants.

Until very recently, much that has been known about Waldhausen's K - theory, $K_*(X) = \pi_* A(X)$ has been rational information, obtained via some form or another of rational homotopy theory, using the techniques for computing the Algebraic K - theory of group algebras over fields of characteristic zero. The basic starting point for these calculations is Waldhausen's observation, that rationally, $K_*(X) \otimes \mathbb{Q}$ is isomorphic to the algebraic K - theory of a certain (simplicial) group ring. Here is how that works.

Recall that given a simplicial set $\mathcal{S}_*$, one can construct the free abelian simplicial group on $\mathcal{S}_*$ denoted $\mathbb{Z}[\mathcal{S}_*]$ which is defined as follows: The n - simplices of $\mathbb{Z}[\mathcal{S}_*]$ is the free abelian group $\mathbb{Z}[S_n]$ on the set of n - simplices, S_n of $\mathcal{S}_*$. The face and degeneracies are determined by the requirements that they be group homomorphisms, and that they extend the face and degeneracy maps of the simplicial set $\mathcal{S}_*$. Thus $\mathbb{Z}[\mathcal{S}_*]$ is a simplicial abelian group, and hence its geometric realization is a topological abelian group, and therefore equivalent to a product of Eilenberg - MacLane spaces. In fact it is well known that there is a homeomorphism

$$|\mathbb{Z}[\mathcal{S}_*]| = SP^{\infty}(|\mathcal{S}_*|_+) \tag{1.2}$$

where $SP^{\infty}(\_)$ is the infinite symmetric product construction. So in particular we have that

$$\pi_*\mathbb{Z}[\mathcal{S}_*] = H_*(|\mathcal{S}_*|)$$

Now let X be a space and let $G_*(X)$ denote Kan's simplicial loop group [29], which is a simplicial group with the homotopy type of the loop space ΩX. Notice then that the induced simplicial set $\mathbb{Z}[G_*(X)]$ has the natural structure of a simplicial group ring. Moreover we can perform the same construction with the rational numbers replacing the integers to construct the rational simplicial group ring $\mathbb{Q}[G_*(X)]$.

Now the algebraic K - theory of simplicial rings is a well understood generalization of Quillen's algebraic K - theory of discrete rings. In the case of interest to us they are defined as the homotopy groups of the space $BGL(\mathbb{Q}[G_*(X)])^+$. The relevence of these groups to Waldhausen's space $A(X)$ is given by the next result which follows from the work of Steinberger [28].

Proposition1.3. The spaces $BGL(\mathbb{Q}[G_*(X)])^+$and $A(X) = BGL(Q(\Omega X_+))^+$ are rationally homotopy equivalent.

This result has been used by several people to carry out successful calculations of $K_*(X) \otimes \mathbb{Q}$. Crucial among all these calculations is the observation that when working rationally, one is reduced to homology calulations. More specifically, since $A(X) = BGL(Q(\Omega X)_+)^+$ is an infinite loop space, its rational homotopy groups are given by the primitives in its rational homology groups, $Prim(H_*(A(X);\mathbb{Q})$. (Recall that for a space Y $H_*(Y)$ has a "coproduct" $\Delta: H_*(Y) \longrightarrow H_*(Y) \otimes H_*(Y)$ which is dual to the cup product multiplication in cohomology. $Prim(H_*(Y)) = \{\alpha \in H_*(Y) : \Delta(\alpha) = \alpha \otimes 1 + 1 \otimes \alpha.)$ Now since the Quillen plus construction does not change homology, we therefore have that

$$K_*(X) \otimes \mathbb{Q} = \pi_*A(X) \otimes \mathbb{Q} = Prim(H_*BGL(Q(\Omega X)_+); \mathbb{Q}) = Prim(H_*BGL(\mathbb{Q}[G_*(X)]); \mathbb{Q})$$

Thus rationally, one is reduced to computing the homology of the classifying space of the general linear group of the rational group ring of the Kan simplicial loop group of X. Now for many years it has been of interest to people working in representation theory, invariant theory, as well as topology, to understand the relationship between the homology of a Lie group, its classifying space, and the Lie algebra homology of the associated Lie algebra. This philosophical point of view has been adapted to this situation with great success by Dwyer, Hsiang, and Staffeldt [17, 25, 26]. In order to describe some of their results, we first recall some of the basics of Lie algebra homology theory.

Let g be a Lie algebra over a field k with Lie bracket denoted by [,]. The Lie algebra homology of g, denoted simply by $H_*(g)$ is defined to be the homology of the universal enveloping algebra U(g); that is

$$H_*(g) = \mathrm{Tor}_{U(g)}(k, k)$$

Less abstractly, one can define $H_*(g)$ as the homology of a specific chain complex, known as the "Koszul complex", which we now describe.

Let $\Lambda^n g$ denote the n - fold exterior product of g over k, and consider the maps $d_{n-1}: \Lambda^n g \longrightarrow \Lambda^{n-1} g$ defined by the formula

$$d_{n-1}(g_1 \wedge g_2 \wedge \ldots \wedge g_n) = \sum_{i<j} (-1)^{i+j} [g_i, g_j]\, g_1 \wedge \ldots \wedge \hat{g}_i \wedge \ldots \wedge \hat{g}_j \wedge \ldots \wedge g_n$$

It is then straightforward to check that one obtains a chain complex

$$\ldots \longrightarrow \Lambda^n g \xrightarrow{d_{n-1}} \Lambda^{n-1} g \xrightarrow{d_{n-2}} \ldots \xrightarrow{d_2} \Lambda^2 g \xrightarrow{d_1} g \xrightarrow{\epsilon} k$$

where $\epsilon : g \longrightarrow k$ is the augmentation. The homology of this complex (the Koszul complex) is the Lie algebra homology $H_*(g)$.

Now let A be an algebra, or a differential graded algebra (DGA) over a field k. Let $g\ell(A)$ be the Lie algebra defined as follows. $g\ell_n(A)$ is the set of $n \times n$ matrices over A which has the Lie algebra structure given by $[X,Y] = XY - YX$. Notice that there is an inclusion of Lie algebras $g\ell_n(A) \subset g\ell_{n+1}(A)$ given by mapping an $n \times n$ matrix to the corresponding $n+1 \times n+1$ matrix whose last column and row consist of zeroes. $g\ell(A)$ is the Lie algebra defined to be the limit of these inclusions. Now $H_*(g\ell(A))$ is a Hopf algebra, and in particular a co - algebra. Its multiplication is induced by Whitney sum of matrices, and its comultiplication is induced by the diagonal. Thus one may consider $\mathrm{Prim}(H_*(g\ell(A)))$. The following is an important result of Dwyer, Hsiang, and Staffeldt [17] which relates $\mathrm{Prim}(H_*(g\ell(A)))$ to rational K - theory.

Theorem 1.4. Let G_* be a simplicial group. Let $\mathcal{S}_*|G|$ denote any differential graded algebra over $\mathbb{Q}$ which models the singular chains on $|G|$. Let $\mathcal{I}_*|G|$ denote its augmentation ideal. Then there is an isomorphism of coalgebras

$$H_*(BGL(\mathbb{Q}[G_*]); \mathbb{Q}) = H_*(g\ell(\mathcal{I}_*|G_*|)$$

and hence when we pass to primitives, we get an isomorphism

$$K_*(\mathbb{Q}[G_*]) \otimes \mathbb{Q} \;=\; \mathrm{Prim}(H_*(g\ell(\mathcal{S}_* \mid G_* \mid))).$$

This theorem was proved using a beautiful application of the classical invariant theory of H. Weyl. The point is that the infinite general linear group $GL(\mathcal{S}_* \mid G_* \mid)$ acts on the Koszul complex and in [17] the invariants of this action are studied. In fact a stronger version of this theorem is proved in which $\mathrm{Prim}(H_*(g\ell(S_* \mid G_* \mid)$ is identified in terms of the invariants of this action. (See theorem 1.2 of [17].) This theorem was used by Hsiang and Staffeldt to construct specific chain complexes from which to compute the rational Waldhausen K - theory for many spaces [25, 26]. Some of these applications of invariant theory were also discovered by Burghelea [3]. Now on the algebraic side, in the late seventies, J. L. Loday announced conjectures concerning the Lie algebra homology of the Lie algebra of matrices, $g\ell_n(A)$ and how it is related to the Lie algebra homology of the stable matrices, $g\ell(A)$. His conjecture was proved by himself and Quillen in [31] by giving a striking application and generalization of an piece of homological algebra invented by A. Connes [13, 14] called "Cyclic homology" theory. The chain complexes of Hsiang and Staffeldt mentioned above in fact turned out to be precursors to this theory. Indeed cyclic homology theory has since proved to be a indispensible tool in all forms of algebraic K - theory. We will describe this theory and some of its applications to Waldhausen's K - theory of spaces, in the next section.

Section 2. Cyclic Homology and the Free Loop Space.

As mentioned earlier, the first bit of evidence that the functors $\tilde{A}(X)$ and $\tilde{B}(X)$ were at all related came from rational calculations making use of the machinery of "Cyclic Homology" theory. This theory was invented by A. Connes [13, 15] to fill the role of DeRham cohomology when doing "noncommutative differential geometry". After Connes' initial work, Tsygan [34] and Loday and Quillen [31] generalized this theory and used it to compute the Lie algebra homology of the matrix Lie algebra, $g\ell(A)$, where A is an algebra over a field of characteristic zero. We will describe some of these results in this section. Our description of cyclic homology follows Loday and Quillen [31].

Let A be an associative algebra with identity over a field k. (Loday and Quillen actually allow k to be any commutative ring.) Following the notation of [31] we let A^n denote the n - fold tensor product (over k) of A with itself. Consider the operators b and b': $A^{n+1} \longrightarrow A^n$ defined by the following formulas:

2.1 $$b(a_0,\dots,a_n) = \sum_{i=0}^{n-1}(-1)^i(a_0,\dots,a_ia_{i+1},\dots,a_n) + (-1)^n(a_na_0,a_1,\dots,a_{n-1})$$

$$b'(a_0,\dots,a_n) = \sum_{i=0}^{n-1}(-1)^i(a_0,\dots,a_ia_{i+1},\dots,a_n)$$

The chain complex

$$\dots \xrightarrow{b'} A^n \xrightarrow{b'} \dots \xrightarrow{b'} A^2 \xrightarrow{b'} A$$

is seen to be acyclic by means of the chain null homotopy $s : A^n \longrightarrow A^{n+1}$ defined by $s(a_0,\dots,a_{n-1}) = (1, a_1,\dots, a_{n-1})$ which satisfies

$$b's + sb' = id.$$

This complex is refered to as the *acyclic Hochschild complex* and forms a free resolution of A as a module over the algebra $A \otimes A^{op}$, where A^{op} is the algebra which is isomorphic to A as a k - vector space, and whose multiplication is given by multiplication in the opposite order in A. That is the product of two elements a and b in A^{op} is given by the product in A of the elements b and a. In this resolution $A \otimes A^{op}$ acts on a module of the form $A \otimes M \otimes A$ by the rule

$$(\alpha, \beta)(a, m, b) = (\alpha a, m, b\beta).$$

(Notice that $A \otimes A^{op}$ acts on the left of A by the rule $(\alpha, \beta)(a) = \alpha a \beta$ and it acts on the right of A by the rule $(a)(\alpha, \beta) = \beta a \alpha$.) Thus if $\mathcal{H}_*$ denotes the acyclic Hochschild complex we may construct the complex $\mathcal{H}_* \otimes_{A \otimes A^{op}} A$ whose homology is the group $\operatorname{Tor}_{A \otimes A^{op}}(A, A)$ which is referred to as the *Hochschild homology of A*, which we denote by $HH_*(A)$. Upon careful but straightforward inspection one sees that the complex $\mathcal{H}_* \otimes_{A \otimes A^{op}} A$ is simply the following complex, known as the "Hochschild complex" for A:

$$\cdots \xrightarrow{b} A^{n+2} \xrightarrow{b} A^{n+1} \xrightarrow{b} \cdots \xrightarrow{b} A^2.$$

Notice that in the above chain complex the grading is such that the nth chain group is A^{n+2}.

Loday and Quillen constructed a double complex out of these chain complexes in the following manner. Consider the action of the cyclic group $\mathbb{Z}_n$ on A^n given by letting the generator act as the operator

$$t(a_1, \dots, a_n) = (-1)^{n-1}(a_n, a_1, \dots, a_{n-1}).$$

Let $N = 1 + t + t^2 + \dots + t^{n-1}$ denote the corresponding norm operator on A^n. In [31] Loday and Quillen defined the following double chain complex, referred to as the *Cyclic complex* for the algebra A, and denoted $\mathcal{C}(A)$:

$$\begin{array}{ccccccc}
\downarrow b & & \downarrow -b' & & \downarrow b & & \\
A^2 & \xleftarrow{1-t} & A^2 & \xleftarrow{N} & A^2 & \xleftarrow{1-t} & \cdots \\
\downarrow b & & \downarrow -b' & & \downarrow b & & \\
A & \xleftarrow{1-t} & A & \xleftarrow{N} & A & \xleftarrow{1-t} & \cdots
\end{array}$$

in which the even degree columns are Hochschild complexes and the odd degree columns are acyclic Hochschild complexes with the sign of the differential changed. Notice that the rows consist of the standard complexes for computing the homology of the cyclic groups $\mathbb{Z}_n$ with (twisted) coefficients in A^n. Loday and Quillen verified that the above diagram forms a double chain complex, and they called its homology the *Cyclic Homology* of A which they denoted by $HC_*(A)$. We note that this double complex was also suggested in Tsygan's work [34]. In [31] Loday and Quillen carried out several calculations of HC_*. We recall those that are most relevant to this discussion. As above let A be an algebra over a field k.

Theorem2.1. Let $A = k$. Then $HC_*(k) = H_*(BS^1; k)$.

Now let A be an augmented algebra over the field k, and define the "reduced" cyclic homology of A, denoted $\tilde{H}C_*(A)$ to be given by

$$\tilde{H}C_*(A) = HC_*(A)/HC_*(k).$$

Theorem 2.2. Let V be a vector space over the field k. Let $A = T(V)$ be the tensor algebra of V. That is, $T(V) = \bigoplus_{n=0}^{\infty} V^n$, where, as above V^n denotes the n - fold tensor product (over k) of V with itself, and where by convention, $V^0 = k$. One then has

$$\tilde{H}C_*(T(V)) = \bigoplus_{n=1}^{\infty} H_*(\mathbb{Z}_n; V^n),$$

where the $\mathbb{Z}_n$ - action on V^n is given by cyclic permutation of coordinates.

For the next theorem we assume that A is a algebra over a field k of characteristic zero. Here we start to see the connection between cyclic homology, Lie algebra homology, and K - theory.

Theorem 2.3. Let A be an algebra over a field of characteristic zero, and let $g\ell = g\ell(A)$ be the Lie algebra of matrices as described in the last section. Then there is an isomorphism

$$HC_*(A) \cong H_{*+1}(g\ell)$$

where $H_*(g\ell)$ denotes the Lie algebra homology as described in section 1.

By comparing theorem 2.3 with theorem 1.4 one can readily see a strong connection between Cyclic homology and Waldhausen's algebraic K - theory of a space. To make this connection precise one has to extend the notion of cyclic homology to cover Differential Graded Algebras, as well as simply discrete algebras over a field. This has been done by Burghelea in [4] where he proved the following:

Theorem 2.4. Let X be a simply connected space and let G_* be a simplicial group of the homotopy type of ΩX. Let $\mathcal{S}_*|G|$ denote any differential graded algebra over $\mathbb{Q}$ that models the rational homotopy type of the singular chains on $|G|$. We then have an isomorphism

$$\mathrm{HC}_{*+1}(\mathcal{S}_*|G|) \cong \pi_*\tilde{A}(X) \otimes \mathbb{Q}.$$

This theorem gives a clear connection between cyclic homology theory and rationalized Waldhausen's algebraic K - theory of a space. The full connection between cyclic homology, Hochschild homology, and algebraic K - theory is now fairly well understood by the deep results of Goodwillie [19] and Bokstedt [2].

Now in the beginning of this section we mentioned that cyclic homology theory was the vehicle in which the first connections between the functors A(X) and B(X) were seen. By the following results of Goodwillie [18] and of Burghelea and Fiedorowicz [5] one sees the relation between cyclic homology and the free loop space, $\Lambda(X)$

Let $\mathcal{S}_*|G|$ denote the singular chains of a simplicial group G_* of the homotopy type of the loop space ΩX of a simply connected space X. (Here the singular chains can be taken over a field k of any characteristic.)

Theorem 2.5. a. $\mathrm{HH}_*(\mathcal{S}_*|G|) \cong H_*(\Lambda(X); k)$

b. $\mathrm{HC}_*(\mathcal{S}_*|G|) \cong H_*(ES^1{}_+ \wedge_{S^1} \Lambda(X); k)$

Now recall that for any space X, $\pi_*Q(X) \otimes \mathbb{Q} \cong H_*(X; \mathbb{Q})$. Thus $\pi_*(\tilde{B}(X)) \otimes \mathbb{Q} = \pi_*Q\Sigma(ES^1{}_+ \wedge_{S^1} \Lambda(X)) \otimes \mathbb{Q} \cong H_*(ES^1{}_+ \wedge_{S^1} \Lambda(X); \mathbb{Q})$. Now since the rational homotopy type of infinite loop spaces is determined by their homotopy groups, then theorems 2.4 and 2.5 imply the following:

Corollary 2.6. $\pi_*\tilde{A}(X) \otimes \mathbb{Q} \cong \pi_*\tilde{B}(X) \otimes \mathbb{Q}$, and therefore $\tilde{A}(X)$ and $\tilde{B}(X)$ are rationally homotopy equivalent.

In order to establish the more global relationship between the functors A(X), the Whitehead space Wh(X), and B(X), as stated in theorem 4 in the introduction, it is helpful to use various combinatorial and simplicial models for the free loop space and its homotopy orbit space. We discuss these models in the next section.

Section 3. Combinatorial Models for the Free Loop Space.

Among other things, the results in the last section yield an effective way of computing the homology of the free loop space and its homotopy orbit space under the S^1 action. That is, one computes the cyclic homology of a differential graded algebra that is chain equivalent to the singular chains $\mathcal{S}_*(\Omega X)$. If one is working over a field k one can use $H_*(\Omega X; k)$ thought of as a differential graded algebra with the trivial differential. One is then naturally lead to try some calculations, and in particular to first look for examples when the calculations are as clean as possible.

When looking at the results described in the last section one sees that perhaps the cleanest nontrivial example of a calculation of cyclic homology comes when the algebra involved is a tensor algebra (thm. 2.2.). To apply this calculation to free loop spaces, one needs to consider spaces X with the property that $H_*(\Omega X; k)$ is a (graded) tensor algebra over k. By the classical calculations of Bott and Samelson this occurs when X is of the homotopy type of the suspension of a connected space Y, $X \simeq \Sigma Y$. In this case one has

$$H_*(\Omega\Sigma Y; k) \cong T(\tilde{H}_*(Y; k))$$

where $\tilde{H}_*$ denotes reduced homology. Using this together with theorems 2.2 and 2.5.b one has the following striking calculation:

Theorem 3.1. $\tilde{H}_*(ES^1{}_+\wedge_{S^1}\Lambda\Sigma Y; k) \cong \bigoplus_{n=1}^{\infty} H_*(\mathbb{Z}_n; \tilde{H}_*(Y; k)^n)$

Now it is well known that the groups appearing in the isomorphism of this theorem can be geometrically realized as follows:

$$\tilde{H}_*(E\mathbb{Z}_{n+}\wedge_{\mathbb{Z}_n} Y^{(n)}; k) \cong H_*(\mathbb{Z}_n; \tilde{H}_*(Y; k)^n)$$

where $Y^{(n)}$ denotes the n - fold smash product of Y with itself which is acted upon by the cyclic group $\mathbb{Z}_n$ by cyclically permuting the coordinates. Thus by the algebraic splitting of theorem 3.1 one is naturally lead to ask whether the space $E^1{}_+\wedge_{S^1}\Lambda\Sigma Y$ (stably) splits into pieces of the form

$E\mathbb{Z}_{n_+} \wedge_{\mathbb{Z}_n} Y^{(n)}$, and whether there is an unstable combinatorial model for this space that is built out of the cyclic groups. These questions were addressed and resolved by the author and Carlsson in [7]. They built a combinatorial complex they called $\mathbb{Z}(Y)$ defined as follows.

Consider the configuration space $F(\mathbb{R}^\infty, n) = \{(t_1, \ldots t_n) \in (\mathbb{R}^\infty)^n$ such that $t_i \neq t_j$ if $i \neq j\}$. It is straightforward to see that $F(\mathbb{R}^\infty, n)$ is a contractible space that is acted upon freely by the group $\mathbb{Z}_n$ by cyclically permuting coordinates. Thus it serves as a model for $E\mathbb{Z}_n$. In [7] the complex $\mathbb{Z}(Y)$ was defined by the rule

$$\mathbb{Z}(Y) = \bigcup_{n\geq 1} F(\mathbb{R}^\infty, n) \times_{\mathbb{Z}_N} Y^n) /\sim$$

where the equivalence relation "~" is generated by setting $(t_1, \ldots, t_{n-1}, t_n) \times_{\mathbb{Z}_n} (y_1, \ldots, y_{n-1}, *) \sim (t_1, \ldots, t_{n-1}) \times_{\mathbb{Z}_{n-1}} (y_1, \ldots, y_{n-1})$, where $* \in Y$ is the basepoint. The following is one of the main results of [7].

Theorem 3.2. Let Y be a connected, nilpotent, based space of the (based) homotopy type of a C.W. complex. (By nilpotent we mean that $\pi_1(Y)$ acts nilpotently on $\pi_*(Y)$.) Then there is a homotopy equivalence

$$h : \mathbb{Z}(Y) \xrightarrow{\simeq} ES^1{}_+ \wedge_{S^1} \Lambda\Sigma Y.$$

The complex $\mathbb{Z}(Y)$ is a naturally filtered complex, and the subquotients of the filtration are the spaces $E\mathbb{Z}_{n_+} \wedge_{\mathbb{Z}_n} Y^{(n)}$. The second result of [7] is that stably this complex splits into a wedge of these subquotients. Combining this with 3.2 we have the following.

Theorem 3.3. There is a stable homotopy equivalence

$$ES^1{}_+ \wedge_{S^1} \Lambda\Sigma(Y) \simeq_s \bigvee_{n\geq 1} E\mathbb{Z}_{n_+} \wedge_{\mathbb{Z}_n} Y^{(n)}$$

where the symbol $\simeq_s$ denotes stable homotopy equivalence; i.e. equivalence of the associated suspension spectra.

One is then naturally lead to ask if there is a similar combinatorial model for the free loop space of a suspension, $\Lambda\Sigma(Y)$, as opposed to its homotopy orbit space. Moreover one would like it to be a clear generalization of the famous James model for the based loops, $\Omega\Sigma(Y)$. Indeed such a model exists and was constructed independantly by the author [11], and by Bodigheimer [1]. The model for $\Lambda\Sigma(Y)$ (again when Y is connected) is a space called L(Y) which is a complex of the form

$$L(Y) = \bigcup_{n\geqslant 1} S^1 \times_{\mathbb{Z}_n} Y^n /\sim.$$

Moreover it was shown in [1, 11] that the complex L(Y) also stably splits. That is we have the following theorem, which was also known to Goodwillie.

Theorem 3.4. For Y a connected space there is a stable homotopy equivalence

$$\Lambda\Sigma(Y) \simeq_s \bigvee_{n\geqslant 1} S^1_+ \wedge_{\mathbb{Z}_n} Y^{(n)}.$$

It turns out that these stable splittings are very important in the proof that $\tilde{A}(\Sigma Y) \simeq \tilde{B}(\Sigma Y)$ in [9]. However before we describe how these splittings are used, we first describe a certain simplicial model for the free loop space which also turns out to be very useful. This model was originally due to Waldhausen [36], which he referred to as the "cyclic bar construction". The idea of this construction is based on the following well known fact.

Let G be a topological group. Consider the homotopy orbit space of the action of G on itself by conjugation: $EG \times_G G$.

Proposition 3.5. $EG \times_G G$ is homotopy equivalent to the free loop space on the classifying space, $\Lambda(BG)$. In fact the fibration sequence

$$G \longrightarrow EG \times_G G \longrightarrow BG$$

induced by mapping G to the basepoint, is fiber homotopy equivalent to the fibration sequence

$$\Omega BG \longrightarrow \Lambda(BG) \longrightarrow BG$$

induced by evaluating a loop in ΛBG at the basepoint of S^1.

Notice that if G is a topological group which is homotopy equivalent as H - spaces to the loop space ΩX, then the homotopy orbit space $EG \times_G G$ is homotopy equivalent to the free loop space ΛX.

The cyclic bar construction on a topological group G which we will denote by $N(G)$ is the simplicial space whose space of n - simplices is G^{n+1} and whose face maps

$$\partial_i : G^{n+1} \longrightarrow G^n, \quad i = 0,1, \ldots, n$$

are given by

$$\partial_i(g_0,\ldots,g_n) = \begin{cases} (g_0, \ldots, g_i g_{i+1}, \ldots, g_n), & 0 \leq i < n. \\ (g_n g_0, g_1, \ldots, g_{n-1}), & i = n \end{cases}$$

The degeneracy maps $s_i : G^n \longrightarrow G^{n+1}$ are defined by

$$s_i(g_0,\ldots, g_{n-1}) = (g_0,\ldots, g_i, 1, g_{i+1},\ldots, g_{n-1}), \quad i = 0,\ldots,n-1.$$

It is easy to see (as is done by Waldhausen [36]) that the geometric realization of the cyclic bar construction is $EG \times_G G$, and hence we have the following result:

Proposition 3.6. Let G be a topological group (or even a monoid) of the H - space homotopy type of the loop space ΩX. Then there is a homotopy equivalence

$$|N(G)| \simeq \Lambda X.$$

Goodwillie actually proved a stronger version of this theorem. To do this he used the notion of "Cyclic objects" originally due to Connes [13, 15]. A cyclic object is a simplicial object with more structure. Namely, a cyclic object X_* in a category $\mathcal{C}$ is a simplicial object in $\mathcal{C}$ together with an action of the cyclic group $\mathbb{Z}_{n+1}$ on the n - simplex object X_n for each $n \geq 0$. Denote a preferred generator of $\mathbb{Z}_{n+1}$ by t_{n+1}. Then the actions of the cyclic groups must satisfy the following compatibility conditions:

$$\partial_i t_{n+1} = \begin{cases} t_n \partial_{i-1}, & 0 < i \leq n \\ \partial_n, & i = 0. \end{cases}$$

$$s_i t_{n+1} = \begin{cases} t_{n+2} s_{i-1}, & 0 < i \leq n \\ (t_{n+2})^2 s_n, & i = 0. \end{cases}$$

As usual it is possible to reinterpret these conditions to say that a cyclic object in a category $\mathcal{C}$ is a contravariant functor from a certain category Λ to $\mathcal{C}$, where Λ is a category defined by Connes in [13, 15] that contains the category Δ of simplicial theory. Now from the definition of Λ it is not difficult to see that if X_* is a cyclic set, then its realization as a simplicial set $|X_*|$ has a natural S^1 - action. So in particular a morphism of cyclic sets realizes to an S^1 - equivariant map. Conversely, if X is any S^1 - space, its singular simplicial set $\mathcal{S}_*(X)$ is in fact a cyclic set. The $\mathbb{Z}_{n+1}$ action on the n - simplices is given by the restriction of the S^1 action. (See [18, 28] for a discussion of these matters.)

Now let G be a topological monoid, and as above let $N(G)$ be the cyclic bar construction. It is straightforward to check that the simplicial structure of $N(G)$ extends to a cyclic structure. The cyclic group $\mathbb{Z}_{n+1}$ acts on the space of n - simplices, G^{n+1} by cyclically permuting coordinates. In [18] Goodwillie proved the following strengthening of 3.6:

Theorem 3.7. There is a map of cyclic spaces

$$\lambda: N(MX) \longrightarrow \mathcal{S}_*(\Lambda X)$$

where MX denotes the Moore loops on X, which induces a homotopy equivalence of their realizations.

Thus we have the following corollary.

Corollary 3.8. There is an S^1 - equivariant map

$$\lambda: |N(MX)| \longrightarrow \Lambda X$$

which is a homotopy equivalence. (Note: This does not mean that λ is an equivariant homotopy equivalence.)

We now describe one last combinatorial model for the free loop space. This is actually a cosimplicial model which was first used in connection with cyclic homology theory by Jones [28]. The point is this. $\Lambda X = \text{Maps}(S^1, X)$ and as with any mapping space, a simplicial decomposition for the source space (in this case S^1) induces a cosimplicial decomposition for the mapping space in the following manner. Let Y_* be a simplicial space, and Z any other space. Consider the cosimplicial space $\text{Maps}(Y_*, Z)$, whose space of n-simplices is the function space $\text{Maps}(Y_n, Z)$. The co-face and co-degeneracy maps of $\text{Maps}(Y_*, Z)$ are induced by the face and the degeneracy maps of Y_* in the obvious way.

Specializing to the case $\Lambda X = \text{Maps}(S^1, X)$, we take as our simplicial model for S^1 the standard model with one zero simplex, one nondegenerate 1-simplex, and all other simplices are degenerate. Notice that in this model, which we denote by S^1_*, there are $n + 1$ n-simplices. Thus $\text{Maps}(S^1, X)$ is a cosimplicial space with the property that its space of n-simplices is X^{n+1}. In [28] Jones took the singular cochains of this cosimplicial space level by level, so that at the nth level he had a complex that is isomorphic to the $n+1$-fold tensor product $\mathcal{S}^*(X)^{n+1}$, and he ended up with a cyclic chain complex, from which he proved the following (compare with theorem2.5.):

Theorem3.9. a. $HH_*(\mathcal{S}^*(X)) \cong H^*(\Lambda X)$

b. $HC^-_*(\mathcal{S}^*(X)) \cong H^*(ES^1 \times_{S^1} \Lambda X)$

where HC^-_* denotes the homology of the Loday - Quillen double complex placed in the 2nd quadrant of the plane rather than in the 1st quadrant.

We end this section with a discussion of how the stable splitting of $ES^1{}_+ \wedge_{S^1} \Lambda\Sigma X$ of theorem 3.3 is used in the proof in [9] that $\tilde{B}(\Sigma X) \simeq \tilde{A}(\Sigma X)$. Let $G = \Omega\Sigma X$. Define a map $\theta_{n,1}: X^n \longrightarrow GL_n(Q(G_+))$ as follows. Given an n-tuple $(x_1, \ldots, x_n)$ in X^n define $\theta_{n,1}(x_1,\ldots,x_n)$ to be the matrix

$$\begin{pmatrix} 1 & (x_1-1) & & & \\ & 1 & (x_2-1) & & 0 \\ & & 1 & \ddots & \\ & 0 & & 1 & (x_{n-1}-1) \\ (x_n-1) & & & & 1 \end{pmatrix}$$

This matrix notation needs some explanation.

We identify $Q(G_+)$ with $Q(S^0) \times Q(G)$, and we think of G as being a subspace of $Q(G_+)$ via the inclusion as $\{1\} \times G \subseteq Q(S^0) \times Q(G)$. Here $\{1\}$ denotes the basepoint of the component of degree one maps in $Q(S^0)$. Finally, via the natural inclusion of X into $\Omega\Sigma X = G$, one can therefore think of elements of X as elements of $Q(G_+)$. This matrix notation should now be more or less clear. See [9] for details. Anyway it is straightforward to see that such matrices actually lie in $GL_n(Q(G_+)$. That is, when viewed as self maps of $Q(G_+)^n$ they have homotopy inverses. This will then define a map

$$\theta_{n,1} : X^n \longrightarrow GL_n(Q(G_+)) \simeq \Omega BGL_n(QG_+) \subseteq \Omega BGL(QG_+) \longrightarrow \Omega\tilde{A}\Sigma X$$

where the last map is accomplished by the plus construction.

By using the $\mathbb{Z}_n$ equivariance of the maps $\theta_{n,1}$ and the infinite loop space structure of $\Omega A\Sigma X$, it was shown in [9] how to construct a map

$$\theta: \Omega\tilde{B}\Sigma X = Q(ES^1_+ \wedge_{S^1} \Lambda\Sigma X) \simeq \coprod_n Q(E\mathbb{Z}_{n_+} \wedge_{\mathbb{Z}_n} X^{(n)}) \longrightarrow \Omega\tilde{A}\Sigma X.$$

The "$\simeq$" in this composition comes from the splitting in theorem 3.3. In [9] it was proved that θ is a homotopy equivalence. In order to do that the authors studied the effect of θ on certain invariants of homotopy functors known as the "Goodwillie derivatives" [20]. In the next section we will review some of the main features of this beautiful theory.

Section 4. The Goodwillie Calculus

In this section we outline some of the constructions and results of Goodwillie's theory of the calculus of homotopy functors [20]. This theory was used heavily in [9] and [8] in order to describe the relationship between the functors $A(X)$ and $\tilde{B}(X)$. In particular this theory was used in [9] to show that the map $\theta : \Omega\tilde{B}(\Sigma X) \longrightarrow \Omega\tilde{A}(\Sigma X)$ described in the last section is a homotopy equivalence. Our description of Goodwillie's calculus follows that in sect. 2 of [9]. The reader is referred to [20] and [9] for details.

Let $\mathcal{C}$ be the category of based topological spaces of the homotopy type of based C.W. complexes, and basepoint preserving maps.

Definition 4.1. A functor $F : \mathcal{C} \longrightarrow \mathcal{C}$ is a *homotopy functor* if it satisfies the following properties:

a. F takes homotopy equivalences to homotopy equivalences.

b. If X is a C.W. complex filtered by subcomplexes $\{X_\alpha\}$, then the natural composition of maps

$$\underset{\alpha}{\underrightarrow{\mathrm{holim}}}\, F(X_\alpha) \longrightarrow \underset{\alpha}{\underrightarrow{\lim}}\, F(X_\alpha) \longrightarrow F(\underset{\alpha}{\underrightarrow{\lim}}\, X_\alpha) = F(X)$$

is a homotopy equivalence.

The derivative of a homotopy functor F is, more or less, a "linear approximation" to F. So we first need the following definition:

Definition 4.2. A homotopy functor $L\ \mathcal{C} \longrightarrow \mathcal{C}$ is called *linear* if it satisfies the following properties:

a. $L(*)$ is contractible, where $*$ is the one point space, and

b. L takes homotopy cocartesian (or pushout) squares to homotopy cartesian (or pullback) squares.

The terminology "linear functor" is motivated from the following easily verified result:

Lemma 4.3. If L is a linear homotopy functor and X and Y are any spaces in $\mathcal{C}$, then there is a natural homotopy equivalence

$$L(X \vee Y) \simeq L(X) \times L(Y).$$

Now given any homotopy functor F with the property that F(*) is contractible, there is a natural "suspension" map

$$\sigma : F(X) \longrightarrow \Omega F \Sigma X.$$

This map has been studied in detail by Waldhausen in [36]. σ is induced by applying F to the homotopy cocartesian square obtained by including X as the equator in both the upper and lower cones lying inside ΣX. Now if L is a linear homotopy functor it is straightforward to check that the suspension map $\sigma : L(X) \longrightarrow \Omega L \Sigma X$ is a homotopy equivalence. In particular by taking X to be a sphere, the maps σ define the structure maps of an Ω - spectrum $\underline{L}$ whose n^{th} space $\underline{L}_n = L(S^n)$. The following classification of linear homotopy functors is essentially the Brown representability theorem phrased in this new language.

Theorem 4.4: If L is a linear homotopy functor and X is a space in $\mathcal{C}$, then there is a homotopy equivalence

$$L(X) \simeq \Omega^\infty(\underline{L} \wedge X)$$

where $\Omega^\infty(\ )$ associates to a spectrum the zeroth space of its Ω - spectrum. Conversely, if E is any spectrum, the homotopy functor defined by the equation

$$E(X) = \Omega^\infty(E \wedge X)$$

is a linear homotopy functor.

We now construct the "derivative of a functor" which in many cases will be a linear homotopy functor.

Definition 4.5. The derivative (at a point) of a homotopy functor F is the functor $DF : \mathcal{C} \longrightarrow \mathcal{C}$ defined by

$$DF(X) = \lim_{\overrightarrow{k}} \Omega^k \tilde{F} \Sigma^k(X)$$

where the functor $\tilde{F}(X)$ is defined to be the homotopy fiber of the natural map $F(X) \longrightarrow F(*)$, and where the direct limit system is taken with respect to the suspension map σ.

Now for "nice" functors F, the derivative functor DF is a linear functor. "Niceness" is a Blakers - Massey - type connectivity condition which is defined in [20, 9]. In particular the functors $F(X) = X$, $Q(X)$, $Q(X^n)$, $Q(X^{(n)})$, $Q(E\mathbb{Z}_{n_+} \wedge_{\mathbb{Z}_n} X^{(n)})$, $A(X)$, and, $B(X)$, are all "nice" homotopy functors.

Given a nice homotopy functor $F(X)$ let $T(F(X))$ be the homotopy pullback for the square

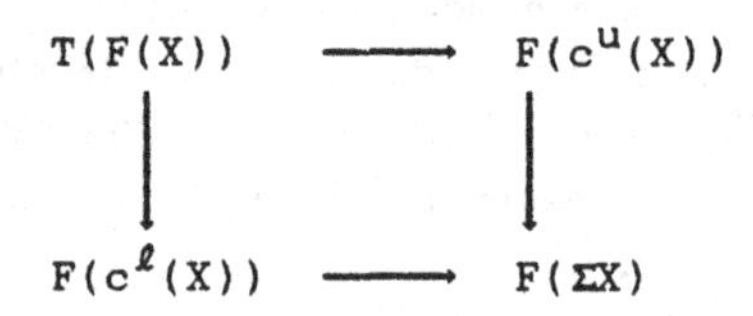

Here $c^u(X)$ and $c^\ell(X)$ are the upper and lower cones of X, thought of as subspaces of ΣX. Let $s_1 : F(X) \longrightarrow T(F(X))$ be the natural map induces by the pullback property from the maps that include X as the equator in $c^u(X)$ and $c^\ell(X)$. Notice that if $F(*)$ is contractible then $T(F(X)) \simeq \Omega F \Sigma X$ and $s_1 = \sigma : F(X) \longrightarrow \Omega F \Sigma X$. Now define

$$P_1F(X) = \lim_{\overrightarrow{k}} T^k(F(X))$$

where T^k denotes the iteration of the transformation T k - times. One can easily check that there is a homotopy fibration sequence $DF(X) \longrightarrow P_1F(X) \longrightarrow F(*)$. Thus the map $s : F(X) \longrightarrow P_1F(X)$ defined to be the composition $F(X) \xrightarrow{s_1} TF(X) \longrightarrow P_1F(X)$, can be viewed as the "first order approximation" to $F(X)$ since s induces an equivalence between F and P_1F when evaluated at a point, and when passing to derivatives at a point. In [20] Goodwillie describes the notion of higher order derivatives and thus of higher order "Taylor approximations" of functors. He also studies to what extent these approximations converge to the original functor.

It is very useful to be able to take derivatives of functors at different spaces Y, $D_Y F$, where the linear functors DF described above correspond to the case when $Y = *$. This is accomplished by mimicking the above constructions in the category $\mathcal{C}_Y$ of "spaces over Y with sections". This category is described in detail in [20] and [9]. In particular there is a notion of homotopy cartesian and cocartesian squares of morphisms in this category, and a linear homotopy functor $L : \mathcal{C}_Y \longrightarrow \mathcal{C}$ is a homotopy functor satisfying the properties

a. L(Y) is contractible, and

b. L takes homotopy cocartesian squares to homotopy cartesian squares.

Given a homotopy functor $F : \mathcal{C} \longrightarrow \mathcal{C}$ one can, in a method analogous to what was done above, define a functor $D_Y F : \mathcal{C}_Y \longrightarrow \mathcal{C}$ which under certain niceness conditions is a linear functor. When two homotopy functors satisfy certain strong niceness conditions, that Goodwillie calls "analyticity", one has the following important theorem, proved in [20].

Theorem 4.6. Let $\eta : F \longrightarrow G$ be a natural transformation between two analytic homotopy functors $F, G: \mathcal{C} \longrightarrow \mathcal{C}$. Then $\eta(X) : F(X) \longrightarrow G(X)$ is a homotopy equivalence for all simply connected X if and only if

a. $\eta : F(*) \longrightarrow G(*)$ is an equivalence, and

b. The induced maps of derivative functors

$$D_Y\eta : D_Y F(X) \longrightarrow D_Y G(X)$$

is an equivalence for all 1 - connected spaces Y and X.

The following was proved by Goodwillie in [20] and supplied the real evidence for the conjectured relationship between the homotopy types of A(X) and B(X).

Theorem 4.7. $D_Y\tilde{A}(X)$ and $D_Y\tilde{B}(X)$ are homotopy equivalent for all simply connected spaces Y and X.

It is easy to verify that the values of the functors A(*) and B(*) are not homotopy equivalent. Thus the best that can be hoped for is that the reduced functors $\tilde{A}(X)$ and $\tilde{B}(X)$ are homotopy equivalent, at least for simply connected X. It was proved in [9] that the map $\theta : \tilde{B}(\Sigma Y) \longrightarrow \tilde{A}(\Sigma X)$ induces an equivalence of derivative functors, and hence is an equivalence. This involved an explicit calculation of the effect of the map θ on derivative functors. Furthermore, it was also proven in [9] that in general $\tilde{B}(X)$ and $\tilde{A}(X)$ are not in general the same, because for example in the case of $X = \mathbb{C}P^2$, the stabilization map $s : \tilde{B}(X) \longrightarrow \tilde{B}^s(X) = D\tilde{B}(X)$ does not have a section, whereas Waldhausen proved in [36] that the stabilization map for $\tilde{A}(X)$ does have a section.

Thus one might conjecture that the homotopy fibers of the stabilization maps for $\tilde{A}(X)$ (which is the reduced Whitehead space $\tilde{Wh}(X)$) and for $\tilde{B}(X)$ are homotopy equivalent for simply connected spaces X. This is what was proven in [8]. This is theorem 4 in the introduction. By theorems 4.6 and 4.7 what one needs to do to prove theorem 4 is to construct a natural transformation between these two functors that induces equivalences on derivative functors. This is what was done. The main tool in constructing this transformation was a sort of "geometric realization" of the Dennis trace map in K - theory constructed by the author and Jones in [12]. We discuss this trace map in the next section.

Section 5. The trace map and the proof of theorem 4.

In this section we outline the proof of theorem 4 (as stated in the introduction). We begin by restating it in terms of the language and constructions described in the last section.

The following is a slightly more precise version of theorem 1.1, which is due to Waldhausen [36, 39]:

Theorem 5.1. The derivative of $\tilde{A}(X)$ is $Q(X)$. The stabilization map

$$s : \tilde{A}(X) \longrightarrow D\tilde{A}(X) = Q(X)$$

is a homotopy trivial fibration with fiber the reduced Whitehead space $\tilde{Wh}(X)$.

The following was proved in [7].

Theorem 5.2: The derivative of $\tilde{B}(X)$ is $Q(X)$.

Let $\tilde{V}(X)$ denote the homotopy fiber of the stabization map

$$s : \tilde{B}(X) \longrightarrow D\tilde{B}(X) = Q(X).$$

The following is the main result of [8], and is essentially a restatement of theorem 4 of the introduction.

Theorem 5.3. For X simply connected there is a natural homotopy equivalence

$$h : \tilde{Wh}(X) \longrightarrow \tilde{V}(X)$$

As a corollary of the results of the last section it is easy to see that the functors $\tilde{Wh}(X)$ and $\tilde{V}(X)$ have the same derivative functors at every space Y. Therefore by theorem 4.6 it is sufficient to construct the natural transformation $h : \tilde{Wh}(\_) \longrightarrow \tilde{V}(\_)$ and then show that it induces and equivalence on derivative functors. As mentioned above the transformation h is constructed using a certain geometric realization of the Dennis trace map due to the author and Jones [12]. We now discuss the construction of this trace map.

Let R be a discrete ring, and let $K_*(R)$ denote its algebraic K - groups, and $HH_*(R)$ its Hochschild homology. The Dennis trace map [16]

$$tr : K_*(R) \longrightarrow HH_*(R)$$

has in recent years proved to be an extremely important invariant in algebraic K - theory. In order to describe the relevant result of [12] concerning this map we first adopt some notation.

Let Y be a space with an S^1 - action. Let Y^{hS^1} denote the homotopy fixed point set:

$$Y^{hS^1} = \mathrm{Maps}_{S^1}(ES^1, Y)$$

where the S^1 - subscript denotes equivariant maps. Let $e : Y^{hS^1} \longrightarrow Y$ denote the map that evaluates an element in $\mathrm{Maps}_{S^1}(ES^1, Y)$ at the basepoint of ES^1.

Theorem 5.4. [12] Let R denote a discrete ring, a simplicial ring, a topological ring, or the ring up to homotopy $Q((\Omega X)_+)$. Then there exists a space $Y(R)$ with an S^1 - action and a map

$$\tau : BGL(R)^+ \longrightarrow Y(R)^{hS^1}$$

satisfying the following properties:

a. If R is a discrete ring or a simplicial ring then $\pi_*Y(R) = HH_*(R)$ and the following composition is the Dennis trace map

$$tr: K_*(R) = \pi_*(BGL(R)^+ \xrightarrow{\tau_*} \pi_*(Y(R)^{hS^1}) \xrightarrow{e_*} \pi_*(Y(R)) = HH_*(R).$$

b. If $R = Q((\Omega X)_+)$ then $Y(R) = Q(\Lambda X)$.

Remarks 1. The spaces $Y(R)$ were built out of variants of Waldhausen's cyclic bar construction as described in section 3 with the resulting S^1- action.

2. Notice that property b. above is related to property a. in that $\pi_* Q(\Lambda X) \otimes \mathbb{Q} = H_*(\Lambda X; \mathbb{Q}) = HH_*(\mathcal{S}_*(\Omega X))$ (see thm. 2.5)

3. When R is a Banach algebra the map τ gives a generalization of the various Chern character maps studied by Connes [14] and Karoubi [30]. This connection is studied in detail in [12].

This theorem therefore supplies us with a map

$$\tau : A(X) \longrightarrow Q(\Lambda X)^{hS^1}.$$

The next step was to use a certain version of the Segal conjecture [6, 32] which analyzes the homotopy type of stable homotopy fixed point sets. Using the results of the Segal conjecture and the map τ, one was then able to construct the natural transformation h of theorem 5.3. As mentioned above theorem 5.3 was then proved by calculating the effect of the transformation h on derivative functors.

Bibliography

1. C.F. Bodigheimer, Stable splittings of mapping spaces, preprint, Gottingen, 1985

2. M. Bokstedt, Topological Hochschild homology, preprint, Bielefeld, 1985.

3. D. Burghelea, Some rational computations of the Waldhausen Algebraic K - theory, Comment. Math. Helv. 54 (1979), 185 - 198

4. D. Burghelea, Cyclic homology and algebraic K - theory of spaces I, Proc. summer inst. on Alg. K-theory, Boulder Colo. 1983

5. D. Burghelea and Z. Fiedorowicz, Cyclic homology and algebraic K-theory of spaces II, Topology 25 (1986), 303-317

6. G. Carlsson, Equivariant stable homotopy and Segal's Burnside ring conjecture, Annals Math. 120 (1984), 189-224

7. G.E. Carlsson and R.L. Cohen, The cyclic groups and the free loop space, to appear in Comment. Math. Helv.

8. G. Carlsson, R. Cohen, and W.-c. Hsiang, The Whitehead space and the free loop space, in preparation

9. G. Carlsson, R. Cohen, T. Goodwillie, and W.-c. Hsiang, The free loop space and the algebraic K - theory of spaces, to appear in K - Theory, 1986

10. J. Cerf, La stratification naturelle des espaces de fonctions differentiable reelle et le theoreme de la pseudo - isotopie, Pub. Math. I.H.E.S. 36 (1970)

11. R.L. Cohen, A model for the free loop space of a suspension, preprint, Stanford, 1985

12. R.L. Cohen, and J.D.S. Jones, Trace maps and Chern characters in K - theory, in preparation

13. A. Connes, Cohomologie cyclique et foncteurs Ext^n, Comptes Rendus Acad. Sc. Paris 296 (1983)m 953-958

14. A. Connes, Noncommutative Differential Geometry, Ch.I, The Chern character in K-homology, preprint, I.H.E.S. 1982

15. A. Connes, Noncommutative Differential Geometry, Ch. II, DeRham homology and noncommutative algebra, preprint, I.H.E.S. 1983

16. K. Dennis, Algebraic K - theory and Hochschild homology, (unpublished)

17. W. Dwyer, W.-c. Hsiang, and R. Staffeldt, Pseudo-isotopy and invariant theory, I, Topology 19 (1980),367-385

18. T. Goodwillie, Cyclic homology, derivations, and the free loop space, Topology 24 (1985), 187-215

19. T. Goodwillie, Algebraic K-theory and cyclic homology, to appear in Annals of Math., 1986

20. T. Goodwillie, The calculus of functors, manuscript, 1985

21. A.E. Hatcher, Higher simple homotopy theory, Annals of Math.,102 (1975), 101-137

22. A. Hatcher, and J. Wagoner, Pseudo - isotopies of compact manifolds, Asterisque 6 (1973)

23. W.-c. Hsiang, Geometric applications of algebraic K - theory, Proc. I.C.M, Warszawa,1983, North-Holland (184), 99-118

24. W.-c. Hsiang, and B. Jahren, The Rigid Handlebody Theory I,

Current trends in Algebraic Topology, CMS conf. proc., 2, part II (1982), 337-385

25. W.-c. Hsiang, and R.E. Staffeldt, A model for computing rational algebraic K-theory of simply connected spaces, Invent. Math. 68 (1982), 227-239

26. W.-c. Hsiang, and R.E. Staffeldt, Rational algebraic K-theory of a product of Eilenberg - MacLane spaces, Contemp. Math. Vol. 19 (1983), 95-114

27. K. Igusa, Stability of pseudo - isotopies, preprint, Brandeis Univ., 1984

28. J.D.S. Jones, Cyclic homology and equivariant homology, preprint, Warwick, 1985

29. D.M. Kan, A combinatorial definition of homotopy groups, Annals of Math., 67 (1958) 282-312

30. M. Karoubi, Homologie cyclique et K - theorie algebraique I., Comptes Rendus Acad. Sc. Paris 297 (1983) 447-450

31. J.L. Loday, and D. Quillen, Cyclic homology and the Lie algebra homology of matrices, Commentarii Math Helv. 59 (1984), 565-591

32. G. Nishida, On the S^1 - Segal conjecture, Pub. Res. Ins. for Math. Sc. of Kyoto Univ. 19 (1983), 1153-1162

33. M. Steinberger, On the equivalence of the definitions of the algebraic K - theory of a topological space, Lect. Notes in Math. 763 (1979), Springer Verlag,317-331

34, B.L. Tsygan, Homology of matrix algebras over rings and the Hochschild homology (in Russian), Uspekhi Mat. Nauk, tom 38 (1983), 217-218

35. F. Waldhausen, Algebraic K-theory of topological spaces I, Proc. Symp. Pure Math 32 (1978), 35-60.

36. F. Waldhausen, Algebraic K-theory of topological spaces II, Lect. Notes in Math., 763, Springer (1979), 356-394

37. F. Waldhausen, Algebraic K-theory of spaces, a manifold approach, Can. Math. Soc. Conf. Proc. vol.2 part I, AMS (1982), 141-184

38. F. Waldhausen, Algebraic K-theory of spaces, Springer Lect. Notes in Math. 1126 (1985), 318-419

39. F. Waldhausen, Algebraic K-theory of spaces, concordance, and stable homotopy theory, to appear, Proc. Conf. in honor of J. Moore, Annals of Math. Studies

Ralph L. Cohen
Dept. of Mathematics
Stanford University
Stanford, California 94305

June, 1986

Global Methods in Homotopy Theory

Michael J. Hopkins†

§1. Introduction and the Main Theorem.

This is a report on joint work with Ethan Devinatz and Jeff Smith concerning results of a global nature in stable homotopy theory. Most of these results were in one way or another conjectured by Doug Ravenel and describe the organization of certain "periodic phenomena." For a general discussion of Ravenel's conjectures see [13] and [14].

The idea that there are periodic phenomena in stable homotopy theory originates in Adams' work [2] on G. Whitehead's J-homomorphism. The J-homomorphism is a map from the homotopy groups of the stable orthogonal group to the stable homotopy groups of spheres. Probably the most striking feature of the homotopy groups of the stable orthogonal group is Bott periodicity. Adams found a geometric manifestation of this periodicity in stable homotopy theory by producing for each prime p a self map $\alpha\colon \Sigma^{k_p} M_p \longrightarrow M_p$ of the mod p Moore spectrum (the cofibre of the degree p map: $S^0 \longrightarrow S^0$). Here $k_p = 2p-2$ if p is odd while $k_2=8$. He showed that the map α induces an isomorphism in K-theory, and that the composites

$$*) \qquad S^{nk_p} \longrightarrow \Sigma^{nk_p} M_p \xrightarrow{\alpha^n} M_p \longrightarrow S^1$$

are (for p odd) precisely the elements of order p in the image of J.

In this way one could, without reference to the J-homomorphism, produce the elements of order p in the image of J and organize them into a family linked by the periodicity operator "multiplication by α ." The operator "multiplication by α" thus plays a role in stable homotopy theory analogous to the operator associated with Bott periodicity in K-theory.

† Supported by an NSF Postdoctoral Fellowship.

Some time later L. Smith launched a program to generalize this. He replaced K-theory with complex cobordism and searched for self maps of finite complexes inducing non-nilpotent endomorphisms in complex cobordism. He and Toda succeeded in producing two important examples [16], [18]. Imitating the construction *) they produced two families of elements in the stable homotopy groups of spheres known as the β and γ families. They were, however, unable to show that the γ family consisted of non-zero elements.

Around 1975, Miller, Ravenel and Wilson [8] introduced the chromatic spectral sequence, ostensibly to demonstrate the non-triviality of the γ-family. They succeeded in doing this and in the process developed a potential framework for organizing the stable homotopy groups of spheres into periodic families associated with the indecomposable elements of the coefficient ring π_*MU.

The general problem therefore emerged of describing the non-nilpotent self maps in the category of finite spectra. Around 1976 Ravenel conjectured that these are precisely the maps inducing non-nilpotent endomorphisms in complex cobordism.

During the next eight years Ravenel formulated the conjectures referred to above. Briefly, these conjectures can be thought of as rendering "geometric" the algebraic organization suggested by the chromatic spectral sequence - much in the same way that Adams' self maps rendered the "algebraic" phenomena of Bott periodicity "geometric."

The only evidence for Ravenel's conjecture about self maps was Nishida's theorem [11] asserting the nilpotence of elements of positive dimension in the stable homotopy groups of spheres. One can think of Nishida's result in three ways:

i) The sphere is a ring spectrum, so it is a result about ring spectra;

ii) The multiplication in the stable homotopy ring comes from the smash product construction, so it is a result about smashing maps;

iii) The multiplication in the stable homotopy ring comes from composing maps, so it is a theorem about iterated composition.

The following result is the backbone of all the global results in stable homotopy theory I am going to discuss. Its three parts are generalization of Nishida's theorem in the directions indicated above.

Theorem 1: (Nilpotence Theorem [4])

i) Let R be a ring spectrum. The kernel of the Hurewicz homomorphism $MU_*: \pi_* R \longrightarrow MU_* R$ consists of nilpotent elements.

ii) Let $f: F \longrightarrow X$ be a map from a finite spectrum F to an arbitrary spectrum X. If the smash product $1_{MU} \wedge f$ is nullhomotopic then f is smash nilpotent (i.e. some $f \wedge \ldots \wedge f$ is null).

iii) Let $\longrightarrow X_n \xrightarrow{f_n} X_{n+1} \longrightarrow \ldots$ be a sequence of spectra. Suppose that X_n is c_n connected and that c_n is bounded below by a linear function of n. If for all n the homomorphism $MU_* f_n$ is zero then $\varinjlim X_n$ is contractible.

Here are some immediate consequences. In part i) if R is connected and $H_*(R;\mathbb{Z})$ is torsion free then the kernel of the MU Hurewicz homomorphism is precisely the torsion in $\pi_* R$.

Corollary. If R is a connected ring spectrum with $H_*(R;\mathbb{Z})$ torsion free then the torsion in $\pi_* R$ is nilpotent.

This applies in particular when $R = S^0$ (Nishida's theorem) and when R=MSp.

The conditions in part iii) are automatically satisfied if the sequence $\longrightarrow X_n \longrightarrow X_{n+1} \longrightarrow$ arises by iterating a self map f of a connected spectrum X. In this case $\varinjlim X_n$ is denoted $f^{-1}X$.

Corollary. Let $f: \Sigma^k X \longrightarrow X$ be a self map of a connected spectrum. If $MU_* f = 0$ then $f^{-1}X$ is contractible. In particular if X is finite then f is nilpotent.

The case of the above corollary with X finite is Ravenel's conjecture.

I'm not going to go into the details of the proof of the nilpotence theorem. I would, however, like to discuss the philosophy which leads to the proof. For simplicity we restrict ourselves to part i) with R connected associative and of finite type (the entire result can be reduced to this case). Suppose also that everything has been localized at 2. We need some sort of upper bound on $\pi_* R$. This is provided by the Adams spectral sequence [1].

In its original guise the Adams spectral sequence starts from $\mathrm{Ext}_A^{s,t}[H^*(R), F_2]$ and abuts to the 2-adic completion of $\pi_{t-s}(R)$. Here A denotes the mod 2 Steenrod algebra and cohomology is taken with F_2 coefficients. This approximation of $\pi_* R$, useful as it is in finite ranges of dimensions, is far too ungainly to yield to global investigation. Matters can be simplified somewhat if we filter $H^*(R)$ in such a way that the Steenrod algebra acts through its augmentation on the associated graded. This increases our upper bound to $\mathrm{Ext}_A[F_2, F_2] \otimes H_*(R)$ and moves us two spectral sequences away from $\pi_* R$.

Computing Hopf algebra cohomology is a lot like computing group cohomology - it's easy to compute the cohomology of abelian Hopf algebras and the real sport lies in wielding an arsenal of spectral sequences to reduce to the abelian case. The natural thing to do, then, is filter the Steenrod algebra by its lower central series and to try and recover its cohomology from that of the (abelian) associated graded Hopf algebra.

I've made it sound as if this puts us infinitely many spectral sequences from $\mathrm{Ext}_A[F_2, F_2]$. In fact the approximation of the cohomology of the Steenrod algebra by that of abelian Hopf algebras can be accomplished in one spectral sequence. This is known as the

May spectral sequence [7] and was one of the first breakthroughs in large scale computation of the cohomology of the Steenrod algebra [17].

At any rate, whether we are 3 or $\infty + 2$ spectral sequences away from π_*R, this is the upper bound we're looking for. The lower central series, as with most algebraic aspects of the Steenrod algebra is more conveniently stated in terms of its dual. The dual Steenrod algebra A_* is isomorphic to $F_2[\xi_1,\xi_2,..]$ with $|\xi_i| = 2^i-1$. The diagonal is given by $\Psi(\xi_n) = \Sigma\, \xi_{n-i}^{2^i} \otimes \xi_i$, and the dual of the lower central series is the filtration by the sub Hopf-algebras

$$F_2 \subset F_2[\xi_1] \subset F_2[\xi_1,\xi_2] \subset \ldots . \subset A_*.$$

The dual of the associated graded of the Steenrod algebra by its lower central series is therefore the Hopf algebra $F_2[\xi_1,\xi_2,..]$ with ξ_i primitive for all i. But this, as a co-algebra, is isomorphic to the exterior algebra $\bigotimes_{ij} \Lambda[\xi_i^{2^j}]$, so its cohomology is isomorphic to $F_2[h_{ij}: i \geq 1,\ j \geq 0]$ with $|h_{ij}| = (2^{i+j}-2^j,1)$.

<u>Proposition</u>: <u>Let</u> R <u>be a connected associative ring spectrum of finite type.</u> <u>The 2-adic completion of</u> π_*R <u>is the abutment of a series of spectral sequences beginning with</u> $F_2[h_{ij}: i \geq 1,\ j \geq 0] \otimes H_*R$.

Note that h_{ij} corresponds to a potential element in π_*R of dimension $2^{i+j} - 2^j-1$.

Next we need to describe the Hurewicz homomorphism $\pi_*R \longrightarrow MU_*R$ in this context. Since we are working 2-locally it is no harm to replace MU with the Brown-Peterson spectrum BP. From the eyes of the Adams spectral sequence the homomorphism $\pi_*R \longrightarrow BP_*R$ is the restriction map $\mathrm{Ext}_A[H^*(R), F_2] \longrightarrow \mathrm{Ext}_E[H^*(R), F_2]$ where $E \subset A$ is the exterior sub-algebra $E[Q_0, Q_1, \ldots]$. In terms of our upper bound, the Hurewicz homomorphism therefore corresponds to the map $F_2[h_{ij}] \otimes H_*R \longrightarrow F_2[h_{i0}] \otimes H_*R$.

<u>Proposition</u>: <u>An "upper bound" for the kernel of the Hurewicz homomorphism</u> $\pi_*R \longrightarrow BP_*R$ <u>is the ideal</u> $(h_{ij}: i, j \geq 1) \subset F_2[h_{ij}] \otimes H_*R$.

Here "upper bound" has to be interpreted rather loosely since objects in question are not necessarily related by spectral sequences.

A typical element of the kernel of the MU Hurewicz homomorphism therefore looks like $\sum_{i,j\geq 1} h_{ij} \otimes a_{ij}$ where almost all of the a_{ij} are zero. We therefore need a good reason for the elements h_{ij} to be nilpotent. Two observations are appropriate. First of all, what distinguishes the elements h_{ij}, $j \geq 1$ from the h_{i0} is that they correspond to (potential) homotopy classes in <u>odd</u> dimensions. Second, from the point of view of our upper bound, the h_{ij} are potential elements of π_*S^0. Thus, if an element h_{ij} ever arises in nature it is "morally" an element of π_*S^0.

Now the proof of Nishida's theorem shows that odd dimensional elements of π_*S^0 are nilpotent for a very good reason - their powers extend over iterated extended powers of an odd sphere. Since these are good approximations to the Eilenberg-MacLane spectrum HF_2, this forces still greater powers to factor through the Hurewicz homomorphism which is zero.

Philosophically then, one should be able to prove the nilpotence theorem by extending iterates of a "universal h_{ij}" over extended powers of an odd sphere.

For the sake of enhanced believability, here are the highlights of the proof of Theorem 1.

Consider the sequence of double loop maps

$$* = \Omega SU(1) \longrightarrow \Omega SU(2) \longrightarrow \cdots \longrightarrow \Omega SU \simeq BU$$

and let

$$S^0 = X(1) \longrightarrow X(2) \longrightarrow X(3) \longrightarrow \cdots \longrightarrow MU$$

be the resulting sequence of bordism theories. The $X(n)$ were originally constructed by Ravenel [14] and generalize the X_k-construction of Barratt and Mahowald [3].

Recall that the ring of co-operations MU_*MU is the polynomial algebra $MU_*[b_1, b_2, \ldots]$, $|b_i| = 2i$.

<u>Proposition. The MU homology of the sequence</u> $X(1) \longrightarrow X(2) \longrightarrow \ldots$ <u>is the sequence of rings</u> $MU_* \longrightarrow MU_*[b_1] \longrightarrow MU_*[b_1, b_2] \longrightarrow \ldots$.

The above proposition portrays the sequence $X(1) \longrightarrow X(2) \longrightarrow \ldots$ as a geometric representation of the filtration $F_2 \longrightarrow F_2[\xi_1] \longrightarrow \ldots$ (or more precisely, of its analogue in complex cobordism).

The program for proving the nilpotence theorem is

<u>Theorem 2</u> [4]. <u>Let</u> R <u>be a connected associative ring spectrum of finite type and let</u> $\alpha \in \pi_* R$. <u>If</u> $X(n+1)_*\alpha = 0$ <u>then</u> $X(n)_*\alpha$ <u>is nilpotent</u>.

Note that this does prove Theorem 1. Indeed let $\alpha \in \ker\{\pi_* R \longrightarrow MU_*R\}$. Since homotopy commutes with directed colimits $X(n+1)_*\,\alpha = 0$ for $n >> 0$. By induction and Theorem 2 this implies that $X(1)_*\alpha$ is nilpotent. But $X(1)_*\,\alpha = \alpha$ since $X(1) = S^0$.

The proof of Theorem 2 requires a means of passing from $X(n)$ to $X(n+1)$. Define a space F_k' by the homotopy cartesian square

$$\begin{array}{ccc} F_k' & \longrightarrow & \Omega SU(n+1) \\ \downarrow & & \downarrow \\ J_k S^{2n} & \longrightarrow & \Omega S^{2n+1} \end{array}$$

where $J_k S^{2n}$ is the kth stage of the James construction. Let $F_k \longrightarrow X(n+1)$ be the resulting map of bordism theories.

<u>Proposition. The MU homology of</u> F_k <u>is isomorphic to the</u> $MU_*X(n) = MU_*[b_1, \ldots, b_{n-1}]$ <u>sub-module of</u> $MU_*X(n+1)$ <u>with basis</u> $1, b_n, \ldots, b_n^k$.

Let G_j be the localization at 2 of F_{2^j-1}. The main step in proving Theorem 2 is

Theorem 3 [4].

i) There are cofibrations $\Sigma^{2n\cdot 2^k-1} G_k \xrightarrow{h} G_k \longrightarrow G_{k+1}$.

ii) $h^{-1}G_k$ is contractible.

The map h is (the analogue in complex cobordism of) a "universal h_{ij}." Part ii) of Theorem 3 is proved by extending iterates of h over the smash product of G_j with extended powers of an odd sphere— a Nishida style argument.

The deduction of Theorem 2 from Theorem 3 is in three steps:

1) $X(n+1)_* \alpha = 0 \Rightarrow \alpha^{-1}R \wedge G_{j+1}$ is contractible for $j >> 0$.

2) $\alpha^{-1}R \wedge G_{k+1} \sim * \Rightarrow \alpha^{-1}R \wedge G_k \sim * \Rightarrow \ldots$

$\ldots \Rightarrow \alpha^{-1}R \wedge X(n) = \alpha^{-1}R \wedge G_0$ is contractible.

3) $\alpha^{-1}R \wedge X(n) \sim * \Leftrightarrow X(n)_*\alpha$ is nilpotent.

Of these steps, 3) is trivial and 2) follows from part ii) of Theorem 1. Step 1) is more difficult and requires use of the Adams spectral sequence based on the homology theory X(n+1).

§2. Refinements of the Nilpotence Theorem.

A natural question at this point is whether the nilpotence theorem remains true with MU replaced by other spectra. It turns out that part iii) is rather special from this point of view.

Definition. A ring spectrum E is said to detect nilpotence if parts i) and ii) of Theorem 1 remain true with MU replaced by E.

In this section we will determine precisely which spectra detect nilpotence. For convenience we assume that everything is localized at a prime p.

Let BP be the Brown-Peterson spectrum and recall that $BP_* \approx \mathbb{Z}_{(p)}[v_1, v_2, \ldots]$ with $|v_n| = 2p^n-2$.

Proposition. For each integer $n > 0$ there is a unique spectrum $K(n)$ with the following properties:

i) $K(n)_*$ is isomorphic to the ring of finite Laurent series $F_p[v_n, v_n^{-1}]$, $|v_n| = 2p^n-2$.

ii) There is a map $f: BP \longrightarrow K(n)$ whose effect in homotopy is the ring homomorphism

$$\pi_* f(v_j) = \begin{cases} 0 & j \neq n \\ v_n & j = n . \end{cases}$$

Moreover, $K(n)$ is a ring spectrum.

The spectra $K(n)$ are known as Morava K-theories. We also define $K(0)$ and $K(\infty)$ to be the rational and mod p Eilenberg-MacLane spectra respectively.

The Morava K-theories enjoy many convenient properties. These stem largely from the fact that the coefficient ring $K(n)_*$ is a graded field in the sense that all of its graded modules are free.

Proposition.

i) If X is any spectrum then $K(n)\wedge X$ is homotopy equivalent to a wedge of suspensions of $K(n)$.

ii) If X and Y are spectra then the external product

$$K(n)_*X \underset{K(n)_*}{\otimes} K(n)_*Y \longrightarrow K(n)_*X\wedge Y$$

is an isomorphism.

The following result is the main refinement of the nilpotence theorem. Its deduction from Theorem 1 is not difficult.

Theorem 4. [6].

i) Let R be a (p-local) ring spectrum and $\alpha\in\pi_*R$. If for all $0\le n\le\infty$ $K(n)_*\alpha = 0$ then α is nilpotent.

ii) A map $f: F\to X$ from a finite spectrum F to any (p-local) spectrum X is smash nilpotent if and only if $K(n)_*f = 0$ for all $0\le n\le\infty$.

Corollary. Let R be a non-contractible ring spectrum. Then for some $0\le n\le\infty$ $K(n)_*R \ne 0$.

Proof: Indeed, the unit $1\in\pi_*R$ is not nilpotent.

We can now give another characterization of Morava K-theories.

By analogy with commutative algebra call a ring spectrum E a (skew) field if for all X, $E\wedge X$ is homotopy equivalent to a wedge of suspensions of E (i.e. all E-modules are free).

Corollary. Let E be a field. Then for some n, E is (non-multiplicativ homotopy equivalent to a wedge of suspensions of $K(n)$.

Proof: (sketch) For some n, $K(n)_*E\ne 0$. This gives an equivalence between a non-trivial wedge of suspensions of E and a non-trivial wedge of suspensions of $K(n)$, banishing the obstructions to an equivalence between E and a wedge of suspensions of $K(n)$.

The above corollary means that the Morava K-theories are the _prime fields_ of the stable category.

Theorem 4 provides the answer to the question raised at the beginning of this section.

Proposition. A (p-local) ring spectrum E _detects (p-local) nilpotence if and only if_ $K(n)_* E \neq 0$ _for all_ $0 \leq n \leq \infty$.

Proof: If $K(n)_* E = 0$ then E does not detect the non-nilpotent element $v_n \in K(n)_*$. Conversely, suppose $K(n)_* E \neq 0$ for all n. Let R be a ring spectrum and suppose $\alpha \in \ker \{\pi_* R \longrightarrow E_* R\}$. Since $K(n) \wedge E$ is a non-trivial wedge of suspensions of $K(n)$ this means that $K(n)_* \alpha = 0$ for all n so α is nilpotent by Theorem 4. The proof that E satisfies part ii) of Theorem 1 is similar.

§3. Applications.

We now restrict our attention to p-local finite spectra. The material in this section is taken from [6].

The following result is easily deduced from Theorem 4.

Theorem 5.

i) Let X, Y and Z be finite and $f: Y \to Z$ a map. Then some $1_X \wedge f\wedge\ldots\wedge f$ is nullhomotopic if and only if $K(m)_*(1_X \wedge f) = 0$ for all $m < \infty$.

ii) Let $f: \Sigma^k X \longrightarrow X$ be a self map of a finite spectrum X. If $K(m)_* f = 0$ for all $m < \infty$ then f is nilpotent.

Part ii) of this result severely restricts the possibility of non-nilpotent self maps in the category of p-local finite spectra. In this section we address the problem of existence of non-nilpotent self maps. Taking part ii) above into account we single out the simplest type.

Definition: Let X be a p-local finite spectrum and $n \geq 1$ an integer. A map $f: \Sigma^k X \longrightarrow X$ is a v_n self map if $K(n)_* f$ is an isomorphism while $K(m)_* f$ is nilpotent for $m \neq n$. A v_0 self map is a map inducing multiplication by p^j for some j in rational homology.

There is a restriction on the existence of v_n self maps. Let C_0 be the category of p-local finite spectra and let $C_n \subset C_0$ be the full subcategory of $K(n-1)$ acyclics. It is a non-trivial result that there are proper inclusions $C_{n+1} \subset C_n$. That $C_{n+1} \subseteq C_n$ has been proved in many guises (see [13] Thm. 2.11 for a version in this form). That the inclusions are proper was first established by Steve Mitchell [9] and later by Jeff Smith [15].

Proposition 1. If $X \notin C_n$ then X does not admit a v_n self map.

Proof: Suppose that $f: \Sigma^k X \longrightarrow X$ is a v_n self map and let Y be the cofibre. Then $K(n-1)_* Y \neq 0$ but $K(n)_* Y = 0$ contradicting the inclusion $C_{n+1} \subset C_n$.

That this is the only restriction will be deduced from the nilpotence theorem and the following result.

Theorem 6. For each $n \geq 0$ there exists a p-local finite spectrum X_n such that

i) $X_n \in C_n$ but $X_n \notin C_{n+1}$

ii) X_n admits a v_n self map.

Theorem 6 is proved by first constructing spectra with very special cohomological properties [15] and then appealing to the Adams spectral sequence.

Clearly v_0 self maps abound. We are going to establish very special properties of v_n self maps. These are all elementary in case n=0 so the only interesting case is $n \geq 1$. Because of this we now work entirely within the category of p-torsion finite spectra.

We will often employ the following device: homotopy classes of self maps $f: \Sigma^k X \longrightarrow X$ are in one to one correspondence with the kth homotopy group of the ring spectrum $X \wedge DX$. This is a simple manifestation in stable homotopy of the isomorphism $\mathrm{End}(V) \simeq V \otimes V^*$ for a vector space V. Some of the assertions we need to make are more conveniently stated in terms of finite ring spectra while others are more conveniently stated in terms of self maps. Any result about finite ring spectra has an analogue for self maps and vice versa. For example, the definition of v_n self map $(n \geq 1)$ becomes

Definition. Let R be a finite ring spectrum. An element $f \in \pi_* R$ is a v_n element if $K(n)_* f$ is a unit while $K(m)_* f$ is nilpotent for $m \neq n$.

From now on translation between these two contexts will be left to the reader.

Lemma 1. Let R be a finite ring spectrum and $f \in \pi_*R$ a v_n element $(n \geq 1)$. There exist integers i, j so that $K(m)_*f^i = 0$ and $K(n)_*f^i = v_n^j$.

Proof: For $m >> 0$ $K(m)_*R \simeq HF_{p*}R \otimes K(m)_*$ and $K(m)_*f = HF_{p*}f \otimes 1$. Raising f to a power if necessary we may arrange that $K(m)_*f = 0$ for $m \neq n$. Now the assertion $K(n)_*f^i = v_n^j$ is equivalent to $K(n)_*f^i \equiv 1$ mod $(v_n - 1)$. But the ring $K(n)_*R/(v_n-1)$ is a finite dimensional F_p-algebra and so has a finite group of units. We may therefore take i to be the order of this group.

Lemma 2. Let $f \in \pi_*R$ be a v_n element. Then there exists $i > 0$ with $f^i \in$ center π_*R.

Proof Raising f to a power if necessary we may (by Lemma 1) assume that $K(m)_* f \in$ center π_*R for all m. Let $ad(f)\colon \Sigma^k R \longrightarrow R$ be the composite

$$S^k \wedge R \xrightarrow{f\wedge 1} R \wedge R \xrightarrow{1-T} R \wedge R \xrightarrow{u} R$$

where $T\colon\ R \wedge R \longrightarrow R \wedge R$ is the flip map. If $\alpha \in \pi_*R$ then $ad(f)_*\,\alpha = [f, \alpha]$. By assumption $K(m)_*\,ad(f) = 0$ for all m so $ad(f)$ is nilpotent by Theorem 5.

Now there is a formula

$$ad(f^m)_*(-) = \sum_{j<m} \binom{m}{j} ad^j(f)_*(-)\cdot f^{m-j}.$$

Taking m to be a large power of p forces in each term of the summation either the binomial coefficient to be divisible by a large power of p or else j to be large. Since R is torsion, every term in the summation is therefore zero. This completes the proof.

Lemma 3. Let f, $g \in \pi_*R$ be v_n elements. Then there exist $i, j > 0$ so that $f^i = g^j$.

Proof: Raising f and g to powers if necessary we may assume that $K(m)_*f = K(m)_*g$ for all m and that f and g commute. Then $K(m)_*(f-g) = 0$ for all m so f-g is nilpotent. It follows that $f^{p^i} \equiv g^{p^i}$ (p) for $i >> 0$. Taking i still larger ensures $f^{p^i} \equiv g^{p^i}$ (p^k) for any fixed k. This completes the proof since R was assumed torsion.

In terms of self maps Lemma 3 asserts that if f and g are two v_n self maps of X then f^i and g^j are homotopic for some i,j.

Lemma 4. Let X and Y be finite spectra with v_n self maps f and g respectively. Then there exist integers i, j > 0 so that for all h: X ⟶ Y the following diagram commutes:

$$\begin{array}{ccc} X & \xrightarrow{h} & Y \\ f^i \downarrow & & \downarrow g^j \\ X & \xrightarrow{h} & Y \end{array} .$$

In the above lemma we have suppressed suspensions from the notation. Thus h could map $\Sigma^k X$ to Y for any k.

Proof. The spectrum $Y \wedge DX$ has two v_n self maps $1 \wedge Df$ and $g \wedge 1$. It follows from Lemma 3 that $1 \wedge Df^i$ and $g^j \wedge 1$ are homotopic for some i, j > 0. The result now follows from a routine manipulation involving Spanier-Whitehead duality.

Call a property P of p-local finite spectra _generic_ if the full subcategory of C_0 consisting of the finite spectra satisfying P is closed under cofibrations and retracts, i.e.

i) a retract of an object with P has P

ii) if two out of three terms in a cofibre sequence have P so does the third.

The following is easily deduced from Lemmas 1-3 and the 5-lemma.

Corollary. The property of admitting a v_n self map is generic.

The next result is equivalent to the nilpotence theorem. It is one of the main tools used in applications of the nilpotence theorem to global questions in stable homotopy theory.

Theorem 7. Let $C \subseteq C_0$ be a full subcategory which is closed under cofibrations and retracts. Then $C = C_n$ for some n.

Proof: For $X \in C_0$ let the support of X, $\mathrm{supp}(X)$, be the set $\{ n: K(n)_* X \neq 0\}$. We must show that if $X \in C$ and $\mathrm{supp}\, Y \subseteq \mathrm{supp}\, X$ then $Y \in C$.

Note that if $X \in C$ then $X \wedge Z \in C$ for $Z \in C_0$ since C is closed under cofibrations. It follows that $Y \wedge X \wedge DX \in C$.

If $f: Z \longrightarrow S^0$ is a map let M_f be the cofibre. There are cofibrations $M_{\underbrace{f\wedge\dots\wedge f}_{(n-1)}} \longrightarrow \Sigma \underbrace{Z\wedge\dots\wedge Z}_{(n-1)} \wedge M_f \longrightarrow M_{\underbrace{f\wedge\dots\wedge f}_{n}}$.

It follows that if $Y \wedge M_f \in C$ then $Y \wedge M_{f\wedge\dots\wedge f} \in C$.

Let $S^0 \longrightarrow X \wedge DX$ be the duality map and $f: Z \longrightarrow S^0$ the fibre. Then with the above notation $X \wedge DX$ is also M_f. It follows that $Y \wedge M_{f\wedge\dots\wedge f} \in C$. But $K(m)_*(1_Y \wedge f) = 0$ for all m (since $\mathrm{supp}\, Y \subseteq \mathrm{supp}(X)$). By Theorem 5 $1_Y \wedge f \wedge \dots \wedge f$ is null so Y is a retract of $Y \wedge M_{f\wedge\dots\wedge f}$. This proves that $Y \in C$.

Theorem 8. A p-local finite spectrum X admits a v_n self map if and only if $X \in C_n$.

Proof. Since the property of admitting a v_n self map is generic the full subcategory of p-local finite spectra admitting a v_n self map is C_j for some j. By Proposition 1, $j \geq n$. On the other hand Theorem 6 gives $j \leq n$. This completes the proof.

Theorem 8 was conjectured by Ravenel (though with a slightly different definition of v_n self map).

Theorem 7 has found many applications. Here is another example.

Definition A map of rings $f: A \longrightarrow B$ is an N-isomorphism if ker f consists of nilpotent elements and if given $b \in B$, $b^j \in \mathrm{im}(f)$ for some j.

Theorem 9. Let X be in C_n *but not* C_{n+1}. *The* $K(n)$ *Hurewicz homomorphism*: $[X,X]_* \longrightarrow \mathrm{End}_{K(n)_*}[K(n)_*X, K(n)_*X]$ *induces an N-isomorphism*

$$\text{center } [X,X]_* \longrightarrow \begin{cases} F_p[v_n] \text{ if } n \geq 1 \\ \mathbb{Z}_{(p)} \text{ if } n=0 \end{cases}$$

Thus the v_n self maps are distinguished by generating the centers of the endomorphism rings of finite spectra modulo nilpotents.

§4. An Algebraic Analogue.

In section 2 the nilpotence theorem was portrayed as determining the prime fields in the category of spectra. In this section we pursue the analogy with commutative algebra and show that Theorem 7 can be thought of as determining the prime spectrum of the sphere spectrum. Here, following Waldhausen we are regarding the sphere as a ring up to homotopy and the category of spectra as a substitute for its category of modules. This, in some sense, justifies the use of the word "generic" in section 3.

Fix a commutative ring R and let C be the category of unbounded chain complexes of R-modules. In many ways the category C is much like the category of spectra and in this section we will formulate and prove analogues of Theorem 1 i), ii) and Theorem 7 for C. To do this we first need to define analogues of the usual constructions in stable homotopy for C. The reader may wish to consult [12] for more details on this sort of thing.

1. Homotopy classes of maps $[X,Y]$: chain homotopy classes of maps from X to Y.

2. The sphere spectrum S^0: R as a chain complex with zero differential.

3. Homotopy groups of $X = [S^0,X]_*$: Homology of the chain complex $X = [R,X]_*$.

4. Smash product $X \wedge Y$: Tensor product $X \otimes_R Y$.

5. CW Spectra: These are the objects of the smallest full subcategory of C which contains the connected (=bounded below), dimensionwise projective chain complexes, and which is closed under the formation of mapping cones, arbitrary direct sums, and retracts.

In stable homotopy theory, the graded abelian group $[X,Y]_*$ is not a functor of the weak homotopy types of X and Y unless we require X to be a CW spectrum. A similar remark goes for the weak homotopy type of the smash product $X \wedge Y$. This is usually dealt with by making the implicit assumption that $[X,Y]_*$ and $X \wedge Y$ denote $[\tilde{X},Y]$ and $\tilde{X}$ is a CW spectrum weakly equivalent to X. We make a similar assumption when doing homotopy theory in C. Thus $[X,Y]$ and $X \wedge Y$ will denote $[\tilde{X},Y]$ and $\tilde{X} \wedge Y$ where $\tilde{X}$ is a "CW complex" weakly equivalent to X.

6. Finite Spectra: These are the small objects. An object X of C is small if for all directed colimites $Y = \varinjlim Y_\alpha$ the natural map $\varinjlim [X,Y_\alpha]_* \longrightarrow [X,Y]_*$ is an isomorphism. This is equivalent to requiring that X be weakly equivalent to a chain complex which is dimensionwise free, and finitely generated when regarded as an (ungraded) R-module.

7. The Spanier-Whitehead dual of X, DX: If X is finite let $\tilde{X}$ be a weakly equivalent complex which is dimensionwise free and finitely generated as an R-module. Then DX is the complex $\hom_R[\tilde{X},R]$. All of the usual properties of Spanier-Whitehead duality continue to hold.

8. Morava K-theories: Fields K which are R-algebras regarded as chain complexes with zero differential.

9. The homology E_*X: As usual this is defined as the homotopy groups of $E \wedge X$.

We can now formulate and prove the analogue of the nilpotence theorem for C.

Theorem 10.

i) A map $f\colon F \longrightarrow X$ from a finite complex to an arbitrary complex is smash nilpotent if and only if $K_*f = 0$ for all fields K.

ii) Let X, Y and Z be finite complexes and $f\colon Y \longrightarrow Z$ a map. Then some $1_X \wedge \underbrace{f \wedge \ldots \wedge f}_{n}$ is null for $n >> 0$ if and only if $K_*(1_X \wedge f) = 0$ for all fields K.

iii) Let $f\colon \Sigma^k X \longrightarrow X$ be a self map of a finite complex. If $K_*f = 0$ for all fields K then f is a nilpotent element of the ring $[X,X]_*$.

Proof. We will only prove part i). The deduction of the other parts from part i) is a formal argument similar to its analogue in stable homotopy theory.

Replacing $f:F \longrightarrow X$ by $Df: S^0 \longrightarrow X \wedge DF$ changes neither the assumptions nor the conclusion so we may assume that $F = S^0$.

For an integer j let X^j be the j-fold smash product of X with itself and let $f^j: S^0 \longrightarrow X^j$ be the j-fold smash product of f with itself. Consider the sequence

$$S^0 \longrightarrow X = S^0 \wedge X \xrightarrow{f\wedge 1} X^2 \xrightarrow{f\wedge 1} X^3 \longrightarrow \ldots \quad .$$

The j-fold composite is f^j. Let X^∞ be the direct limit of this sequence and $f^\infty: S^0 \longrightarrow X^\infty$ the infinite composite of the maps. Since "homotopy commutes with directed colimits" we need to show that $f^\infty = 0 \in \pi_0 X^\infty$.

In general, given a complex Z and a map $g: S^k \longrightarrow Z$ let the support of g, $\operatorname{supp}(g) \subset \operatorname{Spec} R$, be the set of prime ideals p for which the image of g in $\pi_k X_p$ is not zero. Define the annihilator ideal $\operatorname{ann}(g) \subset R$ to be the ideal $\{a \in R: af \text{ is nullhomotopic}\}$. Then $\operatorname{supp}(g)$ is just the set of prime ideals containing $\operatorname{ann}(g)$. In particular $\operatorname{supp}(g)$, if nonempty, must contain minimal elements (provided we use a model of set theory in which Zorn's lemma is true).

It is easy to see that g is null if and only if $\operatorname{supp}(g) = \emptyset$. We will show that f^∞ is null by showing that $\operatorname{supp}(f^\infty) = \emptyset$.

Assume then that $\operatorname{supp}(f^\infty) \neq \emptyset$, that $p \in \operatorname{supp}(f^\infty)$ is minimal, and that everything has been localized at p. Since p is minimal, given $x_1, \ldots, x_n \in p$ the map $x_1^{i_1} \ldots x_n^{i_n} \cdot f$ is null for $i_1 + \ldots + i_n >> 0$. Now let K be the field R/p. By assumption, $K_* f = 0$, so f is homotopic to $\Sigma\, x_i g_i$ for some $x_1, \ldots, x_n \in p$, $g_1, \ldots, g_n: S^0 \longrightarrow X$. Then f^{k+1} is homotopic to $(\Sigma\, x_i g_i)^k \cdot f$ which is null homotopic for $k >> 0$. It follows that $f^\infty = 0$ as an element of the R-module $\pi_0 X_p$ — contradicting $p \in \operatorname{Supp}(f)$. This completes the proof.

The above proof shows that the nilpotence theorem for C is really a manifestation of the simple fact that the nilradical of R consists of nilpotent elements. Thus, the significance of the nilpotence theorem in stable homotopy theory is that it _identifies_ the fields. I don't know if there is an easy proof that the fields detect nilpotence in stable homotopy (without, of course, identifying the fields). Such a result would be interesting.

Now to formulate the analogue of Theorem 7. Let $C_0 \subset C$ be the full subcategory of finite complexes. For $X \in C_0$ let the _support of X_ $\mathrm{supp}(X) \subset \mathrm{Spec}(R)$ be the (Zariski closed) set of primes p for which X_p is not acyclic. We need to study the full subcategories of C_0 which are closed under cofibrations and retracts. Here are some examples.

Let $S \subset \mathrm{Spec}\, R$ be a subset which is stable under specialization [5]. In other words, if $p \in S$ and $p \subset p'$ then $p' \in S$. Let $C_S \subset C_0$ be the full subcategory consisting of complexes X with $\mathrm{supp}(X) \subset S$. Then C_S is closed under cofibrations and retracts.

Theorem 11. Let $C \subseteq C_0$ _be a full subcategory which is closed under cofibrations and retracts. Then_ $C = C_S$ _for some_ $S \subset \mathrm{Spec}\, R$ _which is stable under specialization._

Proof: Let $S = \bigcup\{\mathrm{supp}\, X: X \in C\}$. To show that $C = C_S$ we must show that if $X \in C$ and $\mathrm{supp}\, Y \subseteq \mathrm{supp}\, X$ then $Y \in C$. The proof of this is the same (practically verbatim) as the proof of Theorem 7--only the meanings of the words have been changed.

Knowledge of the full subcategories $C \subseteq C_0$ which are closed under cofibrations and retracts is therefore equivalent to knowing Spec R and its Zariski topology. That one can recover Spec R from the derived category of finite complexes seems to be a new result. In any case, this shows how Theorem 7 can be thought of as determining the prime ideals of the sphere spectrum regarded as a ring up to homotopy.

The phenomena of periodicity in stable homotopy theory began with a geometric manifestation of the "algebraic" phenomena of Bott periodicity. The ideas then progressed in alternate dominions of algebraic and geometric insight. We have now reached a vista where the algebra and geometry are really the same thing: the study of periodicity in stable homotopy is the same as the study of the prime ideals of the sphere spectrum. Unfortunately, we can only understand the ideals of a "ring up to homotopy" indirectly. It would be very useful to have more direct analogues of the concepts of commutative algebra for rings up to homotopy. This would enable one to study "periodic phenomena" in homological and homotopical algebra in general and would undoubtedly spur further developments in global aspects of stable homotopy theory.

Princeton University

References

1. J.F. Adams: On the structure and applications of the Steenrod algebra, Comm. Math. Helv. 32(1958), 180-214.

2. J.F. Adams: On the groups J(X)IV, Topology 5 (1966), 21-71.

3. M.G. Barratt and M.E. Mahowald, private communications.

4. E.S. Devinatz, M.J. Hopkins, and J.H. Smith: Nilpotence and stable homotopy theory I, to appear.

5. R. Hartshorne: Algebraic Geometry, Graduate Texts in Math, Springer-Verlag 1977.

6. M.J. Hopkins and J.H. Smith: Nilpotence and stable homotopy theory II to appear.

7. J.P. May: The cohomology of restricted Lie algebras and of Hopf algebras, J. Algebra 3 (1966), 123-146.

8. H.R. Miller, D.C. Ravenel, and W.S. Wilson: Periodic phenomena in the Adams-Novikov spectral sequence, Ann. of Math. 106 (1977), 469-516.

9. S.A. Mitchell: Finite complexes with A(n)-free cohomology, Topology 24 (1985), 227-246.

10. J. Morava: Noetherian localizations of categories of cobordism co-modules, Ann. of Math. 121 (1985), 1-39.

11. G. Nishida: The nilpotency of elements of the stable homotopy groups of spheres, J. Math. Soc. Japan, 25 (1973), 707-732.

12. D.G. Quillen: Homotopical Algebra, Lecture Notes in Mathematics, Vol. 271, Springer-Verlag, New York, 1972.

13. D.C. Ravenel: Localization with respect to certain periodic homology theories, Am. J. Math., 106 (1984), 351-414.

14. D.C. Ravenel: these proceedings.

15. J.H. Smith: to appear.

16. L. Smith: On realizing complex cobordism modules, IV, applications to the stable homotopy groups of spheres, Am. J. Math. 99 (1971), 418-436.

17. M.C. Tangora: On the cohomology of the Steenrod algebra, Math. Z. 116 (1970), 18-64.

18. H. Toda: On spectra realizing certain exterior parts of the Steenrod algebra, Topology 10 (1971), 53-65.

SUR LA COHOMOLOGIE MODULO p DES p-GROUPES ABELIENS ELEMENTAIRES

J. LANNES
Centre de Mathématiques de l'Ecole Polytechnique
91128 Palaiseau cedex (France) U.A. du CNRS n° 169

0. INTRODUCTION

Soient p un nombre premier, V un p-groupe abélien élémentaire (i.e. un groupe isomorphe à $(\mathbb{Z}/p)^d$ pour un certain d), et BV son classifiant ; on pose $H^*V = H^*BV = H^*(BV;\mathbb{F}_p)$. Nous décrivons dans ce papier certaines propriétés de H^*V à la fois comme module et algèbre instable sur l'algèbre de Steenrod A (voir § 1). A cette fin nous introduisons le formalisme suivant. Soit $\mathcal{U}$ la catégorie des A-modules instables, nous notons $T_V : \mathcal{U} \rightsquigarrow \mathcal{U}$ le foncteur adjoint à gauche du foncteur : $\mathcal{U} \rightsquigarrow \mathcal{U}$, $N \mapsto H^*V \otimes N$ (voir au § 1 la définition du produit tensoriel des deux A-modules instables) ; nous avons donc pour tous A-modules instables M et N :

$$\mathrm{Hom}_{\mathcal{U}}(T_V M, N) = \mathrm{Hom}_{\mathcal{U}}(M, H^*V \otimes N)$$

Voici le résultat algébrique principal de ce papier :

THEOREME 0.1. Le foncteur $T_V : \mathcal{U} \rightsquigarrow \mathcal{U}$ préserve les suites exactes et les produits tensoriels.

L'exactitude de T_V est une propriété de la structure de A-module instable de H^*V. C'est une reformulation du théorème suivant dû à S. Zarati et à l'auteur [13] :

THEOREME 0.2. Soit J un A-module instable injectif (un objet injectif de $\mathcal{U}$, en abrégé $\mathcal{U}$-injectif), alors le produit tensoriel $H^*V \otimes J$ est encore $\mathcal{U}$-injectif.

Ce résultat généralise ceux de G. Carlsson et H.R. Miller qui montrent que H^*V est $\mathcal{U}$-injectif [6] [15] [13].

La préservation des produits tensoriels par T_V est une propriété de la structure de A-algèbre instable de H*V. Ce résultat généralise ceux de l'appendice de [10].

Signalons enfin que les deux propriétés des A-algèbres H*V exprimées par le théorème 0.1 les caractérisent à isomorphisme près.

Les applications homotopiques du théorème 0.1 sont fondées sur la suite spectrale d'Adams instable due à A.K. Bousfield et D.M. Kan [4] [5] et plus précisément sur la généralisation qu'en donne Bousfield [3].

Nous obtenons en particulier l'énoncé suivant :

THEOREME 0.3. Soit Y un espace 1-connexe avec $H^n(Y;\mathbb{F}_p)$ de dimension finie pour tout n. On suppose qu'il existe un espace Z vérifiant les mêmes conditions et une application $\omega : BV \times Z \to Y$ tels que l'application : $T_V H^*(Y;\mathbb{F}_p) \to H^*(Z;\mathbb{F}_p)$, induite par ω, est un isomorphisme. Alors l'application : $Z \to \mathrm{hom}(BV,Y)$, adjointe de ω, est une équivalence à p-complétion près (hom(BV,Y) désigne l'espace des applications de BV dans Y).

Ce théorème (affiné au § 7) montre que l'on peut dans certains cas déterminer le type d'homotopie de l'espace fonctionnel hom(BV,Y) quand on a un bon candidat pour celui-ci. Il implique notamment la conjecture de Sullivan [15] et la conjecture de Segal pour les p-groupes élémentaires [1]. Sous sa forme affinée il implique la conjecture de Sullivan "généralisé" (H. Miller, G. Carlsson).

Nous obtenons également :

THEOREME 0.4. Soit Y un espace nilpotent avec $\pi_1 Y$ fini et $H^n(Y;\mathbb{F}_p)$ de dimension finie pour tout n. L'application naturelle :

$$[BV,Y] \to \mathrm{Hom}_{\mathcal{K}}(H^*(Y;\mathbb{F}_p), H^*(BV;\mathbb{F}_p)) \quad ,$$

[BV,Y] désignant l'ensemble des classes d'homotopie d'applications de BV dans Y et $\mathcal{K}$ la catégorie des A-algèbres instables, est une bijection.

Ce résultat avait été conjecturé par Miller [14]. Il était établi notamment dans les cas suivants. Quand $H^*(Y;\mathbb{F}_p)$ est "very nice" par Miller, dans le papier cité ci-dessus, en utilisant la suite spectrale de Massey et Peterson. Quand Y est un espace de lacets infinis par Zarati et l'auteur, en utilisant la suite spectrale d'Adams stable [11] [12].

Voici un exemple d'application du théorème 0.4 :

COROLLAIRE 0.5. Soient G un groupe de Lie connexe et BG son classifiant. Alors l'application naturelle :

$$_{G}\backslash\mathrm{Hom}(V,G) \to [BV,BG] ,$$

$_{G}\backslash\mathrm{Hom}(V,G)$ désignant le quotient de l'action par conjugaison de G sur Hom(V,G), est une bijection (d'ensembles finis).

1 LES CATEGORIES DES A-MODULES INSTABLES ET DES A-ALGEBRES INSTABLES

Soient Y un espace et p un nombre premier. Nous rappelons dans ce paragraphe les définitions de certaines structures que possède la cohomologie modulo p de Y, notée $H^*(Y;\mathbb{F}_p)$ ou simplement H^*Y.

Tout d'abord H^*Y est un A-module instable. On désigne par A l'algèbre de Steenrod modulo p. Soit $M = \{M^n\}_{n\in\mathbb{Z}}$ un A-module à gauche. On dit que M est instable s'il vérifie la condition suivante :

pour tout x dans M

- $Sq^i x = 0$ si $i > |x|$ quand $p = 2$
- $\beta^e P^i x = 0$ si $2i + e > |x|$, $e = 0, 1$ quand $p > 2$

$|x|$ désignant le degré x.

(On observera que ces conditions impliquent $M^n = 0$ pour $n < 0$). Le produit tensoriel de deux A-modules M et N, noté $M\otimes N$, est le produit tensoriel sur $\mathbb{F}_p$ muni de l'action "diagonale" de A définie à l'aide du coproduit de A ; si M et N sont instables, il en est de même pour $M\otimes N$. On note $\mathcal{U}$ la catégorie dont les objets sont les A-modules instables et dont les morphismes sont les applications A-linéaires de degré zéro.

En outre H^*Y est une A-algèbre instable. On appelle A-algèbre instable un objet M qui possède les propriétés suivantes :

(i) M est une $\mathbb{F}_p$-algèbre graduée, commutative et unitaire ;

(ii) M est un A-module instable et le produit : $M\otimes M \to M$ est A-linéaire ;

(iii) l'élévation à la puissance p-ième dans M est reliée à la structure de A-module par la formule :

$$x^2 = Sq^{|x|}x \quad \text{quand } p = 2$$

$$x^p = P^{\frac{|x|}{2}}x \quad \text{quand } p > 2 \text{ si } |x| \text{ est pair.}$$

On note $\mathcal{K}$ la catégorie dont les objets sont les A-algèbres instables et dont les morphismes sont les applications A-linéaires de degré zéro compatibles avec produit et unité.

On dira qu'un $\mathbb{F}_p$-espace vectoriel gradué $E = \{E^n\}_{n\in\mathbb{Z}}$ est graduellement fini si dim E^n est fini pour tout n. On note respectivement $\mathcal{U}^{g.f.}$ et $\mathcal{K}^{g.f.}$ les sous-catégories pleines de $\mathcal{U}$ et $\mathcal{K}$ dont les objets sont graduellement finis.

2 FORMALISME DE LA DIVISION DANS LES CATEGORIES $\mathcal{U}$ ET $\mathcal{K}$

Nous introduisons dans ce paragraphe le formalisme dont nous aurons besoin au § 3 pour exprimer les propriétés de la cohomologie modulo p des p-groupes abéliens élémentaires que nous avons en vue.

Soient $\mathcal{C}$ l'une des catégories $\mathcal{U}$ ou $\mathcal{K}$ et K, N deux objets de $\mathcal{C}$, alors le produit tensoriel sur $\mathbb{F}_p$, $K\otimes N$, possède une $\mathcal{C}$-structure naturelle.

PROPOSITION-DEFINITION 2.1. Soient $\mathcal{C}$ l'une des catégories $\mathcal{U}$ ou $\mathcal{K}$ et K un objet de $\mathcal{C}^{g.f.}$. Le foncteur : $\mathcal{C} \rightsquigarrow \mathcal{C}$, $N \mapsto K\otimes N$, admet un adjoint à gauche noté : $M \mapsto (M:K)_{\mathcal{C}}$. On a donc pour tous objets M et N de $\mathcal{C}$:

$$\mathrm{Hom}_{\mathcal{C}}(M, K\otimes N) \simeq \mathrm{Hom}_{\mathcal{C}}((M:K)_{\mathcal{C}}, N)$$

Démonstration. On se limitera au cas $\mathcal{C} = \mathcal{K}$; si $\mathcal{C} = \mathcal{U}$ on peut faire une démonstration analogue (en fait plus facile). Avant de commencer la démonstration de la proposition pour $\mathcal{C} = \mathcal{K}$, qui doit beaucoup à L. Illusie, il nous faut rappeler quelques définitions relatives à cette catégorie.

On note $\mathcal{E}$ la catégorie dont les objets sont les $\mathbb{F}_p$-espaces vectoriels $\mathbb{N}$-gradués et dont les morphismes sont les applications linéaires de degré zéro. Le foncteur oubli noté $\mathcal{O} : \mathcal{K} \rightsquigarrow \mathcal{E}$ admet un adjoint à gauche noté $G : \mathcal{E} \rightsquigarrow \mathcal{K}$. On dit qu'une A-algèbre instable M est libre si elle est isomorphe à G(E) pour un certain $\mathbb{F}_p$-espace vectoriel $\mathbb{N}$-gradué E, c'est-à-dire si l'on a pour toute A-algèbre instable N :

$$\mathrm{Hom}_{\mathcal{K}}(M,N) \approx \mathrm{Hom}_{\mathcal{E}}(E, \mathcal{O}N) \quad .$$

Le type même de la A-algèbre instable libre est $H^*K(\mathbb{Z}/p,n) = G(\Sigma^n \mathbb{F}_p)$ ($\Sigma^n \mathbb{F}_p$ désigne le $\mathbb{F}$-espace vectoriel $\mathbb{N}$-gradué E tel que $E^n = \mathbb{F}_p$ et $E^m = 0$ pour $m \neq n$).

DEFINITION 2.2. Soit M une A-algèbre instable. Une $\mathcal{K}$-résolution libre de M est la donnée d'une A-algèbre simpliciale $M_\bullet$ et d'une augmentation $M_\bullet \to M$ telle que :

(i) M_n, $n \in \mathbb{N}$, est libre ;

(ii) la suite de A-modules instables

$$M_n \xrightarrow{\Sigma(-1)^i d_i} M_{n-1} \longrightarrow \cdots \longrightarrow M_o \longrightarrow M \longrightarrow 0$$

est exacte.

Exemple. On obtient une $\mathcal{K}$-résolution libre de M, dite standard, à l'aide du cotriple associé à la paire de foncteurs adjoints $(G,\mathcal{O})$ (voir par exemple l'appendice de [2]). Nous la notons : $G_\bullet M \to M$.

Retour à la démonstration de 2.1 pour $\mathcal{C} = \mathcal{K}$. On remarquera tout d'abord qu'il est facile de définir le foncteur $(\ :K)_{\mathcal{K}}$ sur la sous-catégorie pleine de $\mathcal{K}$, notée $\mathcal{K}^{\ell}$, dont les objets sont les A-algèbres instables libres par la formule :

$$(G(E) : K)_{\mathcal{K}} = G((E : \mathcal{O}K)_{\mathcal{E}})$$

(la définition du bifoncteur $(\ :\ )_{\mathcal{E}}$ est immédiate).
Soit maintenant M un objet quelconque de $\mathcal{K}$ et $M_\bullet \to M$ une $\mathcal{K}$-résolution libre de M. On vérifie que le groupe d'homologie $M' = H_o((M_\bullet:K)_{\mathcal{K}}, \Sigma(-1)^i\, d_i)$ ($(M_\bullet : K)_{\mathcal{K}}$ est un objet simplicial sur $\mathcal{K}^{\ell}$) est naturellement muni d'une $\mathcal{K}$-structure telle que : $\mathrm{Hom}_{\mathcal{K}}(M,K\otimes N) = \mathrm{Hom}_{\mathcal{K}}(M',N)$.

2.3 <u>Relation entre le bifoncteur</u> $(\ :\ )_{\mathcal{K}}$ <u>et la cohomologie des espaces fonctionnels</u>.

Le bifoncteur $(\ :\ )_{\mathcal{K}}$ s'introduit naturellement dans le contexte suivant. Soient X et Y deux espaces (ensembles simpliciaux). On note hom(X,Y) l'espace fonctionnel correspondant (i.e. la version simpliciale de l'espace des applications continues de X dans Y, [5] p. 246). On suppose H*X graduellement fini, alors l'application d'évaluation : $X \times \mathrm{hom}(X,Y) \to Y$ induit en cohomologie un $\mathcal{K}$-morphisme : $H^*Y \to H^*X \otimes H^*\mathrm{hom}(X,Y)$ et donc par par définition un $\mathcal{K}$-morphisme noté e :

$$(H^*Y : H^*X)_{\mathcal{K}} \longrightarrow H^*\mathrm{hom}(X,Y) \quad .$$

L'application e ne donne en général qu'une information "primaire" sur $H^*\mathrm{hom}(X,Y)$, cependant c'est un isomorphisme lorsque Y est un produit fini d'espaces d'Eilenberg-MacLane du type $K(\mathbb{Z}/p,n)$; nous proposons la conjecture "duale" suivante (énoncée si prudemment qu'elle est sûrement vraie !) :

CONJECTURE 2.3. Soient V un p-groupe abélien élémentaire et BV un classifiant de V. Sous certaines conditions sur l'espace Y l'application naturelle e : $(H^*Y : H^*BV)_K \longrightarrow H^*\mathrm{hom}(BV,Y)$ est un isomorphisme (*).

Compte-tenu de 3.5 cette conjecture peut être considérée comme une généralisation de celle faite par Miller dans [14] que nous prouverons en 7.1.1.

3 PROPRIETES DU FONCTEUR $T_V = (\ : H^*V)_{\mathcal{U}}$

On s'intéresse maintenant à la cohomologie modulo p, notée H^*V, d'un p-groupe abélien élémentaire V. On note T_V le foncteur : $\mathcal{U} \rightsquigarrow \mathcal{U}$, $M \rightsquigarrow (M : H^*V)_{\mathcal{U}}$; on pose $H = H^*\mathbb{Z}/p$ et $T = T_{\mathbb{Z}/p} = (\ :H)_{\mathcal{U}}$. Comme H^*V est isomorphe au produit tensoriel $H \otimes H \otimes \ldots \otimes H$, d fois, d désignant la dimension de V sur $\mathbb{F}_p$, le foncteur T_V est équivalent à la composition $T \circ T \circ \ldots \circ T$, d fois. L'étude du foncteur T_V se ramène donc aussitôt à celle du foncteur

THEOREME 3.1. Le foncteur $T_V : \mathcal{U} \rightsquigarrow \mathcal{U}$ est exact.

Démonstration. Soit K un objet de $\mathcal{U}^{g.f.}$, puisque $\mathcal{U}$ a assez d'injectifs, il est clair que le foncteur $(\ :K)_{\mathcal{U}}$ est exact si et seulement si $K \otimes J$ est $\mathcal{U}$-injectif. Il faut donc montrer que $H \otimes J$ est $\mathcal{U}$-injectif dès que J est $\mathcal{U}$-injectif : c'est ce qui est fait dans [13].

Remarque. Il est à noter que les A-modules instables graduellement finis K tels que le foncteur $(\ :K)_{\mathcal{U}}$ est exact sont plutôt rares. Posons $\overline{H} = \widetilde{H}^*(V;\mathbb{F}_p)$, considérons l'ensemble des facteurs directs indécomposables de $\overline{H}^{m\otimes}$, m parcourant $\mathbb{N}$, et choisissons un sous-ensemble L tel que chaque classe d'isomorphisme de ces facteurs directs soit représentée dans L une et une seule

(*) Les récents travaux de Bousfield [20] montrent que la conjecture est vérifiée si l'on suppose que l'algèbre $(H^*Y : H^*BV)_K$ est graduellement finie et nulle en degré un.

fois. Prenons $K = \bigoplus_{L \in \mathcal{L}} L^{a_L}$, a_L désignant des entiers, le théorème 3.1 montre que le foncteur $(\ :K)_{\mathcal{U}}$ est exact. Réciproquement on prouve dans [9] que si le foncteur $(\ :K)_{\mathcal{U}}$ est exact alors K est de cette forme (les a_L sont uniquement déterminés en fonction de K).

3.2 Soient M_i (resp. K_i), $i = 1,2$, deux objets de $\mathcal{U}$ (resp. $\mathcal{U}^{g.f.}$), alors l'adjointe du produit tensoriel des deux applications canoniques : $M_i \to K_i \otimes (M_i : K_i)_{\mathcal{U}}$, $i = 1,2$, est une application : $(M_1 \otimes M_2 : K_1 \otimes K_2)_{\mathcal{U}} \to (M_1 : K_1)_{\mathcal{U}} \otimes (M_2 : K_2)_{\mathcal{U}}$. En particulier si K est un objet de $\mathcal{K}^{g.f.}$ alors on a une application naturelle, notée
$\mu_{M_1,M_2} : (M_1 \otimes M_2 : K)_{\mathcal{U}} \to (M_1 : K)_{\mathcal{U}} \otimes (M_2 : K)_{\mathcal{U}}$.

THEOREME 3.2.1. Pour tous A-modules instables M_1 et M_2 l'application naturelle

$$\mu_{M_1,M_2} : T_V(M_1 \otimes M_2) \longrightarrow T_V M_1 \otimes T_V M_2$$

est un isomorphisme.

Démonstration. Comme nous l'avons déjà remarqué on peut supposer $V = \mathbb{Z}/p$. Nous nous limitons au cas $p = 2$.

On note comme d'habitude F(n) le A-module instable librement engendré par un générateur de degré n ($\mathrm{Hom}_{\mathcal{U}}(F(n),M) = M^n$). Puisque le foncteur T est exact à gauche et préserve les sommes directes (c'est le cas de tous les foncteurs $(\ :K)_{\mathcal{U}}$) il suffit de montrer que $\mu_{F(n_1),F(n_2)}$ est un isomorphisme pour tous n_1 et n_2. En utilisant l'isomorphisme $F(n) \simeq (F(1)^{n\otimes})^{\mathfrak{S}_n}$ et l'exactitude du foncteur T on voit que ceci se ramène à montrer que $\mu_{F(1),F(n)}$ est un isomorphisme pour tout n.

PROPOSITION 3.2.2. Pour tout entier n l'application $\mu_{F(1),F(n)}$ est un isomorphisme.

On procède par récurrence sur n. On considère la suite exacte "naturelle" suivante :

$$0 \longrightarrow \Phi F(n) \longrightarrow \Omega(F(1) \otimes F(n)) \longrightarrow F(1) \otimes F(n-1) \longrightarrow 0$$

$\Phi, \Omega : \mathcal{U} \rightsquigarrow \mathcal{U}$, désignant respectivement le foncteur "double" et l'adjoint à gauche du foncteur suspension $\Sigma : \mathcal{U} \rightsquigarrow \mathcal{U}$. Comme le foncteur T commute avec Ω (c'est évident par définition) et avec Φ (ceci résulte par exemple du point 8.2 de [13]) la suite exacte ci-dessus montre que si $\mu_{F(1),F(n-1)}$ est un isomorphisme alors il en est de même pour $\Omega\mu_{F(1),F(n)}$. Or on a le lemme facile suivant :

LEMME 3.2.3. Soit $f : M \to N$ une application A-linéaire entre A-modules instables ; on suppose que pour tout entier m l'opération $Sq^m = N^m \to N^{2m}$ est injective. Alors les deux conditions suivantes sont équivalentes :

(i) f est un isomorphisme ;

(ii) Ωf est un isomorphisme ainsi que $f : M^o \to N^o$.

Tout se réduit donc à montrer que $\mu_{F(1),F(n)} : T^o(F(1) \otimes F(n)) \to T^oF(1) \otimes T^oF(n)$ est un isomorphisme (on pose $T^oM = (TM)^o$) ou encore que l'application naturelle (voir la remarque 3.3)

$$\mu^*_{F(1),F(n)} : \mathbb{F}_2 = \mathrm{Hom}_{\mathcal{U}}(F(1),H) \otimes \mathrm{Hom}_{\mathcal{U}}(F(n),H) \to \mathrm{Hom}_{\mathcal{U}}(F(1) \otimes F(n),H)$$

est un isomorphisme. Il est clair qu'elle est injective, aussi la démonstration sera-t-elle terminée si l'on prouve l'inégalité $\dim_{\mathbb{F}_2} \mathrm{Hom}_{\mathcal{U}}(F(1) \otimes F(n),H) \leqslant 1$. Cette inégalité résulte de l'exactitude de la suite :

$$0 \longrightarrow F(1) \otimes \Phi F(n) \longrightarrow F(1) \otimes F(n) \longrightarrow \Sigma(F(1) \otimes F(n-1)) \longrightarrow 0$$

et du fait que le A-module $F(1) \otimes \Phi F(n)$ est monogène.

3.3 <u>Remarque</u> ([10], appendice). Considérons le $\mathbb{F}_p$-espace vectoriel $T^o_V M = (T_V M)^o$ formé des éléments de degré zéro de $T_V M$. On a par définition :

$$\mathrm{Hom}_{\mathcal{U}}(M,H^*V) = (T^o_V M)^* \quad \text{et} \quad T^o_V M = (\mathrm{Hom}_{\mathcal{U}}(M,H^*V))'$$

( )* (resp. ( )') désignant le dual d'un $\mathbb{F}_p$-espace vectoriel (resp. le dual "topologique" d'un $\mathbb{F}_p$-espace vectoriel profini). D'après les théorèmes 3.1 et 3.2.1 le foncteur T^o_V est exact et commute au produit tensoriel. Ces propriétés "caractérisent" les A-algèbres instables H^*V. Plus précisément, considérons une A-algèbre instable graduellement fini K et posons $V = (K^1)^*$, alors les deux propriétés suivantes sont équivalentes :

(i) l'application canonique : $H^*V \to K$ est un isomorphisme ;

(ii) le foncteur $(\ :K)^o_{\mathcal{U}}$ est exact et commute au produit tensoriel.

Démonstration de l'implication (ii) ⇒ (i) quand $p = 2$. On a

$$K^n = \mathrm{Hom}_{\mathcal{U}}(F(n),K) = \mathrm{Hom}_{\mathcal{U}}((F(1)^{n\otimes})^{\mathfrak{S}_n},K) = (V^{*n\otimes})_{\mathfrak{S}_n} .$$

3.4 La proposition ci-dessous est essentiellement un corollaire du théorème 3.2.1.

PROPOSITION 3.4. Soit M une A-algèbre instable dont on note $\varphi : M\otimes M \to M$ le produit. La composition $T_V M \otimes T_V M \simeq T_V(M\otimes M) \xrightarrow{T_V\varphi} T_V M$ fait de $T_V M$ une A-algèbre instable. Le foncteur $T_V : \mathcal{U} \rightsquigarrow \mathcal{U}$ induit donc un foncteur que l'on note encore $T_V : \mathcal{K} \rightsquigarrow \mathcal{K}$. Ce foncteur coïncide avec la division par H^*V dans la catégorie $\mathcal{K}$:

$$T_V M = (M : H^*V)_{\mathcal{K}} .$$

Remarque. Soient M_1 et M_2 deux objets de $\mathcal{K}$ alors $M_1 \otimes M_2$ est la somme de M_1 et M_2 dans la catégorie $\mathcal{K}$. Aussi a-t-on, pour tout objet K de $\mathcal{K}^{g.f.}$, un isomorphisme naturel :

$$(M_1 \otimes M_2 : K)_{\mathcal{K}} \simeq (M_1 : K)_{\mathcal{K}} \otimes (M_2 : K)_{\mathcal{K}} .$$

Cet isomorphisme est bien compatible avec le théorème 3.2.1 et la proposition 3.4.

3.5 Nous appelons algèbre p-booléenne une $\mathbb{F}_p$-algèbre commutative (non graduée) dans laquelle $x^p = x$; par exemple les éléments de degré zéro d'une A-algèbre instable forment une algèbre p-booléenne.

COROLLAIRE 3.5 ([10], appendice). Soit M une A-algèbre instable, alors $T^o_V M$ est naturellement muni d'une structure d'algèbre p-booléenne dont le spectre s'identifie à l'ensemble profini $\mathrm{Hom}_{\mathcal{K}}(M,H^*V)$. En particulier si $\mathrm{Hom}_{\mathcal{K}}(M,H^*V)$ est fini alors $\mathrm{Hom}_{\mathcal{U}}(M,H^*V)$ s'identifie au $\mathbb{F}_p$-espace vectoriel de base $\mathrm{Hom}_{\mathcal{K}}(M,H^*V)$.

COROLLAIRE 3.6 ([10], appendice). Soient M une A-algèbre et M' une sous-A-algèbre de M ; alors tout $\mathcal{K}$-morphisme de M' dans H*V se prolonge en un $\mathcal{K}$-morphisme de M dans H*V (H*V est $\mathcal{K}$-injective).

Démonstration. Puisque l'application d'algèbres p-booléennes : $T_V^0 M' \to T_V^0 M$ est injective, l'application induite entre spectres est surjective.

4 EXEMPLES DE CALCULS OÙ INTERVIENT LE FONCTEUR T_V

4.1 Rappelons que nous notons $\overline{H}$ le sous-A-module de H formé des éléments de degré positif ; nous avons $TM = M \oplus (M : \overline{H})_{\mathcal{U}}$ pour tout A-module instable M.

DEFINITION-PROPOSITION 4.1. Nous disons qu'un A-module instable M est ponctuellement fini si pour tout élément x de M le sous-A-module Ax est fini. On a dans ce cas

$$(M : \overline{H})_{\mathcal{U}} = 0 \quad \text{i.e.} \quad TM = M \ .$$

Démonstration. On vérifie que pour tout A-module instable N le seul élément y de $\overline{H} \otimes N$ tel que Ay soit fini est 0.

Il est à noter que la proposition ci-dessus admet une réciproque [9].

4.2 Un résultat de J.F. Adams, J. Gunawardena, et H.R. Miller.

Soient W un p-groupe abélien élémentaire et K une A-algèbre graduellement finie et connexe, le calcul de $(H^*W : K)_{\mathcal{K}}$ est immédiat parce que H*W est libre, on trouve :

$$(H^*W : K)_{\mathcal{K}} = \mathbb{F}_p^E \otimes H^*W = (H^*W)^E$$

avec $E = W \otimes K^1 = \mathrm{Hom}_{\mathcal{K}}(H^*W, K)$. Cette formule correspond à l'équivalence d'homotopie :

$$\mathrm{hom}(X, BW) \simeq H^1(X;W) \times BW \quad ,$$

X désignant un espace connexe avec H*X graduellement fini, voir 2.3.

En prenant $K = H^*V$ et en utilisant l'identité $(H^*W : H^*V)_{\mathcal{K}} = (H^*W : H^*V)_{\mathcal{U}}$ on obtient la proposition suivante qui est un cas particulier du théorème 1.6 de [1].

PROPOSITION 4.2. Soient V, W deux p-groupes abéliens élémentaires, on a un isomorphisme canonique :

$$T_V \; H^*W \simeq \mathbb{F}_p^{L(V,W)} \otimes H^*W = (H^*W)^{L(V,W)}$$

$L(V,W)$ désignant l'ensemble des applications linéaires de V dans W. En d'autres termes, pour tout A-module instable N on a :

$$\mathrm{Hom}_{\mathcal{U}}(H^*W, H^*V \otimes N) \simeq \mathbb{F}_p[L(V,W)] \otimes \mathrm{Hom}_{\mathcal{U}}(H^*W, N) \quad .$$

4.3 <u>Calcul de</u> $\mathrm{Hom}_{\mathcal{K}}(H^*BG, H^*BV)$ <u>quand</u> G <u>est un groupe de Lie compact.</u>

Soient G un groupe topologique et BG son classifiant. L'application (d'ensembles) : $\mathrm{Hom}(V,G) \to \mathrm{Hom}_{\mathcal{K}}(H^*BG, H^*BV)$, $\varphi \mapsto (B\varphi)^*$ se factorise à travers le quotient $_G\backslash\mathrm{Hom}(V,G)$ de l'action de G sur Hom(V,G) induite par l'action par conjugaison de G sur lui-même ; c'est là la définition de l'application : $_G\backslash\mathrm{Hom}(V,G) \to \mathrm{Hom}_{\mathcal{K}}(H^*BG, H^*BV)$ qui apparaît dans l'énoncé ci-dessous :

PROPOSITION 4.3.1. Soient G un groupe de Lie compact et V un p-groupe abélien élémentaire. Alors l'application naturelle :

$$_G\backslash\mathrm{Hom}(V,G) \longrightarrow \mathrm{Hom}_{\mathcal{K}}(H^*BG, H^*BV)$$

est une bijection (d'ensembles finis).

Le résultat ci-dessus est connu de nombreux auteurs (Adams, Miller, Wilkerson ...). La démonstration que nous en donnons montre qu'il est équivalent à la détermination de H^*BG à F-isomorphisme près (D.G. Quillen [17]). Avant d'attaquer la démonstration de la proposition 4.3.1 introduisons la définition suivante :

DEFINITION 4.3.2. On dira qu'un A-module instable M est nilpotent dans les cas suivants.
Quand $p = 2$, si pour tout élément x de M il existe un entier n tel que $Sq^{2^n|x|} Sq^{2^{n-1}|x|} \dots Sq^{|x|} x = 0$.
Quand $p > 2$, si pour tout élément de degré pair x de M il existe un entier n $P^{p^n \frac{|x|}{2}} P^{p^{n-1} \frac{|x|}{2}} \dots P^{\frac{|x|}{2}} x = 0$.

Cette définition posée, nous pouvons énoncer :

LEMME 4.3.2. Soit M un A-module instable, les deux conditions suivantes sont équivalentes :

(i) M est nilpotent ;

(ii) $\mathrm{Hom}_{\mathcal{U}}(M, H^*V) = 0$ pour tout p-groupe abélien élémentaire V.

Démonstration. L'implication (i) ⇒ (ii) est facile (évidente pour $p = 2$, pour $p > 2$ utiliser les méthodes de [19]). Pour l'implication (ii) ⇒ (i) (dont nous n'avons pas en fait besoin pour démontrer 4.3.1) voir [9].

La proposition 4.3.1 résulte de la détermination à F-isomorphisme près de H*BG par Quillen [17] et des deux lemmes suivants :

LEMME 4.3.3. Soit $\rho : M \to L$ un homomorphisme de A-algèbres instables, alors les conditions suivantes sont équivalentes :

(i) le morphisme de $\mathbb{F}_p$-algèbres graduées commutatives unitaires sous-jacent à ρ est un F-isomorphisme ;

(ii) le noyau et le conoyau du $\mathcal{U}$-morphisme sous-jacent à ρ sont des A-modules instables nilpotents ;

(iii) ρ induit un isomorphisme : $\mathrm{Hom}_{\mathcal{U}}(L, H^*V) \xrightarrow{\sim} \mathrm{Hom}_{\mathcal{U}}(M, H^*V)$ pour tout p-groupe abélien élémentaire V ;

(iv) ρ induit une bijection (homéomorphisme) : $\mathrm{Hom}_{\mathcal{K}}(L, H^*V) \xrightarrow{\sim} \mathrm{Hom}_{\mathcal{K}}(M, H^*V)$ pour tout p-groupe abélien élémentaire V.

LEMME 4.3.4. Le foncteur $S_V : M \rightsquigarrow \mathrm{Hom}_{\mathcal{K}}(M, H^*V)$, défini sur la catégorie $\mathcal{K}$ et à valeurs dans la catégorie des ensembles profinis, transforme une limite projective finie en limite inductive. Plus précisément, soient $\mathcal{A}$ une catégorie finie (objets et morphismes en nombre fini) et Θ un foncteur contravariant défini sur $\mathcal{A}$ et à valeurs dans $\mathcal{K}$, alors l'application naturelle :

$$\lim_{\substack{\rightarrow \\ A}} S_V \circ \Theta \longrightarrow S_V(\lim_{\substack{\leftarrow \\ A}} \Theta)$$

est une bijection (homéomorphisme).

Démonstration de 4.3.3. L'équivalence (i) $\Leftrightarrow$ (ii) est claire. L'équivalence (ii) $\Leftrightarrow$ (iii) résulte de la $\mathcal{U}$-injectivité de H^*V (i.e. l'exactitude de T_V^o) et du lemme 4.3.2, l'équivalence (iii) $\Leftrightarrow$ (iv) du corollaire 3.5.

Démonstration de 4.3.4. L'exactitude de T_V^o montre que les deux algèbres p-booléennes $T_V^o(\lim_{\substack{\leftarrow \\ A}} \Theta)$ et $\lim_{\substack{\leftarrow \\ A}} T_V^o \circ \Theta$ coïncident. Il en est donc de même pour leurs spectres.

5 LES FONCTEURS $\mathrm{Ext}_{\mathcal{K}}$ DE BOUSFIELD [3]

Soient $\varphi : M \to N$ un homomorphisme de A-algèbres instables et $t \geq 1$ un entier. On note $\mathrm{Der}^t_{\mathcal{K}}(M,N;\varphi)$ l'image réciproque de φ par l'application :

$$\mathrm{Hom}_{\mathcal{K}}(M,(H^*S^t) \otimes N) \longrightarrow \mathrm{Hom}_{\mathcal{K}}(M,N)$$

induite par l'augmentation : $H^*S^t \to \mathbb{F}_p$; $\mathrm{Der}^t_{\mathcal{K}}(M,N;\varphi)$ s'identifie au $\mathbb{F}_p$-espace vectoriel des applications $\psi : M \to N$, A-linéaire de degré $-t$, qui vérifient

$$\psi(xy) = \psi(x)\ \varphi(y) + (-1)^{t|x|}\varphi(x)\ \psi(y)$$

ce qui justifie la notation (ψ est une φ-dérivation de degré $-t$).

L'objet $\mathrm{Der}^t_{\mathcal{K}}$ apparaît en théorie de l'homotopie de la façon suivante. Soient X, Y deux espaces et $f : X \to Y$ une application. Un élément du groupe d'homotopie $\pi_t(\mathrm{hom}(X,Y);f)$ peut être représenté par une application : $S^t \times X \to Y$ dont la restriction à $* \times X$ est f ; on a donc un "homomorphisme d'Hurewicz" :

$$\pi_t(\mathrm{hom}(X,Y);f) \longrightarrow \mathrm{Der}^t_{\mathcal{K}}(H^*Y,H^*X;H^*f) \quad .$$

Cet homomorphisme est un isomorphisme si Y est un espace $K(\mathbb{Z}/p,n)$ ou plus généralement si Y est un $\mathbb{F}_p$-espace affine simplicial avec H^*Y graduellement fini.

Venons-en à la définition des foncteurs $\mathrm{Ext}_{\mathcal{K}}$. On considère le complexe de cochaînes $(\mathrm{Der}^t_{\mathcal{K}}(G_\bullet M,N;\varphi_\bullet), \Sigma(-1)^i\, d^i)$ associé au $\mathbb{F}_p$-espace vectoriel cosimplicial $Der_{\mathcal{K}}^{t}(G_{\bullet}M,N;\varphi_{\bullet})$, $G_{\bullet}M \to M$ désignant la $\mathcal{K}$-résolution libre standard de M, et $\varphi_n : G_n M \to N$ les applications induites par φ. On pose :

$$Ext_{\mathcal{K}}^{s,t}(M,N;\varphi) = H^s(Der_{\mathcal{K}}^{t}(G_{\bullet}M,N;\varphi_{\bullet}) \quad , \quad \Sigma(-1)^i \, d^i) \quad .$$

On a également $Ext_{\mathcal{K}}^{s,t}(M,N;\varphi) \simeq H^s(Der_{\mathcal{K}}^{t}(M_{\bullet},N;\varphi_{\bullet}),\ \Sigma(-1)^i\ d^i)$ pour toute $\mathcal{K}$-résolution libre $M_{\bullet} \to M$. En fait les foncteurs $Ext_{\mathcal{K}}^{s,t}(\ ,N;\ )$ sont les foncteurs dérivés à gauche du foncteur $Der_{\mathcal{K}}^{t}(\ ,N;\ )$ défini sur la catégorie $\mathcal{K}/N$ des A-algèbres instables au-dessus de N, les objets libres de $\mathcal{K}/N$ étant les morphismes $\varphi : M \to N$ avec M libre (voir par exemple l'appendice de [2]).

6 FORMULES D'ADJONCTION POUR LES FONCTEURS $Ext_{\mathcal{K}}^{s,t}(\ ,H^*V \otimes N;\ )$

THEOREME 6.1. Soient M, N deux A-algèbres instables, $\varphi : M \to H^*V \otimes N$ un homomorphisme de A-algèbres instables, et $\tilde{\varphi} : T_V M \to N$ son adjoint. On a un isomorphisme naturel :

$$Ext_{\mathcal{K}}^{s,t}(T_V M,N;\tilde{\varphi}) \simeq Ext_{\mathcal{K}}^{s,t}(M,H^*V \otimes N;\varphi) \quad .$$

Démonstration. Soit $M_{\bullet} \to M$ une $\mathcal{K}$-résolution libre de M. D'après la définition même du foncteur $Der_{\mathcal{K}}^{t}(\ ,H^*V \otimes N;\ )$ et la proposition 3.4 on a $Der_{\mathcal{K}}^{t}(M_{\bullet},H^*V \otimes N;\varphi) = Der_{\mathcal{K}}^{t}(T_V M_{\bullet},N;\tilde{\varphi})$. On conclut grâce à la proposition suivante

PROPOSITION 6.2. Soit $M_{\bullet} \to M$ une $\mathcal{K}$-résolution libre de M. Alors $T_V M_{\bullet} \to T_V M$ est une $\mathcal{K}$-résolution libre de $T_V M$.

Démonstration. Par définition, pour toute A-algèbre instable graduellement finie K, le foncteur $(\ :K)_{\mathcal{K}} : \mathcal{K} \rightsquigarrow \mathcal{K}$ transforme objet libre en objet libre. Le fait que $T_V M_{\bullet} \to T_V M$ soit une résolution est conséquence de la proposition 3.4 et du théorème 3.1.

COROLLAIRE 6.3. Soit $\varphi : M \to H^*V$ un homomorphisme de A-algèbres instables, alors :

$$Ext_{\mathcal{K}}^{s,t}(M,H^*V;\varphi) = 0 \qquad \text{pour} \quad t - s \leq 0 \quad .$$

Démonstration. Le théorème 6.1 avec $N = \mathbb{F}_p$ permet de se ramener au cas où $V = 0$. On achève alors à l'aide des lemmes suivants :

LEMME 6.4. Soit M une A-algèbre instable munie d'une augmentation ε ; on note M_c la A-algèbre instable connexe $\mathbb{F}_p \otimes_{M^o} M$ ($\mathbb{F}_p$ est un M^o-module via ε). Alors l'application naturelle :

$$\mathrm{Ext}_{\mathcal{K}}^{s,t}(M_c,\mathbb{F}_p\,;\varepsilon) \longrightarrow \mathrm{Ext}_{\mathcal{K}}^{s,t}(M,\mathbb{F}_p\,;\varepsilon)$$

est un isomorphisme.

LEMME 6.5. Soit M une A-algèbre instable augmentée connexe, alors :

$$\mathrm{Ext}_{\mathcal{K}}^{s,t}(M,\mathbb{F}_p\,;\varepsilon) = 0 \quad \text{pour} \quad t-s \leqslant 0 \;.$$

7 APPLICATIONS HOMOTOPIQUES.

Ces applications sont fondées sur la suite spectrale d'Adams instable due à A.K. Bousfield et D.M. Kan [4] [5] et plus précisément sur la généralisation qu'en donne Bousfield [3].

Bousfield et Kan définissent, pour tout espace Y, une application fonctorielle : $Y \to (\mathbb{F}_p)_\infty Y$ qu'ils appellent la $\mathbb{F}_p$-complétion de Y et montrent que si Y est 1-connexe et $H^*(Y;\mathbb{F}_p)$ graduellement fini alors leur $\mathbb{F}_p$-complétion coïncide avec les p-complétions profinies de Quillen et Sullivan [17] [18]. L'espace $(\mathbb{F}_p)_\infty Y$ est la limite inverse d'une tour de fibrations principales "explicites" et c'est cette tour de fibrations qui donne naissance aux suites spectrales évoquées dans ce paragraphe.

On se donne un autre espace X et une application $f : X \to (\mathbb{F}_p)_\infty Y$; on suppose H^*Y graduellement fini ; on note H l'application canonique : $[X,(\mathbb{F}_p)_\infty Y] \to \mathrm{Hom}_{\mathcal{K}}(H^*Y,H^*X)$, $[X,(\mathbb{F}_p)_\infty Y]$ désignant l'ensemble des classes d'homotopie d'applications de X dans Y. Dans [3] Bousfield décrit une suite spectrale dont le terme E^2 est $\mathrm{Ext}_{\mathcal{K}}^{s,t}(H^*Y,H^*X;Hf)$ qui "converge" vers $\pi_{t-s}(\mathrm{hom}(X,(\mathbb{F}_p)_\infty Y);f)$.

7.1 Sur l'ensemble [BV,Y].

Le résultat suivant avait été conjecturé par Miller (voir introduction) :

THEOREME 7.1.1. Soit Y un espace (un ensemble simplicial fibrant) nilpotent avec $\pi_1 Y$ fini et H*Y graduellement finie. Alors l'application naturelle : $[BV,Y] \longrightarrow \mathrm{Hom}_{K}(H^*Y;H^*BV)$ est une bijection .

Démonstration. Un argument de "carré arithmétique" [7] (voir [15], théorème 1.5) montre $[BV,Y] = [BV,(\mathbb{F}_p)_\infty Y]$, et le travail de Bousfield mentionné ci-dessus réduit la démonstration du théorème à vérifier que les $\mathrm{Ext}_K^{s,t}(H^*Y,H^*BV;\varphi)$ sont nuls pour $t-s\leqslant 0$ et tout φ appartenant à $\mathrm{Hom}_K(H^*Y,H^*BV)$, ce que nous donne le corollaire 6.3.

Remarque. La restriction H*Y graduellement finie dans l'énoncé 7.1.1 n'est pas due à ce qu'on travaille en cohomologie plutôt qu'en homologie. En effet l'énoncé homologique correspondant est faux. Par exemple, "l'application fantôme universelle" : $\mathbb{R}P^\infty \to S(\bigvee_{n\in\mathbb{N}} \mathbb{R}P^n)$, qui par construction est nulle en homologie, est stablement non triviale. Il est également nécessaire de faire une hypothèse sur $\pi_1 Y$ (il suffirait de supposer que l'homomorphisme : $\pi_1 Y \to \pi_1(\mathbb{F}_p)_\infty Y$ est surjectif) : Bousfield m'a exhibé un contre-exemple avec $\pi_1 Y = \mathbb{Z}$!

7.1.2 Compte tenu du calcul effectué en 4.3 le théorème 7.1.1 admet le corollaire suivant :

COROLLAIRE 7.1.2. Soit G un groupe de Lie connexe et BG son classifiant. Alors l'application naturelle :

$$_G\backslash\mathrm{Hom}(V,G) \longrightarrow [BV,BG]$$

est une bijection (d'ensembles finis).

7.2 Sur le type d'homotopie de l'espace fonctionnel $\mathrm{hom}(BV,(\mathbb{F}_p)_\infty Y)$.

La formule d'adjonction pour les foncteurs Ext_K du paragraphe 5 la théorie de Bousfield, et le "mapping lemma" de [5] p. 285, conduisent à l'énoncé suivant :

THEOREME 7.2.1. Soit Y un espace avec H*Y graduellement fini. On suppose qu'il existe un espace Z avec H*Z graduellement fini et une application $\omega : BV \times Z \to Y$ tels que la condition suivante soit vérifiée :

(C) l'application : $T_V H^*Y \to H^*Z$, adjointe de $\omega^* : H^*Y \to H^*V \otimes H^*Z$, est un isomorphisme.

Alors l'application :

$$(\mathbb{F}_p)_\infty Z \longrightarrow \hom(BV, (\mathbb{F}_p)_\infty Y)$$

induite par ω est une équivalence d'homotopie.

COROLLAIRE 7.2.2. Si l'on suppose en outre Y et Z 1-connexes (et fibrants) alors l'application $\tilde{\omega}$:

$$Z \longrightarrow \hom(BV, Y)$$

adjointe de ω est une équivalence à p-complétion près. Plus précisément considérons le diagramme

$$\begin{array}{ccc} Z & \xrightarrow{\tilde{\omega}} & \hom(BV,Y) \\ \downarrow & & \downarrow \\ Y & = & Y \end{array}$$

où la flèche de droite est induite par l'inclusion d'un point base dans BV ; alors $\tilde{\omega}$ est une p-complétion fibre à fibre de la "fibration" : $Z \to Y$.

Remarque. Il est clair que la conjecture 2.3 est vérifiée pour l'espace Y qui apparaît dans ce corollaire.

Voici des exemples d'applications du théorème ci-dessus et de son corollaire.

7.2.3 <u>La conjecture de Sullivan.</u>

Soit Y un CW-complexe fini et 1-connexe. On prend pour Z l'espace Y lui-même et pour ω la projection : $B\mathbb{Z}/p \times Y \to Y$; la condition (C) est vérifiée d'après la proposition 4.1.

Le corollaire 7.2.2 montre que l'espace $\hom_*(B\mathbb{Z}/p, Y)$ des applications pointées de $B\mathbb{Z}/p$ dans Y est faiblement contractile (Miller [15]). Compte tenu de l'énoncé 4.1 la conclusion demeure si on remplace l'hypothèse

"Y fini" par "H*Y ponctuellement fini" [8]. Pour une réciproque voir [9].

7.2.4 La conjecture de Segal pour les p-groupes abéliens élémentaires.

Soit $r \geq 2$ un entier. On prend pour Y l'espace QS^r (Q est l'abréviation habituelle de $\Omega^\infty S^\infty$). On note D le bouquet $\bigvee_{W \subset V} B_+(V/W)$, W parcourant les sous-groupes de V ($B_+(V/W)$ désigne la réunion disjointe du classifiant de V/W et d'un point base) et on prend pour Z l'espace $QS^r D$. Il existe une application naturelle telle que la condition (C) est vérifiée ; pour se convaincre de ce dernier point on utilise essentiellement la proposition 4.2.

Le corollaire 7.2.2 montre que les espaces $QS^r D$ et $\hom(BV, QS^r)$ sont équivalents à p-complétion près : il s'agit là de la conjecture de Segal pour les p-groupes abéliens élémentaires [1].

7.2.5 La conjecture de Sullivan "généralisée".

Soit X un $\mathbb{Z}/p$-CW-complexe fini ; on note F le sous-espace des points fixes de l'action $\mathbb{Z}/p$ sur X. On prend pour Y l'espace $E\mathbb{Z}/p \underset{\mathbb{Z}/p}{\times} X$ ($E\mathbb{Z}/p$ désigne un espace contractile sur lequel $\mathbb{Z}/p$ agit librement). Soit $\varphi : \mathbb{Z}/p \to \mathbb{Z}/p$ un homomorphisme de groupe, on pose $Z_\varphi = Y$ si $\varphi = 0$, $Z_\varphi = B\mathbb{Z}/p \times F$ si $\varphi \neq 0$, et on prend pour Z la somme disjointe $\coprod_\varphi Z_\varphi$, φ décrivant l'ensemble $\mathrm{Hom}(\mathbb{Z}/p, \mathbb{Z}/p)$. On définit enfin l'application $\omega : B\mathbb{Z}/p \times Z \to Y$ de la façon suivante : ω est la somme disjointe $\coprod_\varphi \omega_\varphi$ où ω_o est la projection : $B\mathbb{Z}_p \times Y \to Y$ et ω_φ, $\varphi \neq 0$, la composition :

$$B\mathbb{Z}/p \times (B\mathbb{Z}/p \times F) = B(\mathbb{Z}/p \oplus \mathbb{Z}/p) \times F \xrightarrow{B(1\oplus\varphi)\times 1} B\mathbb{Z}/p \times F \xhookrightarrow{i} Y .$$

Pour vérifier la condition (C) on utilise les deux points suivants :

- le noyau et le conoyau de la restriction $i^* : H^*Y \to H^*(B\mathbb{Z}/p \times F)$ sont finis si bien que i^* induit un isomorphisme :

$$(H^*Y : \overline{H})_{\mathcal{U}} \xrightarrow{\sim} (H^*(B\mathbb{Z}/p \times F) : \overline{H})_{\mathcal{U}}$$

(théorème 3.1 et proposition 4.1).

- $T(H \otimes H^*F) = TH \otimes TH^*F = H^{\mathrm{Hom}(\mathbb{Z}/p,\mathbb{Z}/p)} \otimes H^*F$

(théorème 3.2.1, proposition 4.2, et proposition 4.1).

Le théorème 7.2.1 montre que l'espace fonctionnel $\hom(B\mathbb{Z}/p,(\mathbb{F}_p)_\infty Y)$ a le type d'homotopie de $(\mathbb{F}_p)_\infty Z$, c'est-à-dire de la somme disjointe de $(\mathbb{F}_p)_\infty Y$ et de (p-1) copies de $B\mathbb{Z}/p \times (\mathbb{F}_p)_\infty F$. Cet exemple est intimement relié à la conjecture de Sullivan "généralisée" (H.R. Miller, G. Carlsson). Précisons un peu. Notons $\hom_\varphi(E\mathbb{Z}/p,W)$ l'espace des applications φ-équivariantes de $E\mathbb{Z}/p$ dans un $\mathbb{Z}/p$-espace W ; comme le groupe $\mathbb{Z}/p$ est commutatif $\hom_\varphi(E\mathbb{Z}/p,W)$ est encore un $\mathbb{Z}/p$-espace. Posons $E_\varphi\mathbb{Z}/p = \hom_\varphi(E\mathbb{Z}/p,E\mathbb{Z}/p)$; on observera que $E_\varphi\mathbb{Z}/p$ est un espace contractile sur lequel $\mathbb{Z}/p$ agit librement et que la théorie élémentaire des revêtements nous donne la formule :

$$\hom(B\mathbb{Z}/p, E\mathbb{Z}/p \underset{\mathbb{Z}/p}{\times} W) = \coprod_{\varphi\in\mathrm{Hom}(\mathbb{Z}/p,\mathbb{Z}/p)} E_\varphi\mathbb{Z}/p \underset{\mathbb{Z}/p}{\times} \hom_\varphi(E\mathbb{Z}/p,W).$$

En faisant dans cette formule $W = (\mathbb{F}_p)_\infty X$ et en utilisant l'équivalence d'homotopie $(\mathbb{F}_p)_\infty Y \simeq E\mathbb{Z}/p \underset{\mathbb{Z}/p}{\times} (\mathbb{F}_p)_\infty X$ ([5] p. 62), on obtient l'équivalence d'homotopie :

$$\hom(B\mathbb{Z}/p,(\mathbb{F}_p)_\infty Y) \simeq \coprod_\varphi E_\varphi\mathbb{Z}/p \underset{\mathbb{Z}/p}{\times} \hom_\varphi(E\mathbb{Z}/p,(\mathbb{F}_p)_\infty X) .$$

On a donc montré en particulier que l'application naturelle :

$$B\mathbb{Z}/p \times (\mathbb{F}_p)_\infty F \longrightarrow E_1\mathbb{Z}/p \underset{\mathbb{Z}/p}{\times} \hom(E\mathbb{Z}/p,(\mathbb{F}_p)_\infty X)$$

est une équivalence d'homotopie, ou encore que l'application naturelle :

$$(\mathbb{F}_p)_\infty F \longrightarrow \hom_1(E\mathbb{Z}/p,(\mathbb{F}_p)_\infty X)$$

est une équivalence d'homotopie.

*
* *

REFERENCES

[1] J.F. ADAMS, J. GUNAWARDENA and H. MILLER, The Segal conjecture for elementary abelian p-groups, Topology 24 (1985), 435-460.

[2] A.K. BOUSFIELD, Nice homology coalgebras, Trans. of the A.M.S. 148 (1970), 473-489.

[3] A.K. BOUSFIELD, Lettre à J. Neisendorfer (Novembre 1984).

[4] A.K. BOUSFIELD and D.M. KAN, The homotopy spectral sequence of a space with coefficients in a ring, Topology 11 (1972), 79-106.

[5] A.K. BOUSFIELD and D.M. KAN, Homotopy limits, Completion, and Localizations, Springer Lect. Notes in Maths. 304, 1972.

[6] G. CARLSSON, G.B. Segal's Burnside ring conjecture for $(\mathbb{Z}/2)^k$, Topology 22 (1983), 83-103.

[7] E. DROR, W.G. DWYER and D.M. KAN, An arithmetic square for virtually nilpotent space, Ill. J. Math. 21 (1977), 242-254.

[8] J. LANNES et Lionel SCHWARTZ, A propos de conjectures de Serre et Sullivan, Invent. Math. 83 (1986), 593-603.

[9] J. LANNES et Lionel SCHWARTZ, Sur la structure des A-modules instables injectifs, à paraître.

[10] J. LANNES et S. ZARATI, Foncteurs dérivés de la déstabilisation, à paraître.

[11] J. LANNES et S. ZARATI, Invariants de Hopf d'ordre supérieur et suite spectrale d'Adams, Note aux C.R. Acad. Sc. Paris, t. 296 (1983), 695-698.

[12] J. LANNES et S. ZARATI, Invariants de Hopf d'ordre supérieur et suite spectrale d'Adams, à paraître.

[13] J. LANNES et S. ZARATI, Sur les $\mathcal{U}$-injectifs, Ann. Scient. Ec. Norm. Sup. 19 (1986), 1-31.

[14] H.R. MILLER, Massey-Peterson towers and maps from classifying spaces, Algebraic Topology, Aarhus 1982 (proceedings), Springer Lect. Notes in Maths. 1051 (1984), 401-417.

[15] H.R. MILLER, The Sullivan conjecture on maps from classifying spaces, Annals of Maths. 120 (1984), 39-87.

[16] D.G. QUILLEN, An application of simplicial profinite groups, Comm. Math. Helv. 44 (1969), 45-60.

[17] D.G. QUILLEN, The spectrum of an equivariant cohomology ring : I, Ann. of Math. 94 (1971), 549-572.

[18] D. SULLIVAN, Geometric topology, part I : localization, periodicity and Galois symmetry, M.I.T. (1970).

[19] S. ZARATI, Quelques propriétés du foncteur $\mathrm{Hom}_{\mathcal{U}_p}(\ ,H^*V)$, Algebraic Topology Göttingen 1984 (proceedings), Springer Lect. Notes in Maths. 1172 (1985), 204-209.

[20] A.K. BOUSFIELD, On the homology spectral sequence of a cosimplicial space, à paraître.

A View of Some Aspects of Unstable Homotopy Theory since 1950[1]

J. C. Moore and J. A. Neisendorfer[2]

0. Introduction.

The central problem of algebraic topology has been and continues to be that of obtaining algebraically manageable and homotopy invariant information about naturally arising topological spaces[3]. These spaces are often finite complexes or spaces constructed from finite complexes.

Even though homology theory was crude, and homotopy theory almost nonexistent in 1925, Hopf [17] already knew that the homotopy class of a map of the n-sphere, S^n, into itself is determined by its degree. In 1935 homotopy groups were found by Hurewicz [19], he introduced a homomorphism from homotopy groups to homology groups, and proved his celebrated theorem. Thus it became known that for $q, n > 0$ the q-th homotopy group of the n-sphere, $\pi_q(S^n)$ is zero for $q < n$, and isomorphic with the group of integers, $\mathbb{Z}$, for $q = n$. Since S^1 is covered by the line, $\mathbb{R}^1$, which is contractible, one also knew that $\pi_q(S^1) = 0$ for $q > 1$.

---

[1] This article is in part based on lectures given by one of us at the Durham Topology Conference in the summer of 1985.

[2] Both authors were partially supported by the National Science Foundation.

[3] In this article the conventions will usually be those of [45] which is regarded as a general reference.

It was also in 1935 that Hopf [18] exhibited his famous maps $S^3 \longrightarrow S^2$, $S^7 \longrightarrow S^4$, and $S^{15} \longrightarrow S^8$, and showed that $\pi_3(S^2)$ is isomorphic with $\mathbb{Z}$, a generator being given by his map. He also introduced a homotopy invariant for maps $S^{4n-1} \longrightarrow S^{2n}$, and showed that there was always a map of invariant 2, thus showing that $\pi_{4n-1}(S^{2n})$ always has an element of infinite order for $n > 0$. Extensions and generalizations of this invariant later came to be known as Hopf invariants [45], and they have been much studied.

Only two years later the first stability theorem of homotopy appeared. Indeed Freudenthal [13] showed that for $n > 0$ suspension of maps induced a homomorphism $E : \pi_{n+q}(S^n) \longrightarrow \pi_{n+q+1}(S^{n+1})$, and that this homomorphism is an isomorphism for $q < n-1$ and an epimorphism for $q = n-1$. He also showed that $\pi_4(S^3)$ or more generally $\pi_{n+1}(S^n)$ for $n \geq 3$ is a cyclic group of order 2.

For the next fifteen years computations in homotopy theory were relatively isolated, and important general properties of homotopy groups were not found. There were however several important developments in homotopy theory. Products in homotopy groups were introduced by J.H.C. Whitehead [46], and their basic properties derived. He also introduced the notion of CW complex [47], and proved his basic result to the effect that if $f : X \longrightarrow Y$ is a map of connected CW complexes inducing an isomorphism of homotopy groups, then f is a homotopy equivalence. Whitehead's notion of complex generalized earlier ones such as that of Veblen. Its simplicity and generality made it handle sufficiently easily that in concrete situations of interest, constructions could be made giving rise to new CW complexes from old.

In the period 1935-1950 homology theory developed considerably. This was in part due to the introduction of an appropriate singular homology theory by Eilenberg [10], and in part to the fact that CW complexes often afforded a convenient context for explicit calculations. Eilenberg and MacLane [11] introduced $K(\pi, n)$'s, the study of spaces with a single nonvanishing homotopy group π in dimension n. Steenrod [40] introduced squaring operations, and used them in a homotopy classification result. The significance of both of these things was at best poorly understood.

Thus in 1950 homology calculations could often be made fairly readily though not well enough to compute much about the homology of $K(\pi, n)$'s. Homotopy calculations were so difficult that they were often regarded as esoteric. For $n > 2$, an incorrect calculation of $\pi_{n+2}(S^n)$ had appeared, but this had been corrected by G.W. Whitehead [43], who showed this group to have 2 elements.

Chapter 1. The surge of the 1950's

§1. The work of Serre and some related topics.

In 1951 Serre's thesis [37] appeared, and homotopy theory was no longer a theory where most calculations were esoteric, or a theory without general results concerning homotopy groups of a restricted class of spaces. Incidentally a definition of H-space had also appeared.

The principal changes came from Serre's study of the homology of fibrations and the application of these results to homotopy problems. The start was to give a new definition of fibration more convenient for homotopy theory than earlier ones. This was both because of its wide applicability and the ease with which it handled. Today the most commonly used definition is only a small modification of Serre's notion. The next step was to define the singular homology and the singular cohomology spectral sequences of a fibration, and proceed to show that at least for simply connected base spaces these provide much useful information.

Some of the information concerning the homology of fibrations combined to give the first extended Hurewicz theorem. Namely if X is a simply connected space, k is a field, and $q > 0$ then there is a homomorphism $\pi_q(X)\otimes k \longrightarrow H_q(X; k)$ induced by the ordinary Hurewicz homomorphism, such that if $H_q(X; k) = 0$ for $q < n$, then $\pi_q(X)\otimes k = 0$ for $q < n$, and $\pi_n(X)\otimes k \longrightarrow H_n(X; k)$ is an isomorphism.

Some of the other significant results of Serre's thesis are listed below:

1) If X is a simply connected space with integral homology of finite type, then its homotopy is of finite type.

2) The group $\pi_q(S^{2n+1})$ is a finite group for $q > 2n+1$, and for p a prime $\pi_q(S^{2n+1}) \otimes \mathbb{Z}/p\mathbb{Z}$ is zero for $2n+1 < q < 2n+2p-2$ and isomorphic with $\mathbb{Z}/p\mathbb{Z}$ for $q = 2n+2p-2$, and $n > 0$.

3) The group $\pi_q(S^{2n})$ is a finite group for $q > 2n$, $q \neq 4n-1$; and for $n > 0$, $\pi_{4n-1}(S^{2n})$ is the direct sum of a finite group and an infinite cyclic group.

The results concerning the homotopy groups of spheres contain among other things the seeds of rational homotopy theory. We will return to this question after indicating some other results of the next couple of years.

Using the techniques of Serre, the original extended Hurewicz theorem was improved so that a field could be replaced by a principal ideal domain, and so that space could be replaced by pair of simply connected spaces [27]. At the same time Serre placed his original extended Hurewicz theorem in his $\mathcal{C}$-theory framework [38]. The new theorem also applied to the relative case.

A class $\mathcal{C}$ of abelian groups is called a "Serre class"[1] if it is nonempty, and

i) if $\pi' \longrightarrow \pi \longrightarrow \pi''$ is a short exact sequence of abelian groups, then π is in $\mathcal{C}$ if and only if both π' and π'' are in $\mathcal{C}$,

ii) if π is in $\mathcal{C}$ then $H_n(\pi, q)$ is in $\mathcal{C}$ for $n > 0$, $q > 0$.

It is called a "strong Serre class"[1] if for π in $\mathcal{C}$ and A an abelian group $H_n(\pi, q; A)$ is in $\mathcal{C}$ for $n > 0$, $q > 0$.

[1]These conditions are easily seen to be equivalent to the more elegant ones of Serre of class $\mathcal{C}$ satisfying I, II_A, and III, or in the latter case I, II_B, and III [38].

Suppose $\mathcal{C}$ is a Serre class, and $f : A' \longrightarrow A''$ a homomorphism of abelian groups. The homomorphism f is $\mathcal{C}$-injective if its kernel is in $\mathcal{C}$, $\mathcal{C}$-surjective if its cokernel is in $\mathcal{C}$, and a $\mathcal{C}$-isomorphism if both of these conditions are satisfied.

The $\mathcal{C}$-theory Hurewicz theorem now says that if X is a simply connected space, and $\pi_q(X) \in \mathcal{C}$ for $0 < q < n$, then the Hurewicz morphism $\pi_q(X) \longrightarrow H_q(X)$ is a $\mathcal{C}$-isomorphism for $0 < q \leq n$. The relative version makes a similar assertion for a pair of simply connected spaces in the case where the Serre class $\mathcal{C}$ is strong.

Examples of Serre classes are the class of finitely generated abelian groups, the class of torsion groups; and for a prime p the class of p-torsion groups; and the class of torsion groups with torsion prime to p. All of these classes save the first are strong Serre classes.

One application was to show that there is a map $S^{2n+1} \times \Omega S^{4n+3} \longrightarrow \Omega S^{2n+2}$ inducing a homotopy isomorphism modulo 2 torsion groups. This is the first example of a localized splitting theorem.

After noting that the fact that simply connected spaces having integral homology of finite type have integral homotopy of finite type is now a corollary of the $\mathcal{C}$-theory Hurewicz theorem; let us look again at spheres. An easy calculation shows that the double suspension $\Sigma^2 : S^{2n-1} \longrightarrow \Omega^2(S^{2n+1})$ induces a rational homology isomorphism for $n > 0$. Thus a calculation of the rational homotopy of odd spheres follows inductively, and the preceding paragraph takes care of even spheres.

Another basic contribution of Serre in this period was his discovery of the correspondence between cohomology operations and the cohomology of Eilenberg-MacLane spaces [39]. He also computed the cohomology mod 2 of these spaces in case the homotopy group in question is finitely generated. These computations led him to prove also that if X is a simply connected space with homology of finite type, and nontrivial mod 2 homology seeming to be that of a finite complex, then $\pi_q(X) \otimes \mathbb{Z}/2\mathbb{Z}$ is different from zero for infinitely many q's. He conjectured that $\pi_q(X)$ had nontrivial 2 torsion for infinitely many q. Recently this conjecture has been verified [22], and an appropriate extension of his result proved.

The work mentioned above led to both a considerable increase of knowledge concerning $K(\pi, n)$'s, and a much greater appreciation of their significance. Another striking result of this period was the calculation of their homology by H. Cartan [6].

Other striking developments of this period on which we will not dwell include the development of Steenrod operations and algebras, the calculation by Adem of the relations on the squaring operations, and independent calculation of the relations on odd p-th powers by Adem and Cartan [2], [6].

The computations of Cartan mentioned above were not very usable with other than field coefficients. However in the case of a prime field $\mathbb{Z}/p\mathbb{Z}$, they were very usable indeed, and enabled him to show that the mod p Steenrod algebra, and the algebra of stable mod p cohomology operations coincide. Moreover they were basic in his derivation of the relations on Steenrod powers.

By this time a little more was known concerning homotopy groups of spheres [27], [38]. These facts included that for both the first and second time the homotopy groups of an odd sphere have a nontrivial p-primary component it is cyclic of order p. Further for $n > 0$, the groups $\pi_q(\Omega^2 S^{2n+1}, S^{2n-1})$ are torsion prime to p for $q < 2pn-2$.

§2. Hopf invariants-James-Hilton-Milnor.

Various partially defined Hopf invariants had existed before 1955, but it was only in that year that the work of James [20] on reduced products, and the product structure of Hilton [15] on the loop spaces of bouquets of simply connected spheres led to homomorphisms of homotopy groups defined in all degrees and called Hopf invariants. Later work by Milnor [45] on the loop spaces of bouquets of suspensions of connected spaces gave a similar decomposition and gave rise to more generally defined Hopf invariants.

If X is a connected well pointed space, then James' reduced product applied to X is often denoted by $J(X)$. As a set it is the free monoid generated by X with the relation that the base point is the unit 1. Letting $J_r(X)$ denote the subset of words of length $\leq r$, there is a natural surjection from the r-fold product of X with itself to $J_r(X)$ which is given the quotient topology. One topologizes $J(X)$ as the colimit of the subspaces $J_r(X)$. An agreeable property of this construction is that if X is a CW complex, then there is a natural CW structure on $J(X)$. Note $J_0(X) = \{1\}$, $J_1(X) = X$, and if one denotes by $\Lambda_r X$ the r-fold smash product $X \wedge \cdots \wedge \cdots \wedge X$ of X with itself there are cofibration sequences

$$J_{r-1}(X) \longrightarrow J_r(X) \longrightarrow \Lambda_r X$$

for $r > 0$. These sequences induce short split exact sequences in homology. This leads easily to the fact that if R is a ring such that $H_*(X; R)$ is a flat R-module, then $H_*(J(X); R)$ is the tensor algebra generated by the reduced homology of X.

If $\Omega\Sigma X$ denotes the associative loop space [29] of the suspension of X, and one chooses $X \longrightarrow \Omega\Sigma X$ appropriately, a multiplicative map $J(X) \longrightarrow \Omega\Sigma X$ is determined which induces a homology isomorphism first for coefficients as above, and then for all coefficients. Noting that if X has the homotopy type of a CW complex, then so does $\Omega(\Sigma X)$ [25], it follows at once that $J(X)$ has the homotopy type of $\Omega(\Sigma X)$ while in the general case the map above is a weak homotopy equivalence.

Now one may obtain explicit maps $J(X) \longrightarrow J(\Lambda_r X)$, or one may observe that $\Sigma(Y \times Z)$ has canonically the homotopy type of $\Sigma Y \vee \Sigma Z \vee \Sigma Y \wedge Z$; use this fact to obtain a functorial homotopy equivalence between $\Sigma\Omega(\Sigma X)$ and $\bigvee_{r=1}^{\infty}(\Sigma\Lambda_r X)$, and hence maps $\Omega(\Sigma X) \longrightarrow \Omega(\Sigma\Lambda_r X)$ at least in the case where X has the homotopy type of a CW complex. Sometimes James-Hopf invariants are viewed as maps $h_r : J(X) \longrightarrow J(\Lambda_r X)$, or $h_r : \Omega(\Sigma X) \longrightarrow \Omega(\Sigma\Lambda_r X)$ which are functorial up to homotopy, or sometimes they are thought of as the induced homomorphisms of homotopy groups.

The Hilton-Milnor decomposition may be viewed as saying that if X and Y are connected well pointed spaces, then $\Omega(\Sigma(X \vee Y))$ has the weak homotopy type of a weak product of spaces of the forms $\Omega\Sigma(\Lambda_r X \wedge \Lambda_s Y)$, and that upon making some choices these weak equivalences are explicit and functorial up to homotopy. For X and Y CW complexes they are homotopy equivalences.

Thus for a connected CW complex X the comultiplication $\Sigma X \longrightarrow \Sigma(X \vee X)$ induces $\Omega(\Sigma X) \longrightarrow \Omega(\Sigma(X \vee X))$ which followed by projection on a factor gives a map $\Omega\Sigma X \longrightarrow \Omega\Sigma\Lambda_r X$ which is a Hilton-Milnor Hopf invariant. The corresponding homomorphism in homotopy groups is well defined regardless of whether X is CW or not.

Note that if X is a sphere S^n, $\Sigma X = S^{n+1}$, $\Sigma\Lambda_r X = S^{rn+1}$, and a Hopf invariant $h_r : \Omega(S^{n+1}) \longrightarrow \Omega(S^{rn+1})$ induces a homology isomorphism in dimension rn.

Now recall that $H_*(\Omega S^{n+1})$ as a Hopf algebra is $T(i_n)$ the tensor algebra with one primitive generator i_n in degree n. Thus by duality one has for n even $H^*(\Omega S^{n+1})$ is isomorphic with $\Gamma(x_n)$ the divided polynomial algebra with one generator x_n of degree n, while for n odd it is isomorphic with $E(x_n) \otimes \Gamma(y_{2n})$ the tensor product of an exterior algebra and a divided polynomial algebra.

Now for n odd $h_2 : \Omega(S^{n+1}) \longrightarrow \Omega(S^{2n+1})$, and h_2^* takes $H^*(\Omega S^{2n+1})$ isomorphically with the factor $\Gamma(y_{2n})$ in $H^*(\Omega(S^{n+1})$. Thus making h_2 into a fibration, one has with an abuse of notation a fibration sequence

$$S^n \longrightarrow \Omega(S^{n+1}) \longrightarrow \Omega(S^{2n+1})$$

of James whose homotopy sequence is the so-called EHP sequence.

A slightly more complicated calculation shows that for n even $H^*(\Omega S^{n+1}; \mathbb{Z}_{(2)})$ is a free $H^*(\Omega S^{2n+1}; \mathbb{Z}_{(2)})$ module on 2-generators one of degree 0 and one of degree n where $\mathbb{Z}_{(2)}$ denotes the integers localized at 2. Classically this would say that the homotopy fibre had homology looking like that of S^n modulo odd torsion. Localizing in a current way [16], and again abusing notation one has a localized at 2 fibration sequence.

$$S^n \longrightarrow \Omega(S^{n+1}) \longrightarrow \Omega(S^{2n+1})$$

whose homotopy sequence gives the 2-local results of James.

§3. Toda's secondary Hopf invariants and related fibrations.

Localized away from 2 there is a splitting of ΩS^{2n+2} into $S^{2n+1} \times \Omega S^{4n+3}$ well defined up to homotopy. Therefore, working at odd primes one usually studies only the homotopy of odd spheres. Now for convenience for the rest of this paragraph we assume that all spaces are localized at an odd prime p [16]. In this light we look at Toda's work on the double suspension $\Sigma^2 : S^{2n-1} \longrightarrow \Omega^2 S^{2n+1}$ for $n > 0$. It will be factored into a composite, and the terms of the decomposition studied.

First $h_p : \Omega S^{2n+1} \longrightarrow \Omega(S^{2pn+1})$ induces a homology isomorphism in degree $2pn$. A small calculation shows that $H^*(\Omega S^{2n+1}; \mathbb{Z}_{(p)})$ is a free $H^*(\Omega S^{2pn+1}; \mathbb{Z}_{(p)})$ module where $\mathbb{Z}_{(p)}$ denotes the integers localized at p. Now if h_p is made into a fibration one has a map of $J_{p-1}(S^{2n})$ into the fibre, and that $H^*(\Omega S^{2n+1}; \mathbb{Z}_{(p)}) \longrightarrow H^*(J_{p-1}(S^{2n}); \mathbb{Z}_{(p)})$ sends a basis over $H^*(\Omega S^{2pn+1}; \mathbb{Z}_{(p)})$ into a basis over $\mathbb{Z}_{(p)}$. Thus up to homotopy one has Toda's first fibration [41]

$$J_{p-1}S^{2n} \longrightarrow \Omega(S^{2n+1}) \xrightarrow{h_p} \Omega(S^{2pn+1}).$$

In order to construct Toda's second fibration, a "secondary" Hopf invariant $h' : \Omega J_{p-1}(S^{2n}) \longrightarrow \Omega(S^{2pn-1})$ is needed. One starts by computing $H^*(\Omega(J_{p-1}S^{2n}); \mathbb{Z}_{(p)})$, and observing that it is isomorphic with $E(x_{2n-1}) \otimes \Gamma(y_{2pn-2})$ as a Hopf algebra where both x_{2n-1} and y_{2pn-2} are primitive. Dualization gives that $H_*(\Omega J_{p-1}S^{2n}; \mathbb{Z}_{(p)})$ is isomorphic with $E(v_{2n-1}) \otimes S(w_{2pn-2})$ where the second factor is a polynomial algebra with one generator w_{2pn-2} in degree $2pn-2$.

Let K be the subcomplex $S^{2n-1} \cup_\gamma E^{2pn-2}$ of a homology decomposition of $\Omega J_{p-1}S^{2n}$. Indeed one may assume that γ is obtained by taking a generator of $\pi_{2pn-2}(\Omega^2 S^{2n+1}, S^{2n-1})$ which is cyclic of order p, and then choosing a representative in $\pi_{2pn-3}(S^{2n-1})$ of its image under the homotopy connecting morphism. This of course implies that $\Sigma^2\gamma$, the double suspension of γ, is null homotopic. Thus it gives that $\Sigma^2 K$ is equivalent with $S^{2n+1} \vee S^{2pn}$, and that the double suspension of r-fold products of copies of K is equivalent with a bouquet of spheres. The natural maps of r-fold products of K into $\Omega J_{p-1}S^{2n}$ induce surjective morphisms of homology in degrees $\leq r(2pn-2)$. Hence $\Sigma^2\Omega(J_{p-1}S^{2n})$ is homotopy equivalent with a bouquet of spheres.

Now since one may consider $J_{p-1}(S^{2n})$ to be $J_{p-2}(S^{2n}) \cup_\alpha E^{2(p-1)n}$ for an appropriate α, there is a map $J_{p-1}S^{2n} \longrightarrow J_{p-1}(S^{2n}) \vee S^{2(p-1)n}$ sometimes called a coaction map. Now $\Omega(J_{p-1}(S^{2n}) \vee S^{2(p-1)n})$ is homotopy equivalent with

$\Omega(J_{p-1}(S^{2n})) \times \Omega(S^{2(p-1)n}) \times \Omega(\Sigma(\Omega J_{p-1}(S^{2n}) \wedge \Omega(S^{2(p-1)n})))$. Further $\Sigma(\Omega J_{p-1}(S^{2n}) \wedge \Omega(S^{2(p-1)n}))$ is equivalent with $\Omega J_{p-1}(S^{2n}) \wedge \Sigma\Omega(S^{2(p-1)n})$. This shows that the latter space is equivalent with a bouquet of spheres with bottom sphere S^{2pn-1}. Thus there is a map

$h' : \Omega(J_{p-1}(S^{2n})) \longrightarrow \Omega(S^{2pn-1})$ obtained by looping the coaction map and then projecting on a factor. This secondary Hopf invariant induces a homology isomorphism in degree $2pn-2$. There results up to homotopy the second Toda fibration [41]

$$S^{2n-1} \longrightarrow \Omega(J_{p-1}S^{2n}) \longrightarrow \Omega(S^{2pn-1}),$$

and the desired factorization of the double suspension. If $m : S^{2n} \longrightarrow S^{2n}$ is a map of degree m, there is a homotopy commutative diagram

$$\begin{array}{ccc} \Omega(J_{p-1}(S^{2n})) & \xrightarrow{h'} & \Omega S^{2pn-1} \\ \downarrow \Omega(J_{p-1}(m)) & & \downarrow \Omega(m^p) \\ \Omega(J_{p-1}(S^{2n})) & \xrightarrow{h'} & \Omega(S^{2pn-1}) \end{array}$$

where $m^p : S^{2pn-1} \longrightarrow S^{2pn-1}$ is a map of degree m^p.

§4. Applications of the work of James and Toda to exponents for homotopy groups of spheres.

If X is a space with abelian fundamental group, then for p a prime an exponent for the homotopy of X at p is a least positive integer n such that every element of the p-primary component of the torsion subgroup of $\pi_q(X)$ has order dividing n for all q. The first results of this type for odd spheres are due to James [21] for the prime 2, and shortly later to Toda [41] for odd primes. Their result is that $\pi_*(S^{2n+1})$ has exponent at the prime p dividing p^{2n}.

Recall that if Y is a suspension and X is an H-space then the pointed homotopy classes of maps from Y to X, $[Y, X]_*$, form an abelian group with operation coming from the co H-space structure of Y or the H-space structure of X. Any choice of a k-th power map $k : X \longrightarrow X$ induces a k-th power on $[Y, X]_*$ independent of the choice of k or the H-space structure on X.

Localized at an odd prime p, S^{2n+1} is an H-space which is not the case for $p = 2$. Thus in a modern context using localization the result of Toda is easier to obtain than that of James. The modern proofs are in either case more global than the original ones again due to the fact that one works in a localized context rather than using $\mathcal{C}$-theory as originally.

Indeed suppose that $h_p : \Omega S^{2n+1} \longrightarrow \Omega S^{2pn+1}$ is a p-th Hopf invariant localized at p. Then $p \circ \Omega h_p$ is null homotopic, as we will show for odd primes. For $p = 2$, the result due to M.G. Barratt, will appear in course notes of F.R. Cohen from the 1985 Seattle homotopy conference.

Now let p be an odd prime, as remarked earlier $\Omega(p^p) \circ h_p \simeq h_p \circ \Omega(p)$, where p^p and p denote power maps of the appropriate localized sphere, i.e., maps of the stated degree. These induce power maps in the corresponding loop space i.e., $p^p \circ h_p \simeq h_p \circ p$. Since $\Omega(h_p)$ is an H-map, $\Omega(h_p) \circ p \simeq p \circ \Omega(h_p)$, and hence $p^p \circ \Omega(h_p) \simeq p \circ \Omega(h_p)$, and therefore $0 \simeq (p^p - p) \circ \Omega h_p$. However, $p^p - p = p(p^{p-1} - 1)$, $p^{p-1} - 1$ is a unit localized at p, and hence $p^{p-1} - 1$ is a homotopy equivalence. Thus $p \circ \Omega h_p \simeq 0$. Suppose now $\alpha : \Sigma Y \longrightarrow \Omega^2 S^{2n+1}$. Since $p \circ \Omega h_p \simeq 0$, here exists $\gamma : \Sigma Y \longrightarrow \Omega(J_{p-1} S^{2n})$ such that $(\Omega i)\gamma$ is homotopic with $p \circ \alpha$ where

$$J_{p-1}(S^{2n}) \xrightarrow{i} \Omega(S^{2n+1}) \longrightarrow \Omega(S^{2pn+1})$$

is the first Toda fibration. Since $p^p \circ h'$ is homotopic with $h' \circ \Omega J_{p-1}(p)$ where h' is the secondary Hopf invariant, there is a $\beta : \Sigma Y \longrightarrow S^{2n-1}$ such that $j \circ \beta$ is homotopic with $(p^{p-1} - 1)^{-1} \circ (p^p - \Omega J_{p-1}(p)) \circ \gamma$, the second Toda fibration being

$$S^{2n-1} \xrightarrow{j} \Omega(J_{p-1}(S^{2n})) \xrightarrow{h'} \Omega(S^{2pn-1}).$$

Here since ΣY is a suspension $p^p h'\gamma = h'p^p\gamma$. Now

$\Sigma^2 \circ \beta \simeq \Omega(i) \circ j \circ \beta \simeq (p^{p-1}-1)^{-1}\Omega(i)(p^p - \Omega J_{p-1}(p))\gamma \simeq (p^{p-1}-1)^{-1}(p^p-p)\Omega(i)\gamma$ $\simeq p\Omega(i)\gamma \simeq p^2\alpha$. If one lets $\alpha : \Sigma\Omega^3(S^{2n+1}) \longrightarrow \Omega^2 S^{2n+1}$ be the adjoint of the identity, one has $p^2 : \Omega^3 S^{2n+1} \longrightarrow \Omega^3 S^{2n+1}$ factors through $\Omega\Sigma^2 : \Omega S^{2n-1} \longrightarrow \Omega^3 S^{2n+1}$.

Suppose $p = 2$, and $\alpha : \Sigma Y \longrightarrow \Omega^2 S^{2n+1}$. Since $2 \circ \Omega h_2$ is null homotopic there is a $\gamma : \Sigma Y \longrightarrow \Omega S^{2n}$ such that $\Omega\Sigma \circ \gamma$ is homotopic with $2 \circ \alpha$. Since $h_2 \circ \Omega(-1) \simeq h_2$, there is a $\beta : \Sigma Y \longrightarrow S^{2n-1}$ such that $\Sigma \circ \beta \simeq (1-\Omega(-1) \circ \gamma$. Now $\Sigma^2 \circ \beta \simeq (\Omega\Sigma \circ \Sigma) \circ \beta \simeq \Omega\Sigma(1-\Omega(-1))\gamma = \Omega\Sigma \circ 2 \circ \gamma \simeq 4 \circ \alpha$. Proceeding as before $4 : \Omega^3 S^{2n+1} \longrightarrow \Omega^3 S^{2n+1}$ factors through $\Omega\Sigma^2 \Omega S^{2n-1} \longrightarrow \Omega S^{2n+1}$.

Chapter 2. More recent unstable homotopy theory

During the 60's and early 70's some calculations of homotopy groups were made, a considerable amount of homological algebra for use in homotopy theory was developed, and some conjectures were formulated. In particular calculations let Barratt to conjecture in about 1960 that S^3 had exponent p at an odd prime p. Note that this cannot be the case for 2 since $\pi_6(S^3)$ is isomorphic with $\mathbb{Z}/12\mathbb{Z}$.

§1. Exponents for homotopy groups of spheres.

In 1977, Selick [36] proved Barratt's conjecture. This started a period where many problems concerning exponents for homotopy groups were resolved. In order to explain Selick's result more precisely we recall some work of the early 50's.

For a pathwise connected space X, there is a fibration $\mathcal{O}_n : X\langle n\rangle \longrightarrow X$ called the n-connective cover of X such that $\pi_q(X\langle n\rangle) = 0$ for $q \leq n$, and $\pi_q(\mathcal{O}_n)$ is an isomorphism for $q > n$. This was shown independently by G.W. Whitehead, and by Cartan and Serre. In particular there is $\mathcal{O}_3 : S^3\langle 3\rangle \longrightarrow S^3$ with fibre a space $K(\mathbb{Z}, 2)$, and the fibration may be thought of as a principal fibration. An easy calculation [27], [38] shows that $H_q(S^3\langle 3\rangle) = 0$ for q odd, and $H_{2q}(S^3\langle 3\rangle)$ is isomorphic with $\mathbb{Z}/q\mathbb{Z}$ for $q > 0$.

Now let $S^n\{p^r\}$ denote the homotopy theoretic fibre of a degree p^r map $p^r : S^n \longrightarrow S^n$ for $n > 1$. Looking at the fibration sequence $\Omega S^n \longrightarrow S^n\{p^r\} \longrightarrow S^n$ obtained by shifting the original fibration, an easy calculation using the Wang sequence shows that for $q > 0$

$H_q(S^n\{p^r\}) = 0$ if $(n-1)$ does not divide q, and is isomorphic with $\mathbb{Z}/p^r\mathbb{Z}$ for $q \equiv 0 \bmod (n-1)$.

Suppose BS^3 is the classifying space for the group S^3. One may obtain a commutative diagram

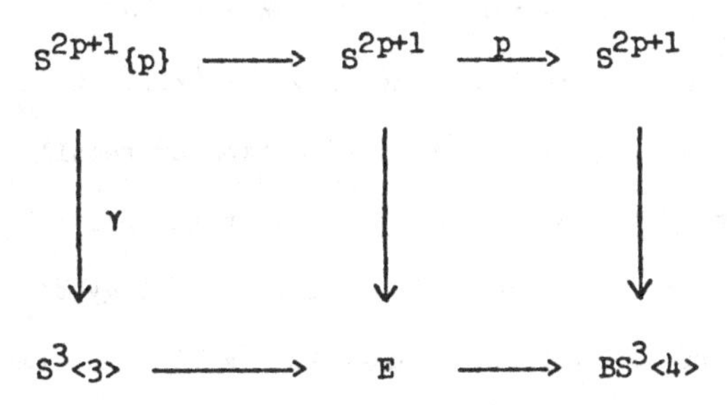

with the rows fibration sequences, E contractible, and γ inducing an injection of homotopy groups in degree $2p$ [27]. Thus localizing at p, γ induces a homology isomorphism in degree strictly less than $2p^2$. In this situation Selick proved that $\Omega^2(\gamma)$ has a cross section for p odd. This proves Barratt's conjecture since $\pi_*(S^{2p+1}\{p\})$ is annihilated by p. Indeed for $n > 3$, the identity map of $\Omega^2(S^n\{p^r\})$ has homotopy order p^r [8]. Observe that Selick's result shows more than that p is an exponent for the homotopy of S^3 at p, for it shows that the p-primary component of the homotopy splits off the mod p homotopy of S^{2p+1} [30].

Shortly after Selick's work the authors of this article together with F.R. Cohen [8] proved a more general conjecture of Barratt. Namely if p is an odd prime and $n > 0$, then p^n is the exponent of $\pi_*(S^{2n+1})$ at p. Since earlier work of Brayton Gray had shown that there were elements of order p^n in these groups, it was only necessary to show that p^n annihilates the p-primary component of $\pi_*(S^{2n+1})$.

As a matter of fact we had not started to attack the problem we solved, but a seemingly quite different one. Indeed suppose $m > 1$, and $P^m(p^r) = S^{m-1} \bigcup_{p^r} E^m$. If $m \geq 3$ and p is an odd prime then $P^m(p^r)$ is a suspension with identity map of homotopy order p^r [28], [30]. We were looking for elements of order p^{r+1} in this space with the idea in mind that they might be detected by a nontrivial $(r+1)$-st Bockstein β_{r+1}.

We begin to go further in this direction by recalling some simple observations about the mod p homology Bockstein spectral sequence of $\Omega P^m(p^r)$. For $s \leq r$ the reduced $\mathbb{Z}/p^s\mathbb{Z}$ homology of $P^{m-1}(p^r)$ is free on two generators u and v of degrees $(m-2)$ and $(m-1)$ respectively. Note here the ground ring is $\mathbb{Z}/p^s\mathbb{Z}$. Applying the Bott-Samelson theorem to the suspension $\Sigma : P^{m-1}(p^r) \longrightarrow \Omega(P^m(p^r))$ shows that $H_*(\Omega(P^m(p^r)); \mathbb{Z}/p^s\mathbb{Z})$ is isomorphic with the primitively generated tensor algebra $T(u, v)$. Hence in the mod p Bockstein spectral sequence $E^1 = E^r$, and $\beta_r v = u$ assuming appropriate choices of u and v. Thus E^r is acyclic and E^{r+1} looks like the homology of a point. Since no nonzero β_{r+1} is possible one proceeds to modify the situation by pinching the bottom cell of $P^m(p^r)$ to a point and making the resulting pinch map part of a fibration sequence

$$F^m\{p^r\} \longrightarrow P^m(p^r) \longrightarrow S^m$$

which defines $F^m\{p^r\}$. The space $\Omega F^m\{p^r\}$ now becomes the one whose homology Bockstein spectral sequence is under study.

One checks easily that $H_*(\Omega P^m(p^r);\ \mathbb{Z}/p\mathbb{Z}) \longrightarrow H_*(\Omega S^m;\ \mathbb{Z}/p\mathbb{Z})$ is a surjective morphism of Hopf algebras. If follows that its Hopf algebra kernel is $H_*(\Omega F^m\{p^r\};\ \mathbb{Z}/p\mathbb{Z})$, and since a sub Hopf algebra of a primitively generated tensor Hopf algebra is again such the next problem is to find information concerning the generators.

For a graded vector space V of finite type, let $\chi(V)$ be its Euler-Poincaré series. For a space X whose mod p homology is of finite type the p-Euler-Poincaré series is $\chi_p(X) = \chi(H_*(X;\ \mathbb{Z}/p\mathbb{Z}))$. Now $\chi_p(\Omega F^{2n+1}\{p^r\})\chi_p(\Omega S^{2n+1}) = \chi_p(\Omega P^{2n+1}(p^r))$, $\chi_p(\Omega S^{2n+1}) = 1/1-t^{2n}$, and $\chi_p(\Omega P^{2n+1}(p^r)) = 1/1-t^{2n-1}-t^{2n}$. Thus $\chi_p(\Omega F^{2n+1}\{p^r\})$ is $1/1-f(t)$ where $f(t) = t^{2n-1}/1-t^{2n}$. In other words $H_*(\Omega F^{2n+1}\{p^r\};\ \mathbb{Z}/p\mathbb{Z})$ is the tensor Hopf algebra on a primitive vector space of finite type having Euler-Poincaré series $f(t)$.

In the homology Bockstein spectral sequence of $\Omega(F^{2n+1}\{p^r\})$ one has $E^1 = E^r$. Looping the fibration sequence of the pinch map, one has a fibration sequence which induces an "algebraic bundle" of E^r terms. Having already remarked that the total space is acyclic a known calculation shows that E^{r+1} is isomorphic which $H_*(\Omega^2 S^{2n+1};\ \mathbb{Z}/p\mathbb{Z})$, i.e. with $E(a_0) \otimes \bigotimes_{j\geq 1} (E(a_j) \otimes S(b_j))$ where a_j and b_j have degrees $2p^j n-1$ and $2p^j n-2$ respectively. An argument in a completely different context enables one to prove $\beta_{r+1} a_j = b_j$ for $j \geq 1$ assuming that appropriate choices have been made. Thus $E^{r+2} = H_*(S^{2n-1};\ \mathbb{Z}/p\mathbb{Z}) = E(a_0) = E^\infty$.

Now fortunately the geometric situation parallels closely the algebraic information provided by the Bockstein spectral sequence. Indeed localized at p, $\Omega F^{2n+1}\{p^r\}$ splits into a product

$$S^{2n-1} \times \prod_{j\geq 1} S^{2p^j n-1}\{p^{r+1}\} \times \Omega(X)$$

where X is a CW complex of finite type which is an infinite bouquet of spaces of the form $P^n(p^r)$. In terms indicated in the preceding paragraph, $H_*(S^{2n-1};\ \mathbb{Z}/p\mathbb{Z}) = E(a_0)$, $H_*(S^{2p^j n-1}\{p^{r+1}\};\ \mathbb{Z}/p\mathbb{Z})$ is isomorphic with $E(a_j) \otimes S(b_j)$, and $H_*(\Omega(X);\ \mathbb{Z}/p\mathbb{Z})$ is something which disappears in the passage from E^r to E^{r+1}.

The problem of finding p^{r+1} torsion in the homotopy of $P^{2n+1}(p^r)$ is solved observing that $\pi_{2p^j n-2}(S^{2p^j n-1}\{p^{r+1}\}) \approx \mathbb{Z}/p^{r+1}\mathbb{Z}$ and this injects into the homotopy of $\Omega P^{2n+1}(p^r)$ in that degree.

Results on the odd primary exponents of the homotopy groups of spheres are obtained by first setting $r = 1$, and then shifting the fibration sequence of the pinch map to obtain a fibration seqeunce

$$\Omega^2 S^{2n+1} \xrightarrow{\partial} \Omega F^{2n+1}\{p\} \longrightarrow \Omega P^{2n+1}(p).$$

Now let $\pi : \Omega^2 S^{2n+1} \longrightarrow S^{2n-1}$ be the composite $\Omega^2 S^{2n+1} \xrightarrow{\partial} \Omega F^{2n+1}\{p\} \xrightarrow{\alpha} S^{2n-1}$ where α is the projection defined using the product decomposition discussed earlier. Assuming that appropriate care has been taken in selecting the splitting one can show that the composite $\Omega^2 S^{2n+1} \xrightarrow{\pi} S^{2n-1} \xrightarrow{\Sigma^2} \Omega^2 S^{2n+1}$ is the p-th power map. Thus $p\pi_* S^{2n+1}$ is contained in the image of the double suspension. Induction starting at S^1 given that p^n is an exponent of the p-primary component of $\pi_*(S^{2n+1})$.

Iterating the factorization $p \simeq \Sigma^2 \circ \pi$ and looping appropriately gives the factorization

$$\Omega^{2n}S^{2n+1} \longrightarrow S^1 \longrightarrow \Omega^{2n}S^{2n+1}$$

of the p-localized p^n power map. In this context passing to universal covering spaces etc. gives the geometric form of the exponent theorem. This says that localized at p the space $\Omega^{2n}(S^{2n+1}\langle 2m+1\rangle)$ has a null homotopic p^n-th power map, i.e. the homotopy order of the identity is p^n.

Given the preceding result one would like to know how many times $S^{2n+1}\langle 2n+1\rangle$ localized at p must be looped in order that the identity have homotopy order p^n. After looping $2n-1$ times the order of the appropriate map is unknown. However Selick and the second author [33] have shown that after looping $2n-2$ times the order is infinite. In the next paragraph we present a short proof due to Mc Gibbon and the second author. The context assumes all spaces localized at p.

For X and Y pointed spaces $map_*(X, Y)$ denotes the pointed space of pointed maps from X to Y. Let G be a finite group and $B(G)$ a classifying space. Applying the functor $map_*(BG, —)$ to the fibration sequence

$$K(\mathbb{Z}, 2n) \longrightarrow S^{2n+1}\langle 2n+1\rangle \longrightarrow S^{2n+1}$$

gives a fibration sequence. Applying Miller's theorem [24],

$\mathrm{map}_*(BG, S^{2n+1})$ is weakly contractible. Hence

$\mathrm{map}_*(BG, K(\mathbb{Z}, 2n)) \longrightarrow \mathrm{map}_*(BG, S^{2n+1}\langle 2n+1\rangle)$ is a weak equivalence,

and upon looping $2n-2$ times

$\mathrm{map}_*(BG, K(\mathbb{Z}, 2)) \longrightarrow \mathrm{map}_*(BG, \Omega^{2n-2}(S^{2n+1}\langle 2n+1\rangle)$ is a weak equivalence,

and moreover it is multiplicative. Now $\pi_0\, \mathrm{map}_*(BG, K(\mathbb{Z}, 2))$ is isomorphic with $H^2(BG; \mathbb{Z})$. If $G = \mathbb{Z}/p^r\mathbb{Z}$ this group is isomorphic with $\mathbb{Z}/p^r\mathbb{Z}$. Since r is arbitrary, the identity map of $\Omega^{2n-2}(S^{2n+1}\langle 2n+1\rangle)$ has infinite order.

§2. Spaces of finite characteristic.

The suspension ΣX of a pointed space X has characteristic m if the identity $1_{\Sigma X}$ has homotopy order m. If the map $1_{\Sigma X}$ does not have finite homotopy order, then ΣX has infinite characteristic. The study of suspension of finite characteristic was initiated by M.G. Barratt [4] at the same time that he introduced the terminology. The first observation concerning spaces ΣX of characteristic m is that m annihilates their reduced homology since $m\,1_{\Sigma X}$ is the composite

$$\Sigma X \xrightarrow{\mathcal{O}_m} \Sigma X \vee \cdots m \cdots \vee \Sigma X \xrightarrow{\alpha_m} \Sigma X$$

where $\mathcal{O}_m$ is an iterated comultiplication and α_m a folding map.

Barratt obtained bounds on the order of elements in the homotopy of a space ΣX of characteristic m by a bootstrap argument which proceeds somewhat as follows below. Let $m = \alpha_m \circ \mathcal{O}_m$, and apply the Hilton-Milnor theorem to obtain a weak product decomposition of $\Omega(\Sigma X \vee \cdots m \cdots \vee \Sigma X)$. In this decomposition there are m terms $\Omega\Sigma X$ and other terms of the form

$\Omega\Sigma(\wedge_r X)$ where $\wedge_r X$ is the r-fold smash product of X with itself and $r \geq 2$. If X is n-1 connected and ΣX has characteristic n, $n \geq 2$, then $\wedge_r X$ is rn-1 connected and $\Sigma\wedge_r X = \Sigma X \wedge \wedge_{r-1} X$ has characteristic dividing m. Since m is null homotopic $\Omega(m)$ is null homotopic, and using the decomposition of the Hilton-Milnor theorem one obtains that multiplication by m in $\pi_q(\Omega\Sigma X)$ is a homomorphism which factors through a finite product of groups of the form $\pi_q(\Omega\Sigma\wedge_r X)$ $r \geq 2$. This implies that m annihilates $\pi_q \Omega\Sigma X$ for $q \leq 2n-1$, which implies m annihilates $\pi_q(\Omega\Sigma\wedge_r X)$ for $q \leq 2rn-1$, and inductively that m^s annihilates $\pi_q(\Sigma X)$ for $q \leq 2^s n$. Thus each group $\pi_q(\Sigma X)$ has an exponent, but the bound for such exponents seems to depend on q. Barratt conjecture is that this is not the case if X is itself a suspension.

More precisely Barratt's exponent conjecture is that if for a prime p the double suspension $\Sigma^2 X$ has finite characteristic p^r, then p^{r+1} is the exponent of the homotopy of $\Sigma^2 X$. More geometrically the homotopy order of the identity of $\Omega^2\Sigma^2 X$ should be p^{r+1}. Very few examples are known where Barratt's conjecture is satisfied. Most general results are of a sliding nature as the results described earlier.

The simplest examples of spaces of finite characteristic are the two cell complexes $P^n(m) = S^{n-1} \cup_m E^n = \Sigma P^{n-1}(m)$, $n \geq 3$. If m is odd or divisible by 4, then $P^n(m)$ has characteristic m. If m is even but not divisible by 4, then $P^n(m)$ has characteristic 2m.

Currently it is not known whether the homotopy of the space $P^n(2^r)$ $r > 0$ has an exponent at 2 or not. On the other hand for an odd prime p, the spaces $P^n(p^r)$ have homotopy groups having exponent p^{r+1} [9], [32] and thus satisfy Barratt's conjecture. Some of the ideas in the proof of this fact will now be indicated.

Recall that homotopy groups with coefficients $\mathbb{Z}/m\mathbb{Z}$ may be defined in the manner of F.P. Peterson. Thus $\pi_n(X; \mathbb{Z}/m\mathbb{Z})$ is the group of pointed homotopy classes of maps $[P^n(m), X]$ [30], and the cofibration sequence $S^{n-1} \longrightarrow P^n(m) \longrightarrow S^n$ induces a universal coefficient exact sequence

$$0 \longrightarrow \pi_n(X) \otimes \mathbb{Z}/m\mathbb{Z} \longrightarrow \pi_n(X; \mathbb{Z}/m\mathbb{Z}) \longrightarrow$$

$$\longrightarrow \mathrm{Tor}(\pi_{n-1}(X), \mathbb{Z}/m\mathbb{Z}) \longrightarrow 0,$$

found by Peterson in his thesis. This sequence is split for m odd, but not generally if $m = 2$.

For $m = p^r$ with p an odd prime, Samelson products are defined in the mod m homotopy of group-like spaces. Indeed let G be a group-like space, and $f : P^i(m) \longrightarrow G$, $g : P^j(m) \longrightarrow G$ represent mod m homotopy classes. The commutator map $(x, y) \longrightarrow f(x)g(x)f(x)^{-1}g(x)^{-1}$ induces a map $P^i(m) \wedge P^j(m) \longrightarrow G$. However $P^i(m) \wedge P^j(m) \simeq P^{i+j}(m) \vee P^{i+j-1}(m)$ canonically, and restriction to $P^{i+j}(m)$ defines the Samelson product $[f, g] : P^{i+j}(m) \longrightarrow G$ which induces the product in the mod m homotopy [30].

If $p > 3$ the Samelson product makes $\pi_*(G; \mathbb{Z}/m\mathbb{Z})$ into a graded Lie algebra. This is not the case for $p = 3$ since the Jacobi identity fails. However the mod m Hurewicz map

$\varphi : \pi_*(G; \mathbb{Z}/m\mathbb{Z}) \longrightarrow H_*(G; \mathbb{Z}/m\mathbb{Z})$ carries Samelson products into commutators, i.e. $\varphi[\alpha, \beta] = [\varphi(\alpha), \varphi(\beta)] = \varphi(\alpha)\varphi(\beta) - (-1)^{ij} \varphi(\beta)\varphi(\alpha)$ for α of degree i and β of degree j. These Samelson products can be used to obtain a weak product decomposition of $\Omega P^{2n+2}(m)$.

Start by recalling that $H_*\Omega P^{2n+2}(m);\ \mathbb{Z}/m\mathbb{Z})$ is the primitively generated tensor algebra $T(u, v)$ where u is of degree $2n$ and v is of degree $2n+1$, and assuming there generators appropriately chosen u is the image of v under an appropriate Bockstein. Further there exist unique mod m homotopy classes μ and ω such that $\varphi(\mu) = u$ and $\varphi(\omega) = v$.

Making $T(u, v)$ commutative gives $A = S(u) \otimes E(v)$ which is precisely $H_*(S^{2n+1}\{m\};\ \mathbb{Z}/m\mathbb{Z})$. Let T' be the Hopf algegra kernel of the natural morphism $T \longrightarrow A$. Now T' is a primitively generated tensor algebra, $\chi(T) = \chi(T')\chi(A)$. Hence $X(T') = 1/1-g(t)$ where $g(t) = (t^{4n+1}+t^{4n+2})/1-t^{2n}$. Now one verifies that the elements $(ad^k(u)[u, v], ad^k(u)[v, v])_{k\geq 0}$ freely generate T', and that the same Bockstein which takes v into u takes $ad^k(u)[v, v]$ into $2\ ad^k(u)[u, v]$ for $k \geq 0$.

The Samelson products $ad^k(\mu)[\omega, \omega]$ combine to give a map $X \longrightarrow \Omega P^{2n+2}(m)$ where $X = \bigvee_{k\geq 0} P^{4n+2kn+2}(m)$. Using the James construction one sees easily that this gives rise to a map $\Omega(\Sigma X) \xrightarrow{\alpha} \Omega(P^{2n+2}(p^r))$ which is multiplicative and in $\mathbb{Z}/m\mathbb{Z}$ homology $H(\alpha;\ \mathbb{Z}/m\mathbb{Z})$ is just the natural inclusion $T' \longrightarrow T$.

Now one can obtain a commutative diagram

$$\begin{array}{ccc} S^{2n+1} & \longrightarrow & E \\ \downarrow{\scriptstyle m} & & \downarrow{\scriptstyle \pi} \\ S^{2n+1} & \xrightarrow{\gamma} & P^{2n+2}(m) \end{array}$$

where γ is the inclusion of the bottom cell and π is the standard acyclic fibration. The resulting map of fibres $\beta : S^{2n+1}\{m\} \longrightarrow \Omega P^{2n+2}(m)$ has the property, that if one composes $T \longrightarrow A$ with $H(\beta; \mathbb{Z}/m\mathbb{Z})$ one obtains an isomorphism $H(S^{2n+1}\{m\}; \mathbb{Z}/m\mathbb{Z}) \xrightarrow{\approx} A$. Now multiplying one has a map $\Omega(\Sigma X) \times S^{2n+1}\{m\} \longrightarrow \Omega P^{2n+2}(m)$ which induces a homology isomorphism, and is thus a homotopy equivalence.

Using the preceding result combined with the Hilton-Milnor theorem one can show that $\Omega P^{2n+2}(m)$ has the homotopy type of a weak product with factors of the type $\Omega(P^{2i+1}(m))$ and $S^{2j+1}\{m\}$ with only a finite number having connectivity less than any given integer. The process of doing this involves an infinite iteration. This result shows that in order to determine whether or not the homotopy order of the identity of $\Omega^2 P^{2n+2}(m)$ is finite or not it suffices to know that the homotopy order of the identity of spaces $\Omega^2 P^{2i+1}(m)$ is bounded independent of i.

A more complicated argument having somewhat the flavor of the preceding gives a weak product decomposition of the spaces $\Omega(P^{2i+1}(m))$ with factors of two types. One type is spaces of the form $S^{2j+1}\{m\}$, and the other is more complicated. These results combine to show that the homotopy order of the identity of $\Omega^2 P^q(m)$ is finite for $q \geq 3$, and that it divides pm^2 [9]. The fact that this order is pm which shows that the spaces $P^q(m)$ verify Barratt's conjecture has been shown by the second author [32].

§3. A general exponent conjecture and a partial solution.

Let K be a simply connected finite complex and p a prime. The first author has conjectured that $\pi_*(K)$ has an exponent at p if and only if the rational homotopy groups of K are bounded, i.e.

$\pi_q(k) \otimes \mathbb{Q}$ is zero for q sufficiently large. If this were the case, then the property of having an exponent at p would depend only on the rational homotopy type of the finite complex K, and would thus be independent of p.

Mc Gibbon and Wilkerson have shown that half of the conjecture is true for almost all primes [23]. More precisely they show that if K is a simply connected finite complex such that $\pi_q(K) \otimes \mathbb{Q}$ is zero for q large, then there exists a finite set of primes $S(K)$ such that after inverting the primes in $S(K)$ the space $\Omega(K)$ has the homotopy type of a finite product of spaces each of which is either an odd dimensional sphere or the loop space of an odd dimensional sphere. Thus for any prime p not in $S(K)$ $\pi_*(K)$ has an exponent at p.

The proof of Mc Gibbon and Wilkerson proceeds by induction on the rank of $\pi_*(K) \otimes \mathbb{Q}$. If the rank is zero, then after inverting finitely many primes K become contractible. If the rank is one, then after inverting finitely many primes K has the homotopy type of an odd dimensional sphere, and ΩK has the homotopy type of the loops in that sphere.

The inductive step splits into two cases depending on the parity of the smallest nonzero degree in $\pi_*(K) \otimes \mathbb{Q}$.

If the lowest degree is $2n+1$, choose a map $K \xrightarrow{\alpha} K(\mathbb{Z}, 2n+1)$ which is surjective in rational homotopy in degree $2n+1$. Let $S^{2n+1} \xrightarrow{i} K(\mathbb{Z}, 2n+1)$ represent a generator of the homotopy in degree $2n+1$. Since K is finite after inverting finitely many primes there exists $K \xrightarrow{\alpha'} S^{2n+1}$ such that $i\alpha' \simeq \alpha$. Making α' into a fibration, one has a fibration sequence

$$E \longrightarrow K \longrightarrow S^{2n+1}$$

such that after inverting perhaps finitely many additional primes the looped sequence splits, i.e. $\Omega(K) \simeq \Omega(E) \times \Omega S^{2n+1}$.

A generalization of a result of Halperin [14], [12] says that after inverting finitely many primes S_1 $H_*(E; S_1^{-1}\mathbb{Z})$ is free of finite rank over $S_1^{-1}\mathbb{Z}$. Clearly E has rational rank one less than that of K, it may be thought of as the localization of an appropriate finite complex, and the inductive step follows in this case.

If the lowest degree is 2n, choose a map $K \xrightarrow{\alpha} K(\mathbb{Z}, 2) \times \cdots \times K(\mathbb{Z}, 2n)$ inducing an epimorphism in rational homotopy in degree 2n. There is a map $BU(n) \xrightarrow{\beta} K(\mathbb{Z}, 2) \times \cdots \times K(\mathbb{Z}, 2n)$ which induces a rational equivalence. Hence inverting finitely many primes there is a map $K \xrightarrow{\alpha'} BU(n)$ such that $\beta\alpha' \simeq \alpha$. Now let $S^{2n-1} \longrightarrow E \longrightarrow K$ be the fibration sequence obtained by as a pull back from the appropriate localization of $S^{2n+1} \longrightarrow BU(n-1) \longrightarrow BU(n)$ using α'. Observe that E has rational rank one less than that of K, and that after inverting possibly finitely many additional primes one has that $\Omega(K) \simeq \Omega(E) \times S^{2m-1}$ and E may be thought of as the localization of a finite complex. Thus the inductive step follows in this case.

Given the result of Mc Gibbon and Wilkerson one can show that given a finite complex K which is simply connected and has finite rational homotopy rank, there is a finite set of primes S Such that $H_*(\Omega K; S^{-1}\mathbb{Z}$ is a free $S^{-1}\mathbb{Z}$ module which is the universal enveloping algebra of the Lie Algebra of its primitive elements. The situation is much more complicated when K does not have finite homotopy rank. Indeed there is an example of Anick such that $H_*(\Omega K)$ has torsion of all orders.

Bibliography

1. Adams, J.F. On the nonexistence of elements of Hopf invariant one, Ann. of Math. (2) 72 (1960), 20-104.

2. Adem, José The relations on Steenrod powers of cohomology classes, Algebraic Geometry and Topology. A symposium in Honor of S. Lefschetz, Princeton, Princeton University Press 1957, 191-238.

3. Anick, D.J. A loop space whose homology has torsion of all orders, to appear.

4. Barratt, M.G. Spaces of finite characteristic, Quart. J. Math., Oxford, Ser. (2) 11 (1960), 124-136.

5. Cartan, Henri Une theorie axiomatiques des cariés de Steenrod, C. R. Acad. Sci., Paris 230 (1950), 425-427.

6. Cartan, Henri Seminaire Henri Cartan 1954/1955, Algèbres d'Eilenberg-MacLane et homotopie, Exposés 1-11, 13-16, New York, Benjamin 1967.

7 Cohen, F.R., Moore, J.C., and Neisendorfer, J.A. Torsion in homotopy groups, Ann. of Math. 109 (1979), 121-168.

8. Cohen, F.R., Moore, J.C., and Neisendorfer, J.A. The double suspension and exponents of the homotopy groups of spheres, Ann. of Math. 110 (1979), 549-565.

9. Cohen, F.R., Moore, J.C., and Neisendorfer, J.A. Exponents in homotopy theory, to appear.

10. Eilenberg, S. Singular homology theory, Ann. of Math. (2) 45 1944, 407-447.

11. Eilenberg, S., and MacLane, S. Relations between homology and homotopy groups of spaces, Ann. of Math. (2) 46 (1945), 480-509.

12. Felix, Y., and Lemaire, J.-M. On the mapping theorem for Lusternik-Schnirelman category, Topology 24 (1985), 41-43.

13. Freudenthal, Hans Uber die Klassen der Spharen abbildungen, Compositio Math. 5 (1937), 299-314.

14. Halperin, S. Finiteness in the mimimal models of Sullivan, Trans. Amer. Math. Soc. 230 (1977), 173-199.

15. Hilton, P.J. On the homotopy groups of the union of spheres, J. London Math. Soc. 30 (1955), 154-172.

16. Hilton, P.J., Mislin, G., and Roitberg, J. Localization of nilpotent groups and spaces, Amsterdam, North Holland, Mathematics Studies 15, 1975.

17. Hopf, Heinz Abbildungen geschlossener Mannigfaltigkeiten auf Kugeln in n Dimensionen, Jahresbericht der Deutschen Mathematika-Vereinigung 34 (1925).

18. Hopf, Heinz Uber die Abbildungen von Spharen auf Spharen niedrigen Dimension, Fund. Math. 25 (1935), 427-440.

19. Hurewicz, Witold Beitrage zur Topologie der Deformationen, I-IV, Nederl. Akad. Wetersch. Proc Ser. A 38 (1935) 112-119, 521-528; 39 (1936), 117-126, 215-224.

20. James, Ioan Reduced product spaces, Ann. of Math. (2) 62 (1955), 170-197.

21. James, Ioan On the suspension sequence, Ann. of Math. 65 (1957), 74-107.

22. Mc Gibbon, C.A., and Neisendorfer, J.A. On the homotopy groups of a finite dimensional space, Comment. Math. Helv. 59 (1984), 253-257.

23. Mc Gibbon, C.A., and Wilkerson, C. The exponent problem for large primes, to appear.

24. Miller, H.R. The Sullivan conjecture on maps from classifying spaces. Ann. of Math. 120 (1984), 39-87.

25. Milnor, J.W. On spaces having the homotopy type of a CW-complex, Trans. Amer. Math. Soc. 90 (1959), 272-280.

26. Milnor, J.W., and Moore, J.C. On the structure of Hopf algebras, Ann. of Math. (2) 81 (1965), 211-264.

27. Moore, J.C. Some applications of homology theory to homotopy problems, Ann. of Math. 58 (1953), 325-350.

28. Moore, J.C. On the homotopy groups of spaces with a single nonvanishing homology group, Ann. of Math. 59 (1954), 549-557.

29. Moore, J.C. Homotopie des complexes monoidaux, I, Seminaire Henri Cartan 1954/1955, Exposé 18, New York, Benjamin 1967.

30. Neisendorfer, J.A. Primary homotopy theory, Memoirs Amer. Math. Soc. Vol. 25 (1980), No. 232.

31. Neisendorfer, J.A. 3-primary exponents, Math. Proc. Camb. Phil. Soc. 90 (1981), 63-83.

32. Neisendorfer, J.A. The exponent of a Moore space, to appear.

33. Neisendorfer, J.A., and Selick, P.S. Some examples of spaces with or without exponents, Can. Math. Soc. Proc. Vol. 2, Part 1 (1982).

34. Samelson, Hans A connection between the Whitehead and the Pontryagin product, Amer. J. Math. 75 (1953), 744-752.

35. Selick, P.S. Odd primary torsion in $\pi_r(S^3)$, Topology 17 (1978), 407-412.

36. Selick, P.S. 2-primary exponents for the homotopy groups of spheres, Topology 23 (1984), 97-99.

37. Serre, J.-P. Homologie singulière des espaces fibrés. Applications, Ann. of Math. (2) 54 (1951), 425-505.

38. Serre, J.-P. Groupes d'homotopie et classes de groupes abeliens, Ann. of Math. (2) 58 (1953), 258-294.

39. Serre, J.-P. Cohomologie modulo 2 des complexes d'Eilenberg-MacLane, Comment. Math. Helv. 27 (1953), 198-232.

40. Steenrod, N.E. Products of cocycles and extensions of mappings, Ann. of Math. (2) 48 (1947), 290-320.

41. Toda, H. On the double suspension E^2, J. Inst. Polytech. Osaka City Univ. Ser. A 7 (1956), 103-145.

42. Whitehead, G.W. On the Freudenthal theorem, Ann. of Math. 57 (1953), 209-228.

43. Whitehead, G.W. The $(n+2)^d$ homotopy group of the n-sphere, Ann. of Math. (2) 52 (1950), 245-247.

44. Whitehead, G.W. On mappings into grouplike spaces, Comment. Math. Helv. 28 (1954), 320-328.

45. Whitehead, G.W. Elements of homotopy theory, New York, Heidelberg, Berlin; Springer-Verlag 1978.

46. Whitehead, J.H.C. On adding relation to homotopy groups, Ann. of Math. (2) 42 (1941), 409-428.

47. Whitehead, J.H.C. Combinatorial homotopy, I, Bull. Amer. Math. Soc. 55 (1949), 213-245.

John C. Moore
Department of Mathematics
Princeton University
Princeton, New Jersey 08544

Joseph A. Neisendorfer
Department of Mathematics
University of Rochester
Rochester, New York 14620

Recent Progress in Stable Splittings

Stewart Priddy[1]

This paper is an exposition of stable splittings for the classifying space of a finite group. These results, obtained over the past half decade, provide p-local splittings $BG = \vee X_i$ for various p-groups G and are distinguished from earlier research by the pervasive use of modular representation theory. Ideally the X_i are indecomposable, thus displaying the homotopy type of BG in simplest terms.

The number of papers in this area has now reached a point that it seems worthwhile to give an overview of some of the basic results, applications, and philosophy. Our goal is twofold. First, we wish to provide a compendium of related splittings which can be used by future workers seeking patterns and general structure theorems. Second, for the neophyte in modular representation theory, we wish to provide enough elementary notions to explain our modest applications of this deep and powerful tool for homotopy theorists. As such, this paper is an outgrowth of lectures presented at the 1985 Durham Symposium on Homotopy Theory. The author wishes to thank the organizers and especially John Jones for his encouragement in writing this paper.

Splittings continue to play an important role in homotopy theory. We assume the reader is aware of some of the classical occurrences, e.g. [Mn], [A], [Sn]. A striking recent example is the use of a finite stable complex free over "half of A_n" in the solution of the Nilpotence Conjecture by Devinatz, Hopkins, and Smith. This complex was split from a product of stunted projective spaces.

[1] Partially supported by N.S.F. Grant DMS-8641403

The Manchester school [C1],[C2],[W] has also made extensive use of splittings via modular representation theory. Their results are applied to study the transfer.

Most of the results described here are from joint work with my collaborators Mark Feshbach, Nick Kuhn, and Steve Mitchell. It is a pleasure to take this opportunity to thank them.

In Section 1 we describe splittings for three related 2-groups: the dihedral, quaternionic, and semi-dihedral groups. These examples are elementary in the sense that only transfer techniques are used. They are complete in the sense that every summand is indecomposable. Section 2 deals with idempotents, the elements of modular representation theory, and its use to produce splittings. Several examples are given to illustrate principal indecomposable modules and blocks for groups related to homotopy theory. The Steinberg idempotent is described along with its summand of $B(F_p{}^k)_+$ in terms of symmetric products of the sphere spectrum. Section 3 deals with application of the Steinberg summand to the solution of the Whitehead Conjecture. Certain 2-groups of symplectic type are discussed in Section 4. These groups form a natural generalization of the dihedral and quaternion groups. We describe the related Steinberg summands of their classifying spaces which are acted on by Chevalley groups.

Finally we note that the material in Section 4 with Feshbach and the splitting of the semi-dihedral group, due to John Martino, are new; details will appear elsewhere. Also the author is preparing a book in which the general subject of stable splittings will be covered more extensively.

<u>Conventions:</u> We work in the category of p-local spectra. All cohomology groups are taken with F_p coefficients.

Section One: Splittings derived using the transfer

Before discussing idempotents in group rings, we consider a more primitive class of splittings for p-groups G derived from the stable transfer $t_H: BG \longrightarrow BH$ for subgroups H<G. These splittings have the form $BG = BG' \vee (\vee X_i)$ where G<G' is a p-Sylow subgroup and $BG \longrightarrow BG'$ is induced by the inclusion. The other requisite maps $BG \longrightarrow X_i$ are obtained as composites

$$BG \xrightarrow{t_H} BH \xrightarrow{g} X_i$$

where H and g are dictated by the situation. The sum of these maps is then shown to be an equivalence in mod-p cohomology. One modular feature of this procedure should be noted: since G is a p-group the index [G:H] is divisible by p, thus one can not use the standard techniques applicable if [G:H] is prime to p. Usually this means making an actual computation of the map in cohomology. In favorable circumstances, e.g. if $H^*(BG)$ is detected on subgroups, one can use the double coset formula to great advantage. The resulting computations are usually quite manageable.

In the following examples p = 2.

Dihedral groups D_{2^n}: We recall

$$D_{2^n} = \langle\, s,t:\ s^{2^{n-1}} = 1 = t^2\ ,\ tst = s^{-1} \,\rangle \quad , \qquad |D_{2^n}| = 2^n \quad , \ n \geq 2$$

For n=2 we are reduced to the special case $D_4 \approx Z/2 \times Z/2$. For n>2, there are exactly two conjugacy classes of maximal elementary abelian subgroups represented by $K = \langle r = s^{2^{n-2}}, t\rangle$ and $T = \langle r, st\rangle$ respectively. These subgroups detect $H^*BD_{2^n}$ in the sense that every nonzero element maps nontrivally to either H^*BK or H^*BT.

A complete splitting of BD_{2^n} is controlled by the subgroups K,T and the projective special linear group.

Theorem 1.1 (Mitchell-Priddy [MP1]) There is a 2-local stable equivalence

$$BD_{2^n} = BPSL_2(F_q) \vee 2(\Sigma^{-2}Sp^4S/Sp^1S) \vee 2(BZ/2)$$

where q is an odd prime power satisfying $n=\nu_2((q^2-1)/2)$.
For n=2, q can be chosen to be 3 in which case $PSL_2(F_3) \approx A_4$.

With this condition on q, D_{2^n} is a 2-Sylow subgroup of $PSL_2(F_q)$.
To describe this splitting we recall that the k-fold symmetric product is defined by $Sp^kX=X^k/\Sigma_k$. An inclusion $Sp^kX \longrightarrow Sp^{k+1}X$ is given by insertion of a basepoint. (By the Dold-Thom theorem colim Sp^kS = HZ for the sphere spectrum S.) There is a pairing $Sp^iS \wedge Sp^jS \xrightarrow{\mu} Sp^{ij}S$ defined unstably by

$$Sp^iS^n \wedge Sp^jS^m \xrightarrow{\mu} Sp^{ij}S^{nm}$$

$$(x_1,\ldots,x_i)\wedge(y_1,\ldots,y_j) \longrightarrow (x_1\wedge y_1,\ldots,x_i\wedge y_j).$$

Moreover $Sp^2S/Sp^1S = \Sigma B(Z/2)$. Let

$$f:BZ/2\times BZ/2 \longrightarrow \Sigma^{-1}Sp^2S/Sp^1S \wedge \Sigma^{-1}Sp^2S/Sp^1S \xrightarrow{\Sigma^{-2}\mu} \Sigma^{-2}Sp^4S/Sp^2S$$

be the resulting map. Then the equivalence in Theorem 1.1 is given by

$$Bi \vee f\circ t_K \vee f\circ t_T \vee \pi_1\circ t_K \vee \pi_1\circ t_T$$

where $i:D_{2^n}<PSL_2(F_q)$ and π_1: $B(Z/2\times Z/2) \longrightarrow BZ/2$ is projection on the first factor. The proof consists of an explicit calculation in cohomology along the lines outlined above. The fact that $H^*BD_{2^n}$ is detected by elementary abelian subgroups simplifies the computation considerably.

The next example is rather different in that the cohomology is not detected on any collection of proper subgroups.

Quaternion groups $Q_{2^{n+1}}$: The (generalized) quaternion groups are defined by

$$Q_{2^{n+1}} = \langle s,t : s^{2^{n-1}} = t^2, s^{2^n}=1, tst^{-1}=s^{-1} \rangle, \quad |Q_{2^{n+1}}| = 2^{n+1}, \quad n\geq 2$$

There is a central extension

$$Z/2 \longrightarrow Q_{2^{n+1}} \longrightarrow D_{2^n}$$

given by factoring out the center $= \langle t^2 \rangle$. Then $i:Q_{2^{n+1}}<SL_2(F_q)$ is a 2-Sylow subgroup lying over $i:D_{2^n}<PSL_2(F_q)$.

Theorem 1.2 (Mitchell-Priddy [MP1]) There is a 2-local stable equivalence

$$BQ_{2^{n+1}} = BSL_2(F_q) \vee 2\Sigma^{-1}(BS^3/BN)$$

where $n = \nu_2((q^2-1)/2)$.

In this splitting N is the normalizer of the maximal torus $S^1<S^3$. Since $S^3/N = RP^2$ has Euler characteristic 1, Becker-Gottlieb transfer for the fibration $RP^2 \longrightarrow BN \longrightarrow BS^3$ yields a splitting

$$BN = BS^3 \vee \Sigma^{-1}(BS^3/BN)$$

The equivalence of Theorem 1.2 is given by

$$Bi \vee \pi'\circ B\psi_{n+1} \vee \pi'\circ B(\psi_3\circ\alpha)\circ t_3$$

where $\psi_{n+1}:Q_{2^{n+1}} \longrightarrow N$ is the usual representation of $Q_{2^{n+1}}$ on S^3, $t_3:BQ_{2^{n+1}} \longrightarrow BQ_{2^3}$ is the transfer, $\pi':BN \longrightarrow \Sigma^{-1}(BS^3)/BN)$ is projection and α is the involution of Q_{2^3} defined by $\alpha(s) = t$.

One similarity between Theorems 1.1 and 1.2 should perhaps be noted. In the former the summands related to the subgroups K and T could equally well have been described in terms of one summand and the involution η of D_{2^n} defined by $\eta(s) = s^{-1}$, $\eta(t) = st$ which sends K to T. In the latter the involution α distinguishes the corresponding summands. (α is defined only for

$Q_{2}3$ which explains the appearance of t_3.)

The next splitting, due to my student John Martino, exhibits interesting features of both the dihedral and quaternionic cases.

Semi-dihedral groups $SD_{2}n$: The semi-dihedral groups are defined by

$$SD_{2}n = \langle s,t:\ s^{2^{n-1}} = 1 = t^2,\ tst^{-1} = s^{2^{n-2}-1} \rangle \ , \quad |SD_{2}n|=2^n \ , \quad n\geq 4$$

Then $H^{*}BSD_{2}n$ is known to be detected by two of its subgroups namely $D_{2}n-1$ and $Q_{2}n-1$ [EP]. This fact has strong implications for the stable homotopy type of $BSD_{2}n$.

Theorem 1.3 (Martino [Mt]) There is a 2-local stable equivalence

$$BSD_{2}n = BSL_3(F_q) \vee \Sigma^{-2}Sp^4S/Sp^2S \vee BZ/2 \vee \Sigma^{-1}(BS^3/BN)$$

where q is an odd prime power with $q\equiv 3 \bmod 4$ and $n = \nu_2((q^3-1)(q^2-1))$.

Under the stated conditions on q, $i:SD_{2}n<SL_3(F_q)$ is a 2-Sylow subgroup. The required equivalence is then given by

$$Bi \vee f\circ t_K \vee \pi''\circ t_Q$$

where $t_K:BSD_{2}n \longrightarrow BK$, $t_Q:BSD_{2}n \longrightarrow BQ_{2}n-1$ are transfer maps and $\pi'':BQ_{2}n-1 \longrightarrow \Sigma^{-1}(BS^3/BN)$ is projection given by Theorem 1.2.

The appearance of SL_3 and not PSL_3 may be somewhat surprising; however, since $q\equiv 3 \bmod 4$, $(q-1,3) = 1$ or 3. Thus the Serre spectral sequence for the extension

$$Z/(q-1,3) \longrightarrow SL_3(F_q) \longrightarrow PSL_3(F_q)$$

yields $H^{*}BSL_3(F_q) \approx H^{*}BPSL_3(F_q)$.

The explicit cohomology rings for all of the groups in this section can be found in the cited references. Martino shows that for $q \equiv 3 \bmod 4$

$$H^*BSL_3(F_q) \approx F_2[\alpha, \beta, \gamma]/(\alpha^2\beta + \gamma^2)$$

where $|\alpha| = 3$, $|\beta| = 4$, $|\gamma| = 5$.

Section Two: Splittings derived from modular representation theory

In this section we review some basic notions from modular representation theory and show how they can be used to obtain various splittings. Use of the Steinberg module provides an important example.

The procedure is quite general; to split BG we seek idempotents in the ring [BG,BG] of stable homotopy classes of self maps of BG localized at p. Since [BG,BG] is a Z_p module (Z_p = p-adic integers) the homomorphism AutG $\longrightarrow$ [BG,BG] given by $f \longrightarrow [Bf]$ extends to a ring homomorphism

$$\sigma: Z_p \mathrm{AutG} \longrightarrow [BG,BG]$$

and so it suffices to find idempotents in the group ring Z_pAutG. We note for later use that σ factors thru Z_pOutG $\longrightarrow$ [BG,BG]. (In many cases it is also useful to replace AutG by EndG , see [HK]; however for our purposes AutG is sufficent.) Now, to reduce the problem still further we consider the reduction epimorphism

$$Z_p \mathrm{AutG} \longrightarrow\!\!\!\!\rightarrow F_p \mathrm{AutG}$$

Then a standard result in algebra [CR] states that an orthogonal idempotent decomposition $e_1,\dots,e_m$ of F_pAutG (i.e. $1 = \Sigma e_i$, $e_i^2 = e_i$, $e_i e_j = 0$ if $i \neq j$, $e_i \neq 0$) can be lifted to an orthogonal idempotent decomposition of Z_pAutG. By abuse of notation we continue to denote by e_i the images of these elements under σ.Let

$$e_i BG = \mathrm{colim}\ (BG \xrightarrow{e_i} BG \xrightarrow{e_i} BG \longrightarrow \dots)$$

the infinite mapping telescope. Then $H_* e_i BG = e_i H_* BG$ and we have an equivalence

$$BG = e_1BG \vee \dots \vee e_mBG$$

If p does not divide $|AutG|$ then the problem of determining orthogonal idempotents for F_pAutG is essentially equivalent to its classical analog for representation theory in characteristic zero. In the modular setting, where p divides $|AutG|$, this theory is deep and rather formidable. However, as we shall show, even some of its elementary aspects have implications for homotopy theory.

Example 2.1 (Holzager [Hz]) $G = Z/p$, $AutG = F_p{}^* \approx Z/p-1$. This is a classic example of a non-modular splitting. Let ζ be a generator of AutG representing multiplication by an element $a \in F_p{}^*$. Then

$$e_i = \prod_{j \neq i} \frac{\zeta - a^i \cdot 1}{a^i - a^j} \qquad i = 0,1,\dots,p-2$$

is an idempotent decomposition by primitive idempotents. We recall that an idempotent e is called primitive if $e \neq e'+e''$ where e', e'' are orthogonal idempotents. Thus

$$BZ/p = e_0BZ/p \vee \dots \vee e_{p-2}BZ/p$$

A transfer argument shows $e_0BZ/p = B\Sigma_p$ and an explicit calculation shows

$$\begin{aligned} H_*e_iBZ/p &= F_p && \text{if } * \equiv 2i,\ 2i-1 \bmod 2(p-1) \\ &= 0 && \text{otherwise} \end{aligned}$$

It is easy to see the e_iBZ/p are indecomposable.

Example 2.2 Let $H < AutG$ with p not dividing $|H|$. Set $e_1 = (1/|H|)(\Sigma\ h\ )$ for $h \in H$, $e_2 = 1-e_1$. Then $BG = e_1BG \vee e_2BG$. To determine e_1BG, let $G \rtimes H$ be the semi-direct product with H acting on G. Then $e_1BG = B(G \rtimes H)$. To see this we note that the Serre spectral sequence for $BG \longrightarrow B(G\rtimes H) \longrightarrow BH$ collapses to $H^*B(G\rtimes H) = H^*(BG)^H$. On the other hand $H^*e_1BG = H^*(BG)e_1 = H^*(BG)^H$ from the definition of e_1. Thus the composite inclusion

$$e_1BG \longrightarrow BG \longrightarrow B(G\rtimes H)$$

is an equivalence.

For $G = D_4 \approx Z/2\times Z/2$, $p=2$, and $H = Z/3$ acting by the matrix $\begin{pmatrix} 0 & 1 \\ 1 & 1 \end{pmatrix}$, we have $D_4\rtimes Z/3 \approx A_4$. Thus $e_1BD_4 = BA_4$; by Theorem 1.1, $e_2BD_4 = 2(\Sigma^{-2}Sp^4S/Sp^2S \vee BZ/2)$. For $G = Q_8$, $p = 2$, and $H = Z/3$ acting by permuting s,t,st cyclically we have $Q_8\rtimes Z/3 \approx SL_2(F_3)$. Thus from Theorem 1.2 we have $e_1BQ_8 = BSL_2(F_3)$ and $e_2BQ_8 = 2(\Sigma^{-1}BS^3/BN)$.

In fact for D_4 and Q_8, e_2 is not primitive. $Out(D_4) = GL_2(F_2) = Out(Q_8)$ and as we shall see in Ex. 2.3, e_2 is the idempotent for the Steinberg block for $F_2GL_2(F_2)$ and thus decomposes into two equivalent primitive idempotents. For Q_8 these correspond to the two summands of $\Sigma^{-1}BS^3/BN$. For D_4 they correspond to the two summands of $\Sigma^{-2}Sp^4S/Sp^2S \vee BZ/2$. In $F_2End(D_4)$, e_2 futher decomposes into the sum of four primitive idempotents accounting for the four summands of e_2BD_4 [HK].

<u>Principal indecomposable modules [CR]</u>: A module M over a ring R is irreducible if it has no non-trivial proper submodules; M is indecomposable if it is not the sum of two non-trivial modules. The Jacobson radical J is the intersection of all maximal right ideals. If $J = 0$, R is semi-simple, i.e., the direct sum of irreducible modules, and these notions agree: R is the direct sum of matrix rings over division algebras. $R = F_pG$ is semi-simple iff p does not divide $|G|$ In the modular setting, however, one can only write

(1) $$R = P_1 \oplus \dots \oplus P_N$$

where the P_i are indecomposable as right R modules (we choose right actions because left actions on spaces are transformed to right actions under cohomology). Since the P_i decompose the group ring they are called <u>principal indecomposable</u> modules, or PIM's. Of course a PIM is projective.
It is a basic fact that

$$\{ P_i \} \longrightarrow \{ P_i/J \}$$

is a 1-1 correspondence between the isomorphism classes of PIM's and the isomorphism classes of irreducible modules.

From (1) we have

(2) $$1 = e_1 + \dots + e_N$$

where $e_i \in P_i$, giving our desired orthogonal decomposition by primitive idempotents. Since the decomposition (1) is unique (up to possible reordering of the factors) by the Krull-Schmidt Theorem, it follows that (2) is also unique up to conjugation by a unit of R.

It is also useful to consider a courser decomposition where actual uniqueness holds. A <u>block</u> $B \subset R$ is a non-zero two sided ideal such that i) $R = B \oplus B'$ for some two sided ideal B' and ii) B is not the sum of two non-zero two sided ideals If R satisfies the D.C.C. (e.g. $R = F_pG$) then R can be uniquely decomposed

(3) $$R = B_1 \oplus \dots \oplus B_M$$

by blocks B_i. Then

(4) $$1 = f_1 + \dots + f_M$$

where the $f_i \in B_i$ are unique central idempotents. The relationship between (1) and (3) is as follows. Two PIM's P,P' are said to be linked if there is a sequence $P = P_1, P_2, \ldots, P_s = P'$ such that for each i, P_i and P_{i+1} have a common composition factor. Being linked is an equivalence relation and a block is just the sum of all elements in an equivalence class of linked PIM's.

Example 2.3 [MP1]: $G = GL_2(F_2)$, p=2. There are two irreducibles: the trivial module $1 \approx F_2$ and the natural representation $V \approx F_2 \oplus F_2$. Then

$$F_2GL_2(F_2) = P_1 \oplus P_2 \oplus P_3$$

where $P_1 = E(\gamma)$, the exterior algebra on $\gamma = \Sigma g$, $g \in G$ and $P_2 = P_3 = V$. Thus

$$F_2GL_2(F_2) = B_1 \oplus B_2$$

with blocks $B_1 = E(\gamma)$ and $B_2 = Mat_2(F_2) = V \oplus V$. Let

$$\tau = \begin{pmatrix} 0 & 1 \\ 1 & 0 \end{pmatrix} \qquad u = \begin{pmatrix} 1 & 1 \\ 0 & 1 \end{pmatrix} \qquad \sigma = \begin{pmatrix} 0 & 1 \\ 1 & 1 \end{pmatrix}$$

Then a primitive idempotent decomposition is given by

$$1 = e_1 + e_2 + e_3$$

where

$$e_1 = 1 + \sigma + \sigma^2$$
$$e_2 = (1+u)(1+\tau)$$
$$e_3 = (1+\tau)(1+u)$$

Note that $E(\gamma) = \langle e_1, e_1\gamma = \gamma \rangle$ is indecomposable but not irreducible since $\langle \gamma \rangle$ is the trivial module. Pictorially

$$P_1 = \left| \begin{matrix} 1 \\ 1 \end{matrix} \right|$$

Example 2.4: $G = \Sigma_4 \approx V \rtimes GL_2(F_2)$. Again we have the two irreducibles 1, V where Σ_4 acts thru the projection $\Sigma_4 \longrightarrow GL_2(F_2)$. Then

$$F_2\Sigma_4 = P_1 \oplus P_2 \oplus P_3$$

where P_1 contains 4 factors of 1 and 2 factors of V

$P_2 \approx P_3$ " 2 " " 1 " 3 " " V .

Since P_1, P_2, P_3 are linked $F_2\Sigma_4$ is itself the unique block.

The following example is central to the Whitehead Conjecture.

Example 2.5 [MP2]: $G = GL_n(F_p)$. Here the complete structure of the PIM's is far from known. However one irreducible PIM was found by R. Steinberg in 1958 [S]. Let $B_n < GL_n(F_p)$ be the Borel subgroup consisting of all upper triangular matrices; let $U_n < B_n$ be the unipotent subgroup consisting of all matrices with 1's on the diagonal. Then U_n is a p-Sylow subgroup of $GL_n(F_p)$. Define

$$e_n = c_n^{-1}\bar{B}_n \cdot \bar{W}_n$$

where $c_n = [GL_n : U_n]$, $\bar{B}_n = \Sigma b$, $b \in B_n$, $\bar{W}_n = \Sigma(-1)^{\sigma}\sigma$, $\sigma \in \Sigma_n$. Then e_n is the Steinberg idempotent and $St_n = e_n F_p GL_n(F_p)$ is the Steinberg module. As a U_n module, St_n is the regular representation, thus $\dim St_n = p^N$, $N = \binom{n}{2}$. As an irreducible PIM, St_n belongs to a block isomorphic to $Mat_{p^N}(F_p)$. For $n = 2$, e_2 is the idempotent e_2 of Example 2.3.

The Steinberg summand [MP2]: To determine the Steinberg summand $e_k B(F_p{}^k)$, we begin by recalling Nakaoka's computation [Nk]

$$H^* Sp^{p^k} S = F_p < \theta^I : I \text{ admissible}, \ l(I) \leq k, \ \theta^I \in A\beta >$$

where θ^I is an admissible sequence of reduced powers and Bocksteins of length$\leq$k. Let

$$D(k) = \text{cofiber}(Sp^{p^{k-1}}S \xrightarrow{\Delta} Sp^{p^k}S)$$

where Δ is the p-fold diagonal map.

Since Δ has degree p on the bottom cell

$$H^*D(k) = F_p\langle \Theta^I: \text{I admissible, } l(I) \leq k \rangle$$

Definition: $M(k) = \Sigma^{-k}D(k)/D(k-1)$.

By convention, $M(0) = S$. Then

$$H^*M(k) = \Sigma^{-k}F_p\langle \Theta^I: \text{I admissible, } l(I) = k \rangle$$

Theorem 2.6 (Mitchell-Priddy [MP2]) Stably $B(F_p{}^k)_+$ contains $p^{\binom{k}{2}}$ summands each equivalent to $M(k)$. Furthermore $e_k(BF_p{}^k)_+ \simeq M(k)$.

Remark: In Theorem 1.1, $M(2) = \Sigma^{-2}(Sp^4S/Sp^2S) \vee B(Z/2)$.

In fact setting

$$L(k) = \Sigma^{-k}Sp^{p^k}S/Sp^{p^{k-1}}S \tag{5}$$

a transfer argument shows

Corollary 2.7: $M(k) = L(k) \vee L(k-1)$.

Sketch proof of 2.6: Let p = 2, the details for p odd are similar. First we observe $\Sigma^{-1}D(1) = P_{-1}$, where P_{-1} is the Thom spectrum of minus the canonical line bundle over $B(Z/2)$; this is clearly true homologically and it is not difficult to construct a map. Then the projection

$$P_{-1} = \Sigma^{-1}D(1) \longrightarrow M(1) = P_0$$

simply pinches out the -1 cell. Hence

$$\begin{array}{ccc} D(1)^{\wedge k} & \xrightarrow{\wedge} & D(k) \\ \downarrow & & \downarrow \\ (\Sigma M(1))^{\wedge k} & \xrightarrow{\wedge} & \Sigma^k M(k) \end{array}$$

commutes, where $\wedge$ is the obvious smash product pairing and the vertical arrows are pinch maps. Now applying Σ^{-k} we obtain the maps f

$$\begin{array}{ccc} (P_{-1})^{\wedge k} & \xrightarrow{f} & \Sigma^{-k}D(k) \\ \downarrow & & \downarrow \\ B(F_2{}^k)_+ = (P_0)^{\wedge k} & \xrightarrow{f} & M(k) \end{array}$$

Let $X_k{}^{-1} = x_1{}^{-1} \ldots x_k{}^{-1}$. Then Nakaoka's calculation and the Cartan formula yield

$$f^*:\ H^*M(k) \xrightarrow{\approx} F_2\langle\ Sq^I(X_k{}^{-1}):\ I \text{ admissible},\ l(I) = k\ \rangle$$

It remains to show $H^*((BF_p{}^k)_+)e_k = \mathrm{Im}(f^*)$. At this stage we invoke a formula relating Steenrod operations to the Steinberg idempotent

$$[x_1{}^{-1}Sq^I(X_{k-1}{}^{-1})]e_k = Sq^I X_k{}^{-1} \tag{6}$$

if $l(I) \le k-1$. Setting $Sq^I = 1$ gives $X_k{}^{-1}e_k = X_k{}^{-1}$, hence $\mathrm{Im}e_k \subset \mathrm{Im}f^*$. The other inclusion also follows from 6) by induction on k.

<u>Remark:</u> Other applications for 6) have been found, (see [Kh]), although variants of the above proof have been given which avoid its use. The proof of 6) is by induction on k directly from the definitions.

<u>Section Three: Appliction to the Whitehead Conjecture</u>

In this section we will be dealing with spectra and the zeroth spaces of their associated Ω-spectra. To avoid confusion let $\Sigma^\infty X$ denote the suspension spectrum of a space X. Let $QX = \Omega^\infty\Sigma^\infty X$.

We recall from 5) that L(k) is defined by the cofibration

$$\Sigma^{-k}Sp^{p^{k-1}}S \longrightarrow \Sigma^{-k}Sp^{p^k}S \longrightarrow L(k)$$

The natural boundary maps of these cofibrations give rise to a sequence of spectra and maps

$$7) \quad \dots \longrightarrow L(k+1) \xrightarrow{\partial_k} L(k) \longrightarrow \dots \xrightarrow{\partial_1} L(1) \xrightarrow{\partial_0} L(0) \xrightarrow{\epsilon} HZ_{(p)}$$

where L(0) = S and ϵ is inclusion of the bottom cell. The original conjecture of G. Whitehead [M1] asserted that this sequence is exact on π_*; however more is true. It can be shown that $L(1) = \Sigma^\infty B\Sigma_p$, hence by the Kahn-Priddy theorem there is a map s_0 (of spaces)

$$QB\Sigma_p \underset{s_0}{\overset{d_0}{\rightleftarrows}} Q_0S^0$$

such that $d_0s_0 \simeq id$ where $d_0 = \Omega^\infty\partial_0$. This suggests the following strengthened form of the conjecture.

<u>Theorem 3.1</u> (Integral Whitehead Conjecture [K], [KP]) Sequence 7) is exact in the sense that there exists maps s_k

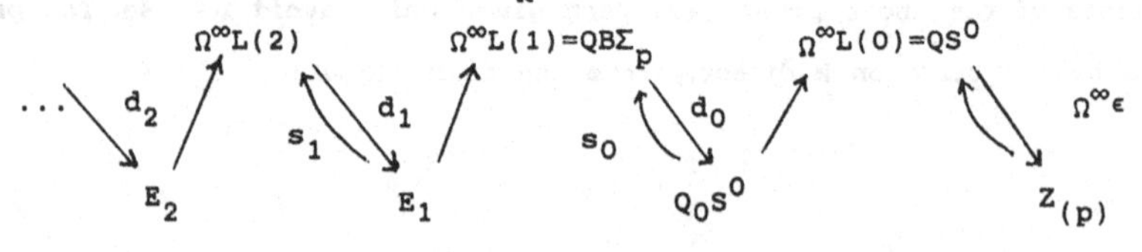

such that $d_k s_k \simeq$ id, $d_k = \Omega^\infty \partial_k$, E_k = homotopy fiber of d_{k-1}.

For p = 2, this was first proved by Kuhn using James-Hopf invariants, see also [K2] for an elaboration of this notion of exactness.

By considering $M(k) = L(k) \vee L(k-1)$ and the connection of M(k) to representation theory via the Steinberg module we are led to

Theorem 3.2 (Mod-p Whitehead Conjecture [KP]) There is an exact sequence of spectra

$$8) \quad \ldots \longrightarrow M(k+1) \xrightarrow{\partial_k} M(k) \ldots \xrightarrow{\partial_1} M(1) \xrightarrow{\partial_0} M(0) \xrightarrow{\epsilon} HZ/p$$

i.e. there exist maps s_k

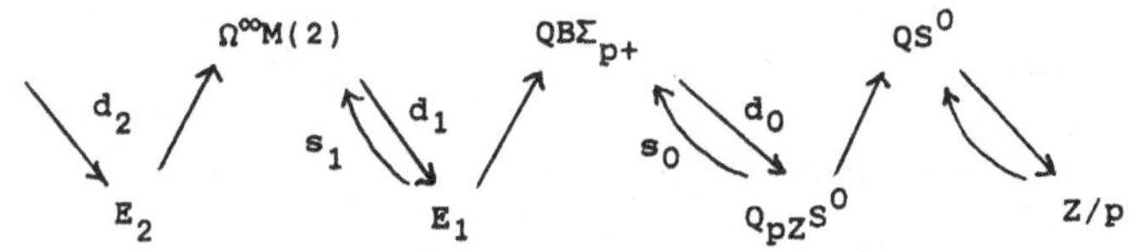

$$\begin{array}{ccccccccc} & & \Omega^\infty M(2) & & QB\Sigma_{p+} & & QS^0 & & \\ \searrow^{d_2} & \nearrow & & {}^{s_1}\nwarrow\searrow^{d_1} & & \nearrow & {}^{s_0}\nwarrow\searrow^{d_0} & \nearrow & \nwarrow\searrow \\ & E_2 & & & E_1 & & Q_{pZ}S^0 & & Z/p \end{array}$$

where $d_k s_k \simeq$ id and d_k, E_k are defined as above. $Q_{pZ}S^0$ consists of maps of degree divisible by p.

It follows from standard methods that Theorem 3.2 implies Theorem 3.1 for any convenient choice of ∂_k. Note that 8) can be considered as a minimal projective resolution of HZ/p in the sense that $\partial_{k*} = 0$ in mod-p homology and that M(k) is a summand of the free (i.e. suspension) spectrum $\Sigma^\infty B\Sigma_p{}^k$.

The projectivity of M(k) can be used to give a remarkably simple construction of ∂_k in terms of the transfer $t:\Sigma^\infty B\Sigma_{p+} \longrightarrow S$

Definition $\partial_k : M(k+1) \xrightarrow{i} \Sigma^\infty BF_{p+}^{k+1} \xrightarrow{t\wedge 1} S\wedge\Sigma^\infty BF_{p+}^{k} \xrightarrow{\pi} M(k)$

Then clearly $\partial_{k*} = 0$. That $\partial_k \partial_{k+1} \simeq 0$ follows from the basic fact that the composite

$$\Sigma^\infty BF_{p+}^2 \xrightarrow{e_2} \Sigma^\infty BF_{p+}^2 \xrightarrow{t} S$$

is null since e_2 belongs to the augmentation ideal of $Z_{(p)}GL_2(F_p)$ [KP].

To construct the splitting maps s_k, we consider $G_k = \Sigma_p \wr \dots \wr \Sigma_p < \Sigma_{p^k}$, the k-fold iterated wreath product. Σ_{p^k} can be viewed as permutations of $F_p{}^k$ and we let $\Delta_k < G_k$ denote the set of translations of $F_p{}^k$. Then $\Delta_k \approx (Z/p)^k$. The inclusion $\Sigma_{p^n} \wr G_{k+1} = \Sigma_{p^n} \wr \Sigma_p \wr G_k \subset \Sigma_{p^{n+1}} \wr G_k$ yields a transfer map

$$t{:}\Sigma^\infty B\Sigma_{p^{n+1}} \wr G_k \longrightarrow \Sigma^\infty B\Sigma_{p^n} \wr G_{k+1}$$

which as $n \to \infty$ defines a map

$$t_k{:}\Sigma^\infty B\Sigma_\infty \wr G_k \longrightarrow \Sigma^\infty B\Sigma_\infty \wr G_{k+1}$$

Looping this map down we define a map tr_k by

$$\begin{array}{ccc}
Q_0BG_{k+} & \xrightarrow{tr_k} & Q_0BG_{k+1+} \\
\downarrow i & & \uparrow \pi \\
QQ_0BG_{k+} & & QQ_0BG_{k+1+} \\
\downarrow \cong & & \uparrow \cong \\
\Omega^\infty\Sigma^\infty B\Sigma_\infty \wr G_k & \xrightarrow{\Omega^\infty t_k} & \Omega^\infty\Sigma^\infty B\Sigma_\infty \wr G_{k+1}
\end{array}$$

where i, π are the standard inclusion and retraction maps. Finally we define s_k by

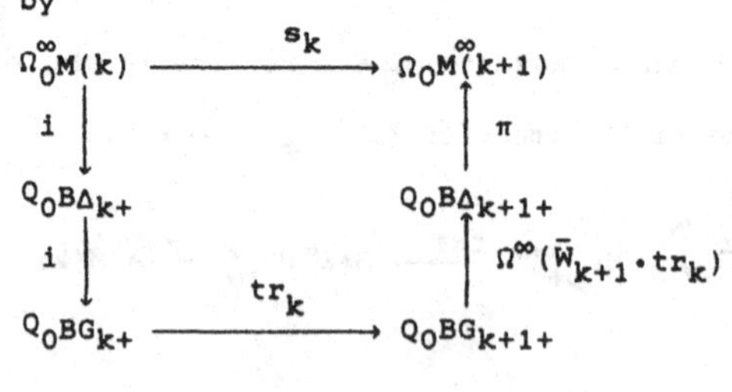

The proof is completed by showing $d_k s_k$ is an equivalence on H_*E_k. This involves showing that the transfer is multiplicative enough to reduce the calculation to generators where the interaction of the transfer and the Steinberg idempotent can be analyzed with the help of the Hecke algebra.

Section Four: Groups of Symplectc Type

The results outlined in this section are joint work with Mark Feshbach; complete details will appear in [FP].

Let V be a vector space of dimension n over F_2. We shall consider group extensions $E = E(n)$ of the form

$$(9) \qquad Z/2 \xrightarrow{i} E \xrightarrow{\pi} V$$

Under certain non-degeneracy assumptions BE is known to be acted upon by a Chevalley group. Our goal here is to describe the summand of BE corresponding to the Steinberg idempotent.

Associated to (9) is the quadratic form $Q:V \longrightarrow F_2$ defined by

$$Q(x) = i^{-1}(\tilde{x}^2) \quad \text{where } \pi(\tilde{x}) = x.$$

Q satisfies

$$Q(x+y) = Q(x) + Q(y) + B(x,y)$$

where $B:V\times V \longrightarrow F_2$ is the bilinear form given by

$$B(x,y) = i^{-1}(\tilde{x}\tilde{y}\tilde{x}^{-1}\tilde{y}^{-1}) \quad \text{where } \pi(\tilde{x}) = x,\ \pi(\tilde{y}) = y.$$

Then B is necessarily symplectic, i.e. $B(x,x) = 0$. Let

$$V_0 = \{ x\in V: B(x,y) = 0 \text{ for all } y\in V \}$$

We say that Q is non-degenerate if $Q(x) \neq 0$ for $x \neq 0$ in V_0. Henceforth we shall assume that Q associated to (9) is non-degenerate; the corresponding groups are said to be of sympletic type.

There are three types of non-degenerate quadratic forms

$$n = 2m \qquad Q = \sum_{i=1}^{m} x_i x_{-i} \qquad \text{(real type)}$$

$$n = 2m \qquad Q = \sum_{i=1}^{m-1} x_i x_{-i} + x_m^2 + x_m x_{-m} + x_{-m}^2 \qquad \text{(quaternion type)}$$

$$n = 2m+1 \qquad Q = x_0^2 + \sum_{i=1}^{m} x_i x_{-i} \qquad \text{(complex type)}$$

for some choice of basis (x_0), $x_1, \ldots, x_m, x_{-1}, \ldots, x_{-m}$ of V^*. Here we follow Quillen's terminology [Q] of real, quaternion, and complex. Classically these forms are shown to be distinguished by their dimension n and Arf invariant $\mathrm{Arf}(Q) \in F_2$. For $n = 2m$ above, $\mathrm{Arf}(Q) = 0,1$ resp. where

$$\mathrm{Arf}(Q) = \begin{cases} 0 & \text{if } |Q^{-1}(0)| > m \\ 1 & \text{if } |Q^{-1}(0)| < m \end{cases}$$

<u>Example 4.1</u>: In the real (resp. quaternion) cases $E(2) = D_8$ (resp. Q_8), the dihedral and quaternion groups of order 8 of Section One. In general for n even, E(n) is an extraspecial 2-group; further it is not hard to show that E(n) is the central product of an appropriate number of copies of D_8 and Q_8.

Let O(V,Q) be the group of automorphisms which preserve Q. These groups are often written

$$O(V,Q) = O^+_{2m}(F_2) \qquad n = 2m \qquad \text{real case}$$

$$= O^-_{2m}(F_2) \qquad n = 2m \qquad \text{quaternion case}$$

$= O_{2m+1}(F_2) \qquad n = 2m+1 \qquad$ complex case

Let $Out_Z(E)$ be the group of outer automorphisms of E which fix the center $Z < E$. Then

Proposition 4.2 $Out_Z(E) \approx O(V,Q)$.

Thus O(V,Q) acts on BE up to homotopy.

The groups O(V,Q) are associated with Chevalley groups [C]. In fact $O_{2m+1}(F_2) \approx B_m(F_2)$ and $\Omega^+_{2m}(F_2) \approx D_m(F_2)$ where $\Omega^+_{2m}(F_2)$ is the commutator subgroup of $O^+_{2m}(F_2)$. The group $O^-_{2m}(F_2)$ is isomorphic to the twisted Chevalley group ${}^2D_{2m}(F_4)$. Thus a Borel subgroup B and a Weyl subgroup W is defined for these groups [C]. In terms of these subgroups the Steinberg idempotent is defined by $e = \Sigma\, b\sigma$ where $b \in B$, $\sigma \in W$.

We now recall Quillen's computation of $H^*BE(n)$ [Q; Th. 4.6].

$$H^*BE = \frac{F_2[(x_0), x_1, \dots, x_m, x_{-1}, \dots, x_{-m}, w_{2^h}]}{(Q, Sq^1Q, \dots, Sq^{2^{h-2}} \dots Sq^1Q)}$$

where h = m, m+1, m+1 for the real, complex, and quaternion cases of dimension n = 2m, 2m+1, 2m resp. Here w_{2^h} denotes the Stiefel-Whitney class of the unique irreducible real representation Δ_n of dimension 2^h. The non-zero positive dimensional Stiefel-Whitney classes of Δ_n are w_{2^h}, $w_{2^h-2^r}$, $w_{2^h-2^{r+1}}$, ..., w_{2^h-1} where r = 0, 1, 2, for the real, complex, and quaternion cases resp. By the uniqueness of Δ_n these classes are O(V,Q) invariants; Quillen has shown that they form a regular sequence in $H^*BE(n)$ and hence generate a polynomial algebra $S_{h,r}$ over which $H^*BE(n)$ is a finitely generated, free module.

We are now in a position to state the main results of this section. Let

$\alpha = \Sigma\, x_{i_1}^{-1} \ldots x_{i_m}^{-1}$, $i_j = \pm j$, with an even no. of minus signs occuring

$\beta =$ " " " " " odd " " " " "

Let $J(m) = \langle\, Sq^J(\alpha+\beta) : J = (i_1, \ldots, i_m)$ admissible, $i_m \geq 2 \,\rangle$.

<u>Theorem 4.3</u>: For n = 2m, the real case

$$H^* eBE = J(m) \oplus F_2[w_{2m}] \oplus S_{m,0}A_{m-2}(Sq^{2^{m-1}} \ldots Sq^1\alpha)$$

For n = 2m+1, the complex case

$$H^* eBE = J(m) \oplus F_2[w_{2m+1}] \oplus S_{m+1,1}A_{m-1}(Sq^{2^m} \ldots Sq^2(\alpha+\beta)x_0)e$$

For n = 2m+2, the quaternion case

$$H^* eBE = J(m) \oplus F_2[w_{2m+2}] \oplus S_{m+2,2}\tilde{A}_m(Sq^{2^{m+1}} \ldots Sq^4(\alpha+\beta)x_{m+1}x_{-m-1}^2)e \quad .$$

Here A_k is the subalgebra of the Steenrod algebra A generated by $Sq^1, \ldots Sq^{2^k}$; $\tilde{A}_k = F_k \cap A_k$ where F_k is the submodule of A spanned by all admissible monomials of length $\leq$ k. Alternately $\tilde{A}_k$ is spanned by Milnor basis elements P^R, $R = (r_1, r_2, \ldots)$, $r_i = 0$ for $i > k$.

<u>Theorem 4.4:</u> For n =2m, the real case, BE contains $2^{m(m-1)}$ wedge summands each equivalent to $eBE(n) \simeq M(m) \vee L(m) \vee eT(\Delta_n)$ where $T(\Delta_n)$ is the Thom spectrum of Δ_n.

These results generalize those of Section 2 for elementary abelian 2-groups. In the notation of the proof of Theorem 2.6

$$H^*M(m) = F_2 < Sq^I(X_m^{-1}) : I = (i_1, \ldots, i_m) \text{ admissible}, i_m \geq 1 >$$

$$= D(m) A_{m-2} Sq^{2^{m-1}} \ldots Sq^1(X_m^{-1})$$

where D(m) is the ring of $GL_m(F_2)$ invariants in $H^*B(F_2{}^m)$ [M]. Proofs of both theorems are obtained by restricting to certain detecting subgroups where the calculations are more tractable.

References

[A] J.F.Adams, Lectures on generalized cohomology, Lecture Notes in Math., no. 99, Springer-Verlag, New York, (1969).

[C1] D.Carlisle, P.Eccles, S.Hilditch, N.Ray, L.Schwartz, G.Walker, R.Wood, Modular representations of GL(n,p), splitting $\Sigma(CP^\infty\times\ldots\times CP^\infty)$ and the β-family as framed hypersurfaces, Math. Z. 189 (1985), 239-261.

[C2] D.Carlisle, The modular representation theory of GL(n,p) and applications, Manchester University Thesis (1985).

[C] R.Carter, Simple Groups of Lie Type, John Wiley and Sons, New York (1972).

[CR] C.Curtis and I.Reiner, Representation Theory of Finite Groups and Associative Algebras, Interscience Publishers, New York (1962).

[EP] L.Evens and S.Priddy, The cohomology of the semi-dihedral group, Proc. of the St. Johns Topology Conference, Contemporary Math. 37 (1985), 61-72.

[FP] M.Feshbach and S.Priddy, Stable splittings associated with Chevalley groups I,II (to appear).

[HK] J.Harris and N.Kuhn, Stable decompositions of classifying spaces of finite abelian p-groups, (to appear).

[Hz] R.Holzager, Stable splitting of K(G,1), Proc. A.M.S. 31 (1972),305-306.

[K] N.Kuhn, A Kahn-Priddy sequence and a conjecture of G.W.Whitehead, Math. Proc. Camb. Phil. Soc. 92 (1982),467-483.

[K2] ______, Spacelike resolutions of spectra, Proc. of the Northwestern University Homotopy Theory Conference, Conteporary Math. 19 (1983), 153-165.

[KP] ______ and S.Priddy, The transfer and Whitehead's conjecture, Math. Proc. Camb. Phil. Soc. 98 (1985), 485-480.

[Kh] J.Kulich, Homotopy models for desuspensions, (to appear).

[Mt] J.Martino, Northwestern University Thesis, (in preparation).

[Ml] J.Milgram (ed.), Problems presented to the 1970 A.M.S. Summer Colloquium in Algebraic Topology, Algebraic Topology, Proc. Symp. Pure Math. XXII, A.M.S. (1971), 187-201.

[Mn] J.Milnor, On the construction FK. In J.F.Adams:Algebraic Topology, a students guide, London Math. Soc. Lecture Note Series 4 (1972),119-136.

[MP1] S. Mitchell and S.Priddy, Symmetric product spectra and splittings of classifying spaces, Amer. J. Math. 106 (1984), 219-232.

[MP2] ____________________________, Stable splittings derived from the Steinberg module, Topology 22 (1983),285-298.

[M] S.Mitchell, Finite complexes with A(n)-free cohomology, Topology 24 (1985), 227-248.

[Nk] M.Nakaoka, Cohomology mod p of symmertric products of spheres, J. Inst. Poly., Osaka City Univ. 9 (1958), 1-18.

[Q] D.Quillen, The mod-2 cohomology rings of the extra-special 2-groups and the spinor groups, Math. Ann. 194 (1971), 197-212.

[Sn] V.Snaith, A stable decomposition of $\Omega^n\Sigma^n X$, J. London Math Soc. 7 (1974), 577-583.

[S] R.Steinberg, Prime power representations of finite linear groups II, Can. J. Math 18 (1956), 580-591.

[W] R.Wood, Modular representations of GL(n,p) and homotopy theory, Lecture Notes in Math. 1172, Springer (1985), 188-203.

Northwestern University

Evanston, IL 60201

Localization and Periodicity in Homotopy Theory

Douglas C. Ravenel

The study of the homotopy groups of spheres can be compared to astronomy. The groups themselves are like distant stars waiting to be discovered by the determined observer, who is constantly building better telescopes to see further into the distant sky. The telescopes are spectral sequences and other algebraic constructions of various sorts. Each time a better instrument is built new discoveries are made and our perspective changes. The more we find the more we see how complicated the problem really is.

We can distinguish three levels of ideas in the subject. The first (comparable to observational astronomy) is the collection of data about homotopy groups by various computational devices (see chapters 1, 3 and 7 of [R1]). While this aspect of the subject is not fashionable and is seldom discussed in public, it is vital to the subject. Without experimental data there can be no valid theories.

Recently attempts have been made to use computers for this purpose by Bruner [Bru], Curtis-Goerss-Mahowald-Milgram [CGMM] and Tangora [Ta]. This subject is wide open and we hope to see more activity in it. There are many mathematical structures that could be exploited in designing software for this purpose. We have not yet seen a dramatic improvement over what was done earlier by hand, but there are grounds for optimism.

The homotopy groups of spheres is as difficult a computational problem as any in algebraic topology. Long hard experience has led to the following.

Mahowald Uncertainly Principle. (i) Any spectral sequence converging to the homotopy groups of spheres (stable or unstable) with a systematically computable E_2-term will have differentials which can be computed only by ad hoc methods.

(ii) No computer program of reasonable length for finding the homotopy groups of spheres can run with reasonable speed without human intervention. □

The second level of ideas in homotopy theory is the identification of certain patterns known as periodic families. This may be compared to the discoveries of Kepler and Halley. The pioneering homotopy theorists in this case were Adams [A1], Smith [S] and Toda [To1 and To2].

The third level (comparable to cosmology) is the formulation of general theories about the mechanisms which produce the observed phenomena. The most recent example of this is the work of Devinatz-Hopkins-Smith described elsewhere in this volume. The earliest results of this sort were Brown's finite computability theorem [Bro] and Nishida's nilpotence theorem [N].

As in theoretical physics one can make various models of the universe based on certain oversimplifications or idealizations. While these constructs have obvious limitations, their study is instructive as it leads to some insight into the nature of the real world. We will discuss several of these models now.

The first of these is the flat earth model. It is based on conjecture that

$$\pi_*(S^n) = H_*(S^n).$$

This is true for $n = 1$, but the discovery of the Hopf map from S^3 to S^2 disproves it for $n > 1$.

The second model is the split EHP sequence theory. At the prime two one has fibrations (James [J])

$$S^{n-1} \to \Omega S^n \to \Omega S^{2n-1}.$$

The resulting long exact sequence of homotopy groups is the EHP sequence. One has a splitting in cohomology,

$$H^*(\Omega S^n) = H^*(S^{n-1}) \otimes H^*(\Omega S^{2n-1}).$$

We assume that the spaces split accordingly, i.e., that

$$\Omega S^n = S^{n-1} \times \Omega S^{2n-1},$$

so that the EHP sequence becomes short exact and split. Using our knowledge of $\pi_*(S^1)$, we can compute $\pi_*(S^n)$ by induction on n and show that it is a certain finitely generated free abelian group. The existence of the Hopf maps is consistent with this theory. In fact it gives us a map from S^{2n-1} to S^n with Hopf invariant one for each $n > 1$.

There are two problems with this theory. The first is that it leads to the conclusion that S^n is parallelizable for all n, and the second is that it contradicts rational homotopy theory.

A model which avoids these difficulties is based on the assumption that the Adams spectral sequence (ASS) collapses. (See [A2] for the stable version and [BC] for the unstable version. See also the first three chapters of [R1].) In this theory the EHP sequence splits iff n is a power of two. S^n is parallelizable only when n is one less than a power of two. More generally, S^n has 2^{k-1} linearly independent tangent vector fields when $n+1$ is an odd multiple of 2^k. There is a map from S^{2n-1} to S^n with Hopf invariant one iff n is a power of 2.

The difficulty with this model is that it contradicts the Adams theorems on Hopf invariant one [A3] and vector fields on spheres [A4]. The latter says that S^n has roughly $2k$ vector fields, with k as above.

Our last model is based on the assumption that the Adams-Novikov spectral sequence (ANSS) collapses. (See Novikov [No] or the last four chapters of [R1]. The unstable version is due to Bendersky-Curtis-Miller [BCM].) This theory is consistent with the Hopf invariant one theorem and almost consistent with the vector field theorem.

By "almost consistent" we mean that the number of vector fields it predicts is also roughly $2k$, but not exactly the number given by Adams. In this sense the model is asymptotically correct. The same can be said of the dimensions of the predicted root invariants of certain periodic families (see [MR1 and 2]).

The most striking feature of this model is the rich array of periodic families that it displays through the chromatic spectral sequence [MRW]. This periodicity can also be found in the ASS (see Mahowald-Shick [MSh]). The question of how much of this structure actually exists in homotopy was studied in [R2], which led to a list of seven conjectures, most of which have been proved recently by Devinatz-Hopkins-Smith. We will describe this in more detail below.

The main shortcoming of this model is that it contradicts Nishida's theorem, which says that some iterate of every positive dimensional element in stable homotopy is null homotopic. There are many elements in the ANSS E_2-term (and also in the ASS E_2-term) which are not nilpotent. Nevertheless, it turns out that nilpotence (which is not apparent in the ANSS) and periodicity (which is) are intimately related. It was suggested in [R2], and proved by Hopkins et al, that the former in some sense implies the latter.

1. The periodic point of view.

Suppose we want to understand the homotopy of a space or spectrum X. If X is finite, its homotopy groups are likely to be intractable. One way to get some good information about them is the following. Suppose we have a finite complex Y which admits a self-map

$$v : \Sigma^d Y \to Y$$

such that all iterates of v are essential. Such a self map will be said to be *periodic*.

We now know, thanks to Hopkins et al, that any finite complex Y, after a suitable suspension, admits a periodic self map and that it is essentially unique. Its existence was conjectured in [R2], but the uniqueness result caught us by surprise.

Let $\pi_*(X;Y)$ denote the graded group of homotopy classes of maps of various suspensions of Y into X. Then v acts on this group by composition on the left, making it a module over $Z[v]$.

1.1. Periodicity Problem. With X, Y and v as above, compute

$$v^{-1}\pi_*(X;Y). \qquad \square$$

Experience has shown that this is a good problem. If X is finite then this object tends to be finitely generated as a module over $Z[v, v^{-1}]$. There are some powerful methods for computing it which will be described in the next section. In some sense $\pi_*(X)$ is determined by these periodic groups.

The first theorem of this sort was Serre's identification of the rational homotopy groups of spheres. If X and Y are spheres localized at a prime p and v is the degree p map on Y. Then inverting v is the same thing as tensoring $\pi_* X$ with the rationals. The following result is well known.

1.2. Serre Finiteness Theorem.

$$\pi_k(S^0) \otimes Q = \begin{cases} Q & \text{if } k = n \\ Q & \text{if } k = 2n-1 \quad \text{and} \quad n \quad \text{is even} \\ 0 & \text{otherwise}. \end{cases} \qquad \square$$

1.3 Corollary. *There is a map*

$$S^{2n+1} \to K(Z, 2n+1)$$

which is a rational equivalence, i.e., it induces an isomorphism of rational homotopy groups.

□

Now let Y be a mod p Moore space which is 2-connected. Then Adams showed in [A1] that there is a self map

$$v : \Sigma^q Y \to Y,$$

where $q = 8$ when $p = 2$ and $q = 2p - 2$ when p is odd, having the desired property. He showed that all of its iterates are essential by showing that it induces an isomorphism in K-theory.

The group

$$v^{-1} \pi_*(S^{2n+1}; Y)$$

was determined for $p = 2$ by Mahowald in [M1]. The odd primary case was recently settled by my student, R. Thompson. Their results are best described in the language of Corollary

1.3. We will replace $K(Z, 2n+1)$ by a certain infinite loop space K_n such that there is a map

$$\Omega^{2n+1}S^{2n+1} \to K_n$$

which induces an isomorphism after applying the functor $v^{-1}\pi_*(: Y)$.

We will describe the 2-primary case first. Recall the J spectrum is the fiber of a certain map

$$bo \to \Sigma^4 bsp$$

defined in terms of an Adams operation. For a space X let $J(X)$ denote the infinite loop space associated with the smash product of J with the suspension spectrum of X.

Snaith [Sn] constructed a map

$$\Omega^n S^n \to QRP^{n-1},$$

where RP^{n-1} denotes $(n-1)$-dimensional real projective space and QX denotes

$$\varinjlim \Omega^k \Sigma^k X.$$

Inclusion of the bottom cell in J leads to a map from QX to $J(X)$. Then we have

1.4. Theorem [M1]. *(a) The Snaith map*

$$\Omega^{2n+1}S^{2n+1} \to QRP^{2n}$$

induces an isomorphism in $v^{-1}\pi_*(; Y)$.

(b) The map

$$QRP^{2n} \to J(RP^{2n})$$

induces a similar isomorphism, and the group

$$v^{-1}\pi_*(J(RP^{2n}); Y)$$

is explicitly computable. □

For $p > 2$, J is the fiber of a map

$$bu \rightarrow \Sigma^q bu$$

and RP^{2n} is replaced by the $2n(p-1)$-skeleton of $B\Sigma_p$, the classifying space for the symmetric group on p letters.

Presumably the Snaith map induces an isomorphism in K-theory, but we do not know how to compute the K-theory of the source. That of the target was computed by Miller-Snaith [MS].

This result identifies a localization of the mod p homotopy of S^{2n+1}. A more delicate question is to determine the precise image of the mod p homotopy in it. This is done in [M1] up to an ambiguity that is related to the Kervaire invariant problem. This image includes that of the mod p J homomorphism and nearly all elements whose Hopf invariants lie in its image.

Thus 1.3 enables us to read off the behavior of every such element in the EHP sequence, i.e., its sphere of origin and sphere of death if it is unstable. Empirical evidence indicates that through any given range of dimensions, most elements in the homotopy groups of spheres are of this type. None of the machine computations referred to above exploits this fact in any way.

Notice that in 1.2 we took Y to be the sphere and v to be the degree p map. In 1.4 Y was the cofiber of the map used in 1.2. Experience has shown that if we were to replace the mod p Moore space Y by the mod p^i Moore space, there would still be a map v which is a K-theory equivalence, and the statement of the theorem would be the same. The actual groups involved would be different since they would be localized mod p^i homotopy groups instead of mod p homotopy groups.

The next theorem of this sort one could hope for would have Y being the cofiber of the Adams self map of the Moore space. This is the *v_2-periodicity* problem. This complex has long been known to admit a periodic self map, but its iterates are not detectable by K-theory since Y itself is K-theoretically acyclic, being the cofiber of a K-theory equivalence.

Such a theorem has not been proved because we do not know what the infinite loop space should be. We do not know what to substitute for either RP^{2n} or J. For $p \geq 5$ the homotopy elements that would be detected by such a result include the β-family of Smith [S] and Toda [T1]. For $p = 2$ some results on such elements were obtained by Davis-Mahowald in [DM].

2. BP-theory and Morava K-theory.

BP-theory is a generalized homology theory that is useful for understanding the periodicity described in the last section. For each prime p there is a spectrum BP with

$$\pi_*(BP) = Z_{(p)}[v_1, v_2, \ldots]$$

with $\dim v_n = 2p^n - 2$. This coefficient ring will be denoted by BP_*.

If $V(0)$ is a mod p Moore space then the reduced BP homology of $V(0)$ is a suitable suspension of $BP_*/(p)$. The Adams self map induces multiplication by v_1^4 if $p = 2$ and by v_1 if p is odd. Hence BP-theory detects all iterates of it.

Fix an odd prime p and let $V(1)$ denote the cofiber of the Adams self map. Then we have

$$BP_*(V(1)) = BP_*/(p, v_1).$$

Smith and Toda showed that for $p \geq 5$ there is a periodic self map

$$\beta : \Sigma^d V(1) \to V(1),$$

with $d = 2p^2 - 2$, which induces multiplication by v_2 in BP-theory.

Let $V(2)$ denote the cofiber of β. Then

$$BP_*(V(2)) = BP_*/(p, v_1, v_2).$$

Smith and Toda showed that for $p \geq 7$ there is a periodic self map

$$\gamma : \Sigma^d V(2) \to V(2),$$

with $d = 2p^3 - 2$, which induces multiplication by v_3 in BP-theory. It is not known whether the cofiber of this map admits a self map inducing multiplication by v_4.

These were the earliest examples of periodic self maps. Each of them can be used in 1.1. Now we will describe a method for computing the groups in 1.1 in the world where the ANSS collapses. In section 4 we will discuss the problem of carrying this analysis over to the real world of homotopy theory.

In the ANSS model of the universe, the set of homotopy classes of maps from Y to X is

$$\operatorname{Ext}(BP_*(Y), BP_*(X)).$$

This Ext is in the category of comodules over $BP_*(BP)$, the BP-theoretic analog of the dual Steenrod algebra. If Y is finite then we have (in honest homotopy theory)

$$[Y, X] = [S^0, DY \wedge X],$$

where DY is the Spanier-Whitehead dual of Y. Thus we can replace the Ext group above by

$$\operatorname{Ext}(BP_*,\ BP_*(DY \wedge X)).$$

Now suppose Y is one of the complexes discussed above, i.e., $V(n+1)$ for $n = -1, 0$, or 1, and that X has torsion free homology (e.g., X could be the sphere spectrum). Each $V(n)$ is self-dual and our Ext group is

$$\operatorname{Ext}(BP_*, BP_*(X)/I_n)$$

where I_n is the ideal $(p, v_1, \ldots v_{n-1})$. The periodic group of 1.1 in this case is

$$\operatorname{Ext}(BP_*, v_n^{-1} BP_*(X)/I_n). \tag{2.1}$$

In this algebraic setting $V(n)$ exists for all n and all primes p. The Ext group 2.1 is surprisingly accessible thanks to some insights of Jack Morava.

At this point it is convenient to translate to the language of MU-theory. MU is the Thom spectrum associated with BU. After localization at p, MU splits into a wedge of suspensions of BP. The E_2-term is an Ext group defined in the category of comodules over $MU_*(MU)$, but more interestingly it can be identified as the cohomology of a certain group Γ acting on $MU_*(X)$.

To define this group Γ we need to discuss the theory of formal group laws. The connection between this theory and complex cobordism was first discovered by Quillen [Q], and has been essential to much of the recent progress in the subject. A complete account with references can be found in Appendix 2 of [R1].

A *formal group law* over a commutative ring R is a power series $F(x,y)$ over R satisfying three conditions:

(2.2)

(i) $F(x,0) = F(0,x) = x$,

(ii) $F(y,x) = F(x,y)$ and

(iii) $F(x, F(y,z)) = F(F(x,y),z)$.

These allow us to define a binary operation on the set of power series in one variable over R with vanishing constant term, by setting the product of $f(x)$ and $g(x)$ to be $F(f(x), g(x))$. The three conditions say that this operation has an identity element (zero), is commutative an associative. We do not need an inverse axiom as the existence of an inverse follows easily by solving the appropriate equation.

There is a universal formal group law G defined over a certain ring L such that given any F as above, there is a unique ring homomorphism θ from L to R such that

$$F(x,y) = \theta(G(x,y)).$$

This is easily constructed in the following way. Let

$$F(x,y) = \Sigma a_{i,j} x^i y^j.$$

Regard the $a_{i,j}$ as indeterminates and set

$$L = Z[a_{i,j}]/I$$

where I is the ideal generated by the relations among the $a_{i,j}$ forced by the definition 2.2.

The explicit structure of L was determined by Lazard, and Quillen showed that it is naturally isomorphic to $\pi_*(MU)$.

Now we are ready for the group Γ. It is the set of power series $g(x)$ over the integers with leading term x, with the group operation being formal composition. This group acts on L as follows. If $G(x,y)$ is the universal formal group law, then $g^{-1}(G(g(x),g(y))$ is another formal group law and is therefore induced by an automorphism of L.

By Quillen's theorem, $L = \pi_*(MU)$. It turns out that the Adams-Novikov E_2-term for the homotopy of the sphere spectrum is

$$H^*(\Gamma; L).$$

Moreover for any spectrum X there is an action of Γ on $MU_*(X)$ such that

$$E_2 = H^*(\Gamma; MU_*(X)). \tag{2.3}$$

The Ext group in (2.1) is isomorphic to

$$H^*(\Gamma; v_n^{-1}MU_*(X)/I_n).$$

Here we are regarding v_n as an element of MU_*. Now there is a change of rings isomorphism between this group and one that is much easier to compute, namely

$$H^*(\Gamma; v_n^{-1}MU_*(X)/I_n) = H^*(S_n; K(n)_*(X)), \tag{2.4}$$

where S_n is a certain pro-p-group and $K(n)$ is the nth Morava K-theory. This is a generalized homology theory with

$$\pi_*(K(n)) = Z/(p)[v_n, v_n^{-1}].$$

Very briefly, this isomorphism can be explained as follows. Recall that a formal group law F over a ring R is induced by a homomorphism θ from L to R. If R is a finite field of characteristic p, then we can define the ***height*** of F to be the smallest n such that $\theta(v_n)$ is nonzero. Equivalently, F has height n iff θ factors through $v_n^{-1}L/I_n$. It is known that this height determines such a formal group law up to isomorphism over the algebraic closure. S_n is the automorphism group of a height n formal group law F over an algebraically closed field.

The action of Γ on

$$v_n^{-1}MU_*(X)/I_n$$

is $K(n)_*$-linear. S_n can be thought of as a subgroup of Γ and the above Γ-module is induced from the S_n-module $K(n)_*(X)$, hence the isomorphism 2.4.

The alert reader may recognize this explanation as the swindle that it is, but it gives the right idea. A more accurate (and necessarily longwinded) account can be found in [MRW] or in Chapter 6 of [R1]. These references also give an explicit arithmetic description of the group S_n.

For $n = 1$ it is the group of units in the p-adic integers. If n is not divisible by $p-1$ then it has cohomological dimension n^2, and if n is divisible by $p-1$ then it has periodic cohomology. The group of units of the ring of integers of any number field of degree N can be embedded in it. Cohomological computations with it are manageable.

3. The chromatic spectral sequence.

In the last section we described a method for computing the algebraic analog of the group 1.1 in the case when Y is one of the $V(n)$'s. In this section we will explain how to generalize this to other finite Y and how all of this information fits together and determines the algebraic analog of $\pi_*(X)$. In the next section we will explain how all of this can be carried over to the stable homotopy category.

We say that a finite complex Y has ***type*** n if n is the smallest integer such that $v_n^{-1}BP_*(Y)$ (or equivalently $K(n)_*(Y)$) is nontrivial. This number is known to be finite for any finite Y. The internal properties of BP-theory tell us that multiplication by a suitable power of v_n is a comodule-endomorphism of $BP_*(Y)$. Moreover any periodic endomorphism must have this form. If X has torsion free homology then the same is true of $BP_*(DY \wedge X)$.

Thus 1.1 reduces to computing

$$\mathrm{Ext}(BP_*, v_n^{-1}BP_*(DY \wedge X)). \tag{3.1}$$

The main difference between this and the Ext group of 2.1 is that the second variable is not annihilated by I_n, but by some finite power of it. Filtering this comodule by powers of that ideal leads to a spectral sequence whose E_2-term can be computed with the methods of the previous section.

Now what does this have to do with $\pi_*(X)$, or in the algebraic setting, with

$$\mathrm{Ext}(BP_*, BP_*(X))?$$

The answer is provided by the chromatic spectral sequence of [MRW]. This is a spectral sequence converging to this Ext group, whose E_1-term consists of Ext groups similar to those in 3.1. It is like a spectrum in the astronomical sense of the term in that it resolves $\pi_*(X)$ into its v_n-periodic components, i.e., into various frequencies.

This spectral sequence is derived from a long exact sequence called the *chromatic resolution* of $BP_*(X)$, obtained by tensoring $BP_*(X)$ with

$$0 \to BP_* \to M^0 \to M^1 \to \dots \tag{3.2}$$

This is obtained by splicing together certain short exact sequences

$$0 \to N^n \to M^n \to N^{n+1} \to 0 \tag{3.3}$$

defined inductively by

$$N^0 = BP_* \quad \text{and}$$

$$M^n = v_n^{-1}N^n.$$

Each element in N^n (but not N^n itself) is annihilated by a finite power of I_n. Thus the group

$$\mathrm{Ext}(BP_*, M^n),$$

which is part of the chromatic E_1-term, is similar to that of 3.1 and is accessible through the methods of the previous section. In particular it has cohomological dimension n^2 when $p-1$ does not divide n.

By composing the connecting homomorphisms of the short exact sequence 3.3, we get a map

$$\mathrm{Ext}(BP_*, N^n) \to \mathrm{Ext}(BP_*, BP_*)$$

which raises degree by n and leads to a decreasing filtration of

$$\mathrm{Ext}(BP_*, BP_*(X))$$

which we call the *chromatic filtration.*

This whole construction can be carried out without the assumption that X has torsion free homology, but the details are more complicated.

4. The chromatic filtration in homotopy theory.

The algebraic constructions of the last two sections lead us to wonder if the chromatic apparatus can be carried over to the homotopy category. We studied this problem in [R2], and were led to the following seven conjectures, which were first announced in 1977.

4.1 Nilpotence conjecture. *If*

$$f : \Sigma^d X :\to X$$

is a self map of a finite spectrum inducing a nilpotent map in MU_*-theory, then some iterate of f is null homotopic. □

4.2 Periodicity conjecture. *Let X be a p-local finite spectrum and let n be the smallest integer such that $K(n)_*(X)$ is nontrivial. Then X admits a self map (not homotopic to the identity) which is a $K(n)$-equivalence.* □

4.3 Class invariance conjecture. *Let X be as in 4.2. The Bousfield equivalence class (this notion will be defined below) of X depends only on n.* □

4.4 Telescope conjecture. *Let X be as in 4.2 and let $\widehat{X}$ denote the direct limit obtained by iterating the periodic self map. Then $\widehat{X}$ is Bousfield equivalent to $K(n)$.* □

4.5 Smash product conjecture. *Let L_n denote Bousfield localization (this will also be defined below) with respect to $v_n^{-1}BP$. Then*

$$L_n(X) = X \wedge L_n(S^0). \quad \square$$

4.6 Localization conjecture. *With L_n as in 4.5,*

$$BP \wedge L_n(X) = X \wedge L_n(BP). \quad \square$$

4.7 Boolean algebra conjecture. *The Boolean algebra of Bousfield equivalence classes generated by finite spectra and their complements is the same as that generated by Morava K-theories and their complements, which all exist.* □

All of these but the telescope conjecture have now been proved. 4.6 was proved by the author [R3] in 1983, and the others (excluding 4.4) were proved by M. Hopkins and various collaborators including E. Devinatz, J. Smith and the author.

We will now describe how these results relate to the program described in the previous sections.

We began by discussing periodic self maps. All the known examples of such maps were known to be detected by BP-theory. so we could ask if there are any that are not. The nilpotence conjecture says that any self map which does not appear to be periodic by BP-theory is in fact nilpotent. Thus there is no periodicity other than the sort readily detected by BP-theory.

In [R2] we suggested programs for deriving most of the other conjectures from 4.1. While these have not worked out as proposed in detail, the proofs of 4.2, 4.3, 4.5 and 4.7 have depended on that of 4.1 as expected.

BP-theory predicts that a finite p-torsion spectrum X will have a periodic self map inducing multiplication by some power of v_n, where n is the smallest integer such that $K(n)_*(X)$ is nontrivial. The periodicity conjecture says that such a self map actually exists. Hopkins and Smith show that this map is unique in the sense that if f and g are two such maps then some iterate of f is homotopic to some iterate of g. This leads to the conclusion that the homotopy endomorphism ring of X has Krull dimension one. Moreover some iterate of f lies in the center of this ring, that is it commutes with any other self map (periodic or nilpotent) of X.

They show even more. These self maps are asymptotically central in the category of finite spectra of type n in the following sense. Suppose X and Y are two such spectra with periodic self maps f and g. Then these maps have iterates f' and g' such that for any map

$$h : X \to Y,$$
$$hf' = g'h.$$

Before proceeding further we need to digress and recall some definitions and results of Bousfield. Precise references and more details can be found in [R2]. Suppose we have a generalized homology theory represented by a spectrum E. A spectrum Y is said to be *E_*-local* if whenever

$$E_*(X) = 0,$$

then

$$[X, Y] = 0.$$

An *E_*-localization* of a spectrum X is a map from X to an E_*-local spectrum which is an E-equivalence, i.e., a map inducing an isomorphism in E-theory. Bousfield proved that this localization always exists. It follows easily from the definition that it is unique, and we denote it by $L_E(X)$.

If E and X are both connective then $L_E(X)$ is merely an arithmetic localization or completion of X. But if either E or X fails to be connective then $L_E(X)$ is much harder to predict. For example, if K denotes complex K-theory, then $L_K(S^0)$ is not connective and we find that

$$\pi_{-2}(L_K(S^0)) = Q/Z.$$

Two spectra E and F are said to be *Bousfield equivalent* when they give the same localization functor, or equivalently when $E_*(X) = 0$ iff $F_*(X) = 0$. The equivalence class of E is denoted by $< E >$. The class invariance conjecture 4.3 characterizes the Bousfield class of a finite complex.

There is a partial ordering on the set of Bousfield classes. We say that $< E > \geq < F >$ if $E_*(X) = 0$ implies that $F_*(X) = 0$. Thus $< S^0 >$ as the biggest class and $< pt. >$ is the smallest. Smash products and wedges are well defined on Bousfield classes. A class $< F >$ is the *complement* of $< E >$ if

$$< E > \vee < F > = < S^0 > \quad \text{and}$$

$$< E > \wedge < F > = < pt. > .$$

A class may or may not have a complement. Bousfield shows that the class of any wedge of finite spectra has a complement. It is easy to find examples of classes (e.g., that of an integer Eilenberg-Mac Lane spectrum) that do not. Those that do form a Boolean algebra denoted by BA. Our last conjecture, 4.7, concerns the structure of part of it.

Now we can return to our discussion of the chromatic theory. The chromatic resolution 3.2 was built out of short exact sequences 3.3. We can realize the former geometrically if we can realize the latter by cofibrations. Suppose inductively that we have a spectrum N_n satisfying

$$BP_*(N_n) = N^n.$$

Then one could hope that

$$BP_*(L_n(N_n)) = M^n, \tag{4.8}$$

so we would have a cofiber sequence

$$N_n \to M_n \to N_{n+1}$$

realizing 3.3.

Now 4.8 follows from 4.6, which tells us how to compute the BP-homology of $L_n(X)$, the spectrum $L_n(BP)$ having been analyzed in [R2]. Thus we can realize the chromatic filtration geometrically.

We have an inverse system of spectra

$$L_0(X) \leftarrow L_1(X) \leftarrow L_2(X) \leftarrow \ldots$$

Hopkins has recently shown that the inverse limit of this tower is X when X is finite and p-local. The fiber of the nth map in this tower is a suspension of $M_n \wedge X$. In order to complete the picture we need to know that the ANSS for this spectrum converges. In other words we need to know how to compute the homotopy of a localization. This convergence question is closely related to the smash product conjecture 4.5. The details of this are too technical to be given here and will appear in [HR].

Finally we return to the original periodicity problem 1.1. Stably this amounts to computing the homotopy of a telescope as in 4.4. Again it needs to be shown that the ANSS converges. This will follow from what we already know if we can identity the telescope with the localization of the spectrum $(DY \wedge X)$ with the periodic self map. The telescope has the right BP-homology, but it is not clear that it is local. The telescope conjecture 4.4, which is still open, assures that it is.

The telescope conjecture has another useful corollary in the study of v_n-periodicity. Suppose Y_1 and Y_2 are finite spectra of type n. Each has a periodic self map which we denote abusively by v_n. The following result shows that the information about X given by the periodic homotopy groups $v_n^{-1}\pi_*(X;Y)$ depends only on n and not on the choice of the finite type n spectrum Y.

4.9 Corollary. *Let f be a map*

$$f : X_1 \to X_2$$

and let Y_1 and Y_2 be as above. then the following three statements are equivalent if 4.4 is true.

(i) f induces an isomorphism in $v_n^{-1}\pi_*(\ ;Y_1)$,

(ii) f induces an isomorphism in $v_n^{-1}\pi_*(\ ;Y_2)$ and

(iii) f induces an isomorphism in $K(n)_*(\ )$.

Proof. (Assuming 4.4): The functor

$$v_n^{-1}\pi_*(\ ;Y_1)$$

is the same thing as the homology theory represented by the telescope associated with the self map of DY_1. (It is known that Spanier-Whitehead duality preserves the Bousfield class.) This telescope, along with the one similarly derived from Y_2, is Bousfield equivalent to $K(n)$ by 4.4. It follows easily from the definitions that an equivalence in one homology theory is also an equivalence in any Bousfield equivalent theory. □

References

[A1] J.F. Adams, On the groups $J(X)$ IV, *Topology* **5** (1966), 21–71.

[A2] J.F. Adams, On the structure and applications of the Steenrod algebra, *Comm. Math. Helv.* **32** (1958), 180–214.

[A3] J.F. Adams, The nonexistence of elements of Hopf invariant one, *Ann. of Math.* **2 72** (1960), 20–104.

[A4] J.F. Adams, Vector fields on spheres, *Ann. of Math.* **75** (1962), 603–632.

[BCM] M. Bendersky, E.B. Curtis, and H.R. Miller, The unstable Adams spectral sequence for generalized homology, *Topology* **17** (1978), 229–248.

[BC] A.K. Bousfield, E.B. Curtis, D.M. Kan, D.G. Quillen, D.L. Rector and J.W. Schlesinger, The mod p lower central series and the Adams spectral sequence, *Topology* **5** (1966), 331–342.

[Bro] E.H. Brown, Finite computability of Postnikov complexes, *Ann. of Math.* **65** (1957), 1–20.

[Bru] R.R. Bruner, Machine calculation of the cohomology of the Steenrod algebra, to appear.

[CGMM] E.B. Curtis, P. Goerss, M.E. Mahowald and R.J. Milgram, Some calculations of unstable Adams E_2-terms for spheres, to appear in *Proceedings of the Seattle Emphasis Year in Homotopy Theory*, Springer-Verlag.

[DM] D.M. Davis and M.E. Mahowald, v_1- and v_2-periodicity in stable homotopy theory, *Amer. J. Math.* **103** (1981), 615–659.

[HR] M. Hopkins and D.C. Ravenel, A proof of the smash product conjecture, to appear.

[J] I.M. James, Reduced product spaces, *Ann. of Math.* **62** (1953), 170–197.

[M1] M.E. Mahowald, The image of J in the EHP sequence, *Ann. of Math.* **116** (1982) 65–112.

[MR1] M.E. Mahowald and D.C. Ravenel, Toward a global understanding of the homotopy groups of spheres, to appear in *Proceedings of the Lefschetz conference*, Mexico, 1984

[MR2] M.E. Mahowald and D.C. Ravenel, Implications of Lin's theorem in stable and unstable homotopy, to appear in *Proceedings of the Seattle Emphasis Year in Homotopy Theory*, Springer-Verlag.

[MRW] H.R. Miller, D.C. Ravenel and W.S. Wilson, Periodic phenomena in the Adams-Novikov spectral sequence, *Ann. of Math.* **106** (1977), 469–516.

[MS] H.R. Miller and V.P. Snaith, On the K-theory of the Kahn-Priddy map, *J. London Math. Soc.* **20** (1979), 339–342.

[MSh] M.E. Mahowald and P. Shick, Periodic phenomena in the classical Adams spectral sequence, to appear in *Trans. A.M.S.*

[N] G. Nishida, The nilpotence of elements of the stable homotopy groups of spheres, *J. Math. Soc. Japan* **25** (1973), 707–732.

[No] S.P. Novikov, The methods of algebraic topology from the viewpoint of cobordism theories, *Math. USSR Izv.* (1976), 827–913.

[Q] D.G. Quillen, On the formal group laws of unoriented and complex cobordism theory, *Bull. A. M. S.* **75** (1969), 1293–1298.

[R1] D.C. Ravenel, *Complex cobordism and stable homotopy groups of spheres*, Academic Press, 1986.

[R2] D.C. Ravenel, Localization with respect to certain periodic homology theories, *Amer. J. Math.* **106** (1984), 351–414.

[R3] D.C. Ravenel, On the geometric realization of the chromatic resolution, to appear in *Proceedings of the J.C. Moore Conference*, Princeton, 1983.

[S] L. Smith, On realizing bordism modules, *Amer. J. Math.* **92** (1970), 793–856.

[Sn] V.P. Snaith, Stable decomposition of $\Omega^n \Sigma^n X$, *J. London-Math. Soc.* **7** (1974), 577–583.

[Ta] M.C. Tangora, Computing the homology of the Lambda algebra, *Mem. A. M. S.* **337**, 1985.

[To1] H. Toda, On realizing exterior parts of the Steenrod algebra, *Topology* **10** (1971) 53–65.

[To2] H. Toda, p-primary compositions of homotopy groups, IV, *Mem. Coll. Sci. Univ Kyoto* **32** (1959), 288–332.

University of Washington
Seattle, Washington 98195

Partially supported by NSF

SYMBOLIC CALCULUS: A 19TH CENTURY APPROACH TO MU AND BP

Nigel Ray
Mathematics Department, The University, Manchester, M13 9PL, England.

INTRODUCTION

In the latter part of the last century, two rather different types of symbolic calculus emerged onto the mathematical scene, both with English roots. These were the umbral calculus, or "representative notation", of John Blissard (1803-1875), and the operator calculus of Oliver Heaviside (1850-1925). The umbral calculus was intended to simplify and extend various types of algebraic manipulation in classical analysis, whilst the operator calculus was designed to facilitate the solution of certain kinds of differential equation. For many years, the only contact between the two seems to have been in cases of mistaken identity!

Typical and instructive examples of the early pioneering work can be found in Blissard (1861) and Heaviside (1893), although we should note that the term "umbra" was coined by J.J. Sylvester (1851), and apparantly first used in this connection by E.T. Bell (1929).

Remarkably, neither of the two inventors was a professional mathematician, Blissard being a country vicar and Heaviside an electrical engineer. Perhaps this helps to explain the strikingly similar histories of their respective methods.

Both at first appeared to be a success. But then the mathematical powers of the day began to level accusations of insufficient rigour, and the rejection that followed was sometimes accompanied by considerable disdain. The theories then lay dormant for a while, with only a few authors attempting to revive them for their own purposes. For example the umbral calculus was championed by Bell (1927), and the operator calculus by B. Van der Pol (1929). As a further indignity, other mathematicians who did find such symbolic methods useful often contrived to attribute them to incorrect sources!

For more detailed (and emotive) historical details, see Bell (1938) on the work of Blissard and I.Z. Shtokalo (1976) on Heaviside.

Only since the last war have the calculuses truly re-entered the mathematical fold, and simultaneously the areas of their applicability have begun tentatively to overlap. Thus in the late 60's G-C. Rota finally set umbral calculus on a sound footing in terms of linear operators and duality, and eventually coalgebras, a theme which he subsequently developed with several co-authors, e.g. in S. Roman & G-C. Rota (1978). This programme culminated in the recent book of Roman (1984).

Somewhat earlier, the work of J. Mikusinski (1950; 1959) on generalised functions and convolution quotients rigorously justified Heaviside's intuitive procedures without resort to the Laplace transform. His ideas have been cogently and concisely expounded by A. Erdelyi (1962), and a wide variety of alternative approaches are now available.

It is our purpose in this article to suggest that symbolic calculus of both ilks can provide an elegant and illuminating framework for certain recent computations in MU theory. We make special use of the Roman-Rota umbral calculus over a graded ring, rather than a field, an extension which we have investigated fully in N. Ray (1986), and subsequently discovered was hinted at by Bell (1938)! It is a more recent observation that the concept of generalised function can provide a valuable signpost at a certain crucial point, and that its usage allows the two methods to become neatly, if briefly, intertwined.

We must emphasise that neither calculus is indispensible for understanding most results and computations described here. However, both can assist in getting to the heart of the matter, and in suggesting a variety of developments and conjectures which might have remained obscure in a more conventional approach. The umbral calculus in particular provides a language for simplifying several subtle algebraic concepts, and for offering unexpected and powerful combinatorial interpretations.

We would like to think that such applications are entirely in keeping with the philosophies of Blissard and Heaviside!

From an algebraic topologist's viewpoint, the relevance of symbolic calculus to bordism theory is its intimate relationship with complex oriented Thom isomorphisms. As witnessed elsewhere in this volume Thom complexes and their accompanying paraphernalia are of the utmost importance in homotopy theory, and an algebraic category which not only offers the usual comodule and coalgebra structures, but is also equipped with the precursors of Thom isomorphisms, cannot fail to be potentially useful. Our contention is that modern umbral calculus provides such a category.

Moreover, when we enrich our calculus to admit a product, or Leibniz formula, we then obtain an algebraic theory closely linked to that of formal groups, thus providing further evidence of its relevance.

Our chapters are organised as follows:

1. Umbral calculus over a graded ring
2. Thom classes & Δ-operators
3. Umbrally numerical polynomials & $E_*(CP_0)$
4. Combinatorial aspects
5. Stunted projective space & Dirac's δ
6. Transfer & the Kummer congruences
7. Von Staudt's theorem for Morava K-theory
8. Universal Bernoulli numbers & Fermat's theorem
9. Several variables & symmetrisation
10. $E_*(BG)$ for certain Lie groups G.

Several of our results are presented here without full proofs, although we have tried to be reasonably self-contained for the adventurous reader willing regularly to leap the crevasses. Our minimal aim is to encourage other topologists to translate their own problems into symbolic language.

Special thanks are due to A. Baker, who first brought umbral calculus to our attention, and to F. Clarke, for a host of relevant and enjoyable discussions. E.K. Lloyd also assisted with his extensive knowledge of combinatorial literature. Elmer Rees, G-C. Rota and C.T.C. Wall all offered encouragement in diverse ways, and at crucial times, of which they may be unaware!

We are particularly indebted to Ro Horton, for her speedy and beautiful typing.

Throughout we write E_* for a graded ring, and for coefficients such as $\mathbb{Q}$ we abbreviate $E_* \otimes \mathbb{Q}$ to $E\mathbb{Q}_*$. Many of our E_*-modules M_* split as $M_* \cong E_* \oplus \tilde{M}_*$ into a copy of the scalars and a reduced part.

Finally, we use and elaborate on the notation

$$[x]^n = x(x+1)\dots(x+n-1), \quad [x]_n = x(x-1)\dots(x-n+1)$$

for the <u>rising</u> and <u>falling factorial polynomials</u>.

1. UMBRAL CALCULUS OVER A GRADED RING

In this section we give a brief outline of our umbral background, summarised from Ray (1986).

Let E_* be a commutative ring with 1, graded by an integer-valued dimension function, and assumed for convenience to be torsion free. We write E^* for its dual (over E_*), graded so that $E^i \cong E_{-i}$ for all i.

Let $E_*[x]$ denote the binomial coalgebra over E_*, in an indeterminate x of dimension 2. This is the free E_*-module on $\{1,x,x^2,\ldots\}$ with coproduct

$$x^n \longmapsto \sum_{j=0}^{n} \binom{n}{j} x^{n-j} \otimes x^j .$$

Then $D = \frac{d}{dx}$ operates linearly on $E_*[x]$, and is itself assigned dimension 2. This permits us to identify the continuous dual of $E_*[x]$ with the ring $E^*((D))$ of multiplicative formal divided power series, free over E^* on $\{1,D,\frac{D^2}{2!},\ldots\}$, and known as the umbral algebra. The duality is expressed by

$$\begin{aligned} \left\langle \frac{D^m}{m!} \,\middle|\, x^n \right\rangle &= \left\langle 1 \,\middle|\, \frac{D^m}{m!} x^n \right\rangle \\ &= \frac{D^m}{m!} x^n \Big|_{x=0} \\ &= \delta_{m,n} . \end{aligned}$$

We now select an umbra, or sequence

$$\theta = (\theta_0,\theta_1,\theta_2,\ldots), \quad \theta_i \in E_{2i}$$

and write

$$e^{\theta D} \in E^2((D)) \qquad\qquad \theta^{i+1} \equiv \theta_i$$

for the operator

$$1 + \theta_0 D + \theta_1 \frac{D^2}{2!} + \ldots + \theta_i \frac{D^{i+1}}{(i+1)!} + \ldots .$$

This is Blissard's original "representative notation", as utilised, for example, by J. Riordan (1958).

Remark that we have

$$e^{\theta D}x^n = (x+\theta)^n \qquad \theta^{i+1} \equiv \theta_i$$

which may be linearly extended to all $p(x) \in E_*[x]$. Hence

$$<e^{\theta D} \mid p(x)> = p(x+\theta)\Big|_{x=0} = p(\theta), \qquad \theta^{i+1} \equiv \theta_i$$

which we construe as umbral substitution of x by θ.

Let us now always suppose that $\theta_0 = 1$.

(1.1) Definition. We call an operator of the form

$$D + \theta_1\frac{D^2}{2!} + \ldots + \theta_i\frac{D^{i+1}}{(i+1)!} + \ldots \qquad \in E^2((D))$$

a Δ-operator, and write it in any one of the equivalent forms $\Delta = \theta(D) = e^{\theta D}-1$. □

Observe that

$$\Delta p(x) = p(x+\theta) - p(x), \qquad \theta^{i+1} \equiv \theta_i$$

so Δ represents the umbral difference operator. Such Δ-operators are the essence of the Roman-Rota umbral calculus, and clearly we may consider several distinct Δ-operators for a particular ring E_*. Thus we often write $E = (E_*, \Delta)$ to denote a specific choice.

Two examples are especially illustrative.

(1.2) $E = K$. Here $K_* = \mathbb{Z}[u,u^{-1}]$, where $\dim u = 2$, and $\Delta = u^{-1}(e^{uD}-1)$, so that $\theta_i = u^i$. Thus Δ is the discrete derivative

$$\Delta p(x) = \frac{p(x+u) - p(x)}{u}.$$

(1.3) $E = \Phi$. Here $\Phi_* = \mathbb{Z}[\phi_1, \phi_2, \ldots]$, where the ϕ_i are independent variables and $\dim \phi_i = 2i$. Then $\Delta = e^{\phi D}-1$, so that $\theta_i = \phi_i$. This is the generic, or universal, example. □

To every Δ-operator E we may assign the (normalised) associated sequence of polynomials $(b_n^E(x))$, or simply $(b_n(x))$, where

$$b_n^E(x) \in E\mathbb{Q}_{2n}[x], \qquad n \geq 0.$$

These are characterised by the property

$$b_0(x) = 1 \; ; \; \Delta b_n(x) = b_{n-1}(x) \text{ and } b_n(0) = 0 \; \forall n > 0, \tag{1.4}$$

and always form a <u>divided power sequence</u> in the sense that

$$b_n(x+y) = \sum_{j=0}^{n} b_{n-j}(x) b_j(y) \qquad \forall n > 0 \tag{1.5}$$

in $\widetilde{E\mathbb{Q}}_*[x,y]$.

In a seminal work, R. Mullin & G-C. Rota (1970) established the converse; that any such divided power sequence is associated to a particular Δ-operator.

They simultaneously developed and codified several vital computational tools, such as the <u>recurrence</u> and <u>transfer</u> formulae

$$nb_n(x) = x\dot{\Delta}^{-1} b_{n-1}(x)$$

$$n!b_n(x) = \begin{cases} \dot{\Delta} P^{-n-1} x^n \\ \\ xP^{-n} x^{n-1} \end{cases} \tag{1.6}$$

where $\dot{\Delta} = \frac{d\Delta}{dD}$ and $P = \frac{\Delta}{D}$ are multiplicatively invertible series in $E^o((D))$.

<u>(1.7) Examples</u>. For $E = K$, the associated sequence consists of the normalised falling factorial, or binomial coefficient, polynomials

$$b_n^K(x) = \frac{1}{n!} x(x-u)\ldots(x-(n-1)u) = u^n \binom{u^{-1}x}{n},$$

also expressible as $\frac{1}{n!}[x]_{nu}$.

For $E = \Phi$, the situation is far more intricate, because $(b_n^\Phi(x))$ is the <u>universal associated sequence</u>. This and several other examples will be discussed in detail in later sections. Suffice it here t give

$$b_1^{\Phi}(x) = x, \quad b_2^{\Phi}(x) = \frac{1}{2!}(x^2 - \phi_1 x)$$
$$b_3^{\Phi}(x) = \frac{1}{3!}(x^3 - 3\phi_1 x^2 + (3\phi_1^2 - \phi_2)x)$$

Observe that to recover $b_n^K(x)$ from $b_n^{\Phi}(x)$, we substitute u^i for ϕ_i □

We may now describe the penumbral coalgebra $\Pi(E)_*$ of E, which is the smallest E_*-module

$$E_*[x] \hookrightarrow \Pi(E)_* \hookrightarrow EQ_*[x]$$

with the property that $\Delta\colon \widetilde{\Pi(E)}_* \longrightarrow \Pi(E)_*$ is an isomorphism.

(1.8) Proposition

$$\Pi(E)_* = E_*\{1, b_1(x), b_2(x), \ldots\}$$

with coalgebra structure furnished by (1.5). □

The proof, of course, follows from (1.4). Again, we shall see several examples, and an alternative characterisation of penumbral coalgebras, in later sections. The fact that $\Delta^{-1}\colon \Pi(E)_* \longrightarrow \widetilde{\Pi(E)}_*$ exists is the germ of the Heaviside calculus. Clearly, $\Pi(\Phi)_*$ is universal.

The continuous dual of $\Pi(E)_*$ is the formal power series algebra $E^*[[\Delta]]$, where the duality can be expressed in terms of the action of Δ on $EQ_*[x]$ precisely as before.

As adumbrated above, $\Pi(E)_*$ is in general not an algebra. In fact it has such a multiplicative structure exactly when $E^*[[\Delta]]$ is a coalgebra, a condition which is guaranteed by a suitable coproduct formula

$$\Delta \longmapsto \Delta \otimes 1 + 1 \otimes \Delta + \sum_{i,j} e_{i,j} \Delta^i \otimes \Delta^j \tag{1.9}$$

where $e_{i,j} \in E_{2(i+j-1)}$. Since (1.9) yields a rule

$$\Delta(p(x)q(x)) = (\Delta p(x))q(x) + p(x)\Delta q(x) + \sum_{i,j} e_{i,j}(\Delta^i p(x))(\Delta^j q(x))$$

for evaluating Δ on products, we label such a Δ as a Leibniz Δ-operator.

As we shall explain, K is Leibniz whereas Φ is not. However, every E has a Leibniz extension ${}^L E = ({}^L E_*, \Delta)$, unique in a suitable sense, with $E_* \subset {}^L E_* \subset E\mathbb{Q}_*$. In particular, ${}^L\Phi_*$ is a polynomial algebra U_* of the form

$$U_* \cong \mathbb{Z}[w_1, w_2, \ldots], \ \dim w_i = 2i \ . \tag{1.10}$$

This is established by Lazard's theorem on the universal formal group, the relevance of which is manifestly suggested by (1.9).

In spite of the importance of this theory, we suggest that several concepts and formulae concerning Δ-operators are independent of the Leibniz property, whose introduction can sometimes obscure the basic simplicity of the underlying phenomena. We examine a case in point in §7.

A further central idea is the fact that the Δ-operators over a chosen ring E_* form a group under composition of power series, the identity being D. Thus

$$\Delta = \theta(D) = e^{\theta D} - 1 \ \in E^2((D))$$

has an inverse Δ-operator

$$\bar{\Delta} = \bar{\theta}(D) = e^{\bar{\theta} D} - 1 \in E^2((D))$$

satisfying $\theta(\bar{\theta}(D)) = \bar{\theta}(\theta(D)) = D$.

<u>(1.11) Definitions.</u> We call $\bar{\Delta}$ the <u>conjugate Δ-operator</u> to Δ, and its associated sequence $(\bar{b}_n^E(x))$ the <u>conjugate sequence</u> of Δ. □

We may write $(E_*, \bar{\Delta})$ as $\bar{E}$.

<u>(1.12) Examples.</u> $\bar{K}$ is specified by

$$\bar{\Delta} = u^{-1}\log(1 + uD),$$

and the polynomials $\bar{b}_n^K(x)$ are the <u>exponential polynomials</u>.

Once more, Φ is subtler by far, and the polynomials $\bar{b}_n^\Phi(x)$ are a version of the <u>Bell polynomials</u>, as documented, for example, by Riordan (1958). □

Following S. Joni & G-C. Rota (1979), we can now enunciate the

<u>fundamental fact of umbral calculus</u> as

$$\sum b_n(x)D^n = e^{x\bar{\Delta}} \quad \text{in} \quad \pi(E)_* \hat{\otimes} E^*((D)) \tag{1.13}$$

Essentially, this is a simple restatement of duality. Even so, in tandem with the transfer formulae (1.6) it can afford surprisingly powerful results.

A striking example, and a linchpin of our thesis, is a purely formal proof of the various <u>Lagrange inversion</u> principles for the compositional inverse of a power series. Until relatively recently, these were established by analytic methods such as the calculus of residues (e.g. see E.T. Whittaker & G.N. Watson (1927)). Nowadays , we can state and prove them in a more conceptual and illuminating fashion, as in Roman (1984). The two we most require refer to a Δ-operator $e^{\theta D}-1$, and an arbitrary power series $g(t) \in E^*[[t]]$, with derivative $\dot{g}(t)$. They are LIP - The Lagrange inversion principle

$$\bar{\theta}_{i-1} = \langle (\tfrac{\Delta}{D})^{-i} \mid x^{i-1} \rangle \tag{1.14}$$

and

FLIP - The full Lagrange inversion principle

$$\langle g(\bar{\Delta}) \mid x^i \rangle = \langle \dot{g}(D)(\tfrac{\Delta}{D})^{-i} \mid x^{i-1} \rangle \tag{1.15}$$

Visibly, LIP is FLIP with $g(t) = t$.

2. <u>THOM CLASSES & Δ-OPERATORS</u>

In this section, we establish the basic link between algebraic topology and umbral calculus.

Wherever possible, and to dovetail in with future usage, we write CP_0 for CP_+^∞ and $E^*(\ )$ for a multiplicative (torsion free) cohomology theory, with complex orientation $t^E \in E^2(CP_0)$. Thus we may denote the pair $(E^*(\ ), t^E)$ by E. We have the dual structure theorems

$$\begin{aligned} E^*(CP_0) &\cong E^*[[t^E]] \\ E_*(CP_0) &\cong E_*\{1, b_1^E, b_2^E, \ldots\} \end{aligned} \tag{2.1}$$

e.g. see J.F. Adams (1974). Often, t^E and b_n^E can be abbreviated to t and b_n without ambiguity.

The simplest example is $E^*(\ ) = H^*(\ )$, integral cohomology, with $t^H = c_1(\eta)$ where η is the Hopf bundle. So adopting $x \in H_2(CP_0)$ as the standard spherical generator, taking Pontrjagin powers gives $x^n = n!b_n^H$, and the $\cap$ product is fully described by

$$t^m \cap x^n = [n]_m x^{n-m}. \tag{2.2}$$

This formula is the quintessence of the symbolic approach, since it allows us to write $t^H = \frac{d}{dx} = D$ and $H^*(CP_0) \cong \mathbb{Z}[D]$.

Other notation we require is provided by viewing the Thom class as a map $t : CP_0 \longrightarrow E$ into the second stage of the spectrum E, and labelling $t_* b_{n+1}^H$ as b_n in $H_{2n}(E)$. This notation is less confusing, and more useful, than might appear at first sight! For example, since $x^{n+1} \in H_{2n+2}(CP_0)$ is spherical, we deduce a simple result.

<u>(2.3) Lemma</u> For each $n \geqslant 0$, there is an element $\theta_n \in E_{2n}$ with Hurewicz image $(n+1)!b_n \in H_{2n}(E)$. □

Of course, $\theta_0 = b_0 = 1$.

We must now recall, e.g. from Switzer (1975), the general Hurewicz and Boardman maps. The Hurewicz map

$$\underline{e} : E_*(CP_0) \longrightarrow H_*(E\mathbb{Q}) \otimes H_*(CP_0) \cong E\mathbb{Q}_*[x], \tag{2.4}$$

with x as above, is monic, and so allows us to regard b_n as a polynomial $b_n(x) \in E\mathbb{Q}_{2n}[x]$. Similarly, the Boardman map

$$\bar{e} : E^*(CP_0) \longrightarrow H_*(E\mathbb{Q}) \hat{\otimes} H^*(CP_0) \cong E\mathbb{Q}^*[[D]], \tag{2.5}$$

with D as above, is also monic.

<u>(2.6) Lemma</u> - The 1st divided power principle: the map $\bar{e}$ embeds $E^*(CP_0)$ in the ring of divided power series $E^*((D))$.

Proof Since $t_*(b^H_{n+1}) = b_n$,

$$\bar{e}(t) = \sum_{i\geq 0} b_i D^{i+1}$$

$$= \sum_{i\geq 0} \theta_i \frac{D^{i+1}}{(i+1)!} \in E^*((D)).$$

Hence $\bar{e}(t^m) \in E^*((D))$ for each m, and (2.1) yields the result. □

(2.7) Corollary - The 2nd divided power principle: under $\bar{e}$, $\frac{t^m}{m!} \in E^*((D))$, and $E^*((t)) \cong E^*((D))$. □

Henceforth, we think of $\underline{e}$ and $\bar{e}$ as embeddings of submodules and omit them from our formulae, so leading to the pivotal idea.

(2.8) Theorem. The $\cap$ product action of $t \in E^2(CP_0) \subset E^2((D))$ on $E_*(CP_0) \subset E\mathbb{Q}_*[x]$ is that of the Δ-operator $e^{\theta D} - 1$.

Proof. This follows from (2.1), (2.2) and the proof of lemma (2.6), together with the product properties of $\underline{e}$ and $\bar{e}$. □

Thus given a complex oriented theory $E = (E^*(\ ),t)$, we may construe it as a Δ-operator $E = (E_*, e^{\theta D} - 1)$. In addition, the H-space structure on CP_0 ensures that $E^*[[\Delta]]$ has a coproduct as in (1.9), whence E is Leibniz. This extra information can sometimes cloud the eal issues, as we have already remarked!

The LIP of (1.14) provides a formula

$$D = e^{\bar{\theta}\Delta} - 1 \in E^2((\Delta)),$$

which often proves to be extremely useful, in spite of the fact that, in general, D does not lie in $E^2(CP_0)$.

Now for some important examples.

(2.9) Complex K-theory. Here $E = K$, with $K_* = \mathbb{Z}[u,u^{-1}]$ and $\Delta = ch(\eta - 1) = u^{-1}(e^{uD}-1)$. So $\theta_i = u^i$, and

$$D = u^{-1}\log(1+u\Delta) = \sum_{i\geq 0} (-1)^i i! \frac{u^i \Delta^{i+1}}{(i+1)!}.$$ □

(2.10) Adams G-theory, constructed, for example, in Adams (1969). Here E is the p-local spectrum G, p an odd prime,

$$G_* = \mathbb{Z}_{(p)}[u^{p-1}, u^{1-p}] \subset K_{(p)*},$$

and Δ is given by applying the LIP to

$$D = \Delta + (p-1)!u^{p-1}\frac{\Delta^p}{p!} + \ldots + (p^i-1)!u^{p^i-1}\frac{\Delta^{p^i}}{p^i!} + \ldots \qquad \square$$

This particular Δ might be called the p typical orientation class in $G^2(CP_0)$. There are others, such as described by Jankowski (1979).

(2.11) Complex cobordism. Here $E = MU$, with $MU_* \cong U_*$ of (1.10), and $\Delta = cf_1(\eta) = e^{\phi D} - 1$. The LIP immediately supplies Miscenko's theorem

$$D = \sum_{i \geq 0} i!CP^i \frac{\Delta^{i+1}}{(i+1)!}.$$

To examine the ϕ's of (2.11) in more detail, we hark back to Lemma (2.3). This shows that a manifold Fi^n representing $\phi_n \in MU_{2n}$ is constructed by making x^n transverse regular to $CP^{\infty-1}$ for the diagram

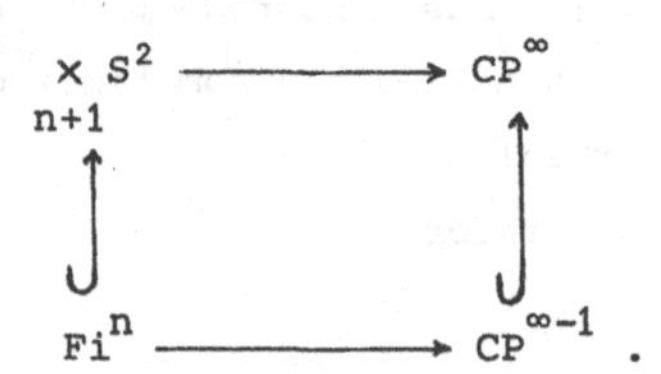

In fact the stable normal bundle of Fi^n factors through ΩS^3.

Note that the Todd genus provides the usual link $td : MU \longrightarrow K$, via $td\ \phi_n = u^n$.

Of course, MU is the universal Leibniz Δ-operator over a graded ring, and it is a splendid feature that the 'derivative' Δ can be realised on a subvariety of CP^N by intersection with a suitable CP^{N-1}.

We can manufacture a less concrete, but equally interesting, Δ-operator out of MU.

(2.12) Let $\Phi_* = \mathbb{Z}[\phi_1, \phi_2, \ldots] \subset MU_*$, and $\Delta = e^{\phi D} - 1$ as in (2.11), so that $\Phi = (\Phi_*, \Delta)$ is the universal non-Leibniz Δ-operator. $\square$

Although there is no homology theory $\Phi_*(\ )$, for any X with free MU homology we can define the Φ_*- modules $\Phi_*(X)$ and $\Phi^*(X)$. Observe in particular that if we write

$$!CP_* = \mathbb{Z}[CP^1, 2!CP^2, \ldots, i!CP^i, \ldots] \subset MU_*$$

then by the LIP we have

$$!CP_* = \Phi_*. \tag{2.13}$$

(2.14) The p-local BP theory, E = BP. This is the Leibniz extension of the Δ-operator given by applying the LIP to

$$D = \Delta + (p-1)!CP^{p-1}\frac{\Delta^p}{p!} + \ldots + (p^i-1)!CP^{p^i-1}\frac{\Delta^{p^i}}{p^i!} + \ldots$$

over

$$\mathbb{Z}_{(p)}[(p-1)!CP^{p-1}, \ldots, (p^i-1)!CP^{p^i-1}, \ldots] \subset MU_{(p)*}. \qquad \square$$

Actually, (2.10) and (2.14) both arise naturally as integral theories, and localisation is only required for certain more sophisticated applications.

3. UMBRALLY NUMERICAL POLYNOMIALS & $E_*(CP_0)$.

We are now ready to describe $E_*(CP_0)$ in umbral terms, using the notation of §§1 and 2.

(3.1) Proposition. As a subcoalgebra of $E\mathbb{Q}_*[x]$,

$$E_*(CP_0) \cong \Pi(E)_*.$$

Proof. This follows directly from combining (1.5), Proposition (1.8), (2.1) and Theorem (2.8). $\square$

The proof identifies the Thom isomorphism with the action of Δ, and the generators b_n^E of (2.1) with the associated sequence $b_n^E(x)$ for Δ, so elaborating on (2.4). We therefore have access to a store of intriguing umbral formulae for these generators, by appealing to (1.6) and (1.13).

We can also obtain an alternative and equally important characterisation of $E_*(CP_0)$.

(3.2) Definitions. We call a polynomial $q(x) \in E\mathbb{Q}_*[x]$ umbrally numerical if, for each $m = 0,1,2,\ldots$, the element $q(m\theta) \in E\mathbb{Q}_*$ actually lies in E_*. Here

$$(m\theta)^j = (\underbrace{\theta+\ldots+\theta}_{m})^j \qquad \theta^{i+1} \equiv \theta_i.$$

We construe this evaluation of $q(m\theta)$ as umbral substitution of x by $m\theta$, or as substitution of x by the umbral integer m. □

(3.3) Theorem. The coalgebra $E_*(CP_0) \subset E\mathbb{Q}_*[x]$ consists of all umbrally numerical polynomials.

Proof. Clearly $q(x) \in E_*(CP_0)$ iff the Kronecker pairing satisfies

$$<\Delta^m |q(x)> \in E_*, \quad m = 0,1,2,\ldots,$$

or equivalently,

$$E_* \ni <(\Delta + 1)^m|q(x)>$$

$$= <(e^{\theta D})^m|q(x)> = <e^{m\theta D}|q(x)> = q(m\theta).$$ □

Several examples follow below. But first we comment that each $E_*(CP_0)$ is multiplicative, indeed a Hopf algebra, since E is Leibniz. This may not be transparent from the penumbral description!

Furthermore, in the style of (2.12), each $E_*(CP_0)$ has a precursor $\Theta_*(CP_0)$ which either agrees with $E_*(CP_0)$ or else fails to be an algebra. For $(E^*(\ ), e^{\theta D}-1)$ this arises from choosing

$$\Theta_* = \mathbb{Z}[\theta_1,\theta_2,\ldots] \subset E_*,$$

whence $\Theta_*(CP_0)$ is the corresponding penumbral coalgebra, representable by polynomials which are umbrally numerical over $\Theta_* \subset \Theta\mathbb{Q}_*$. Of course, a basis for $\Theta_*(CP_0)$ over Θ_* is still provided by the associated sequence $(b_n^E(x))$, because $^L\Theta_* \subseteq E_*$.

We turn to our examples.

(3.4) $H_*(CP_0)$. Since $\Delta = D$ over $\mathbb{Z}$, the penumbral coalgebra is the divided power algebra, with $b_n^H(x) = \frac{x^n}{n!}$. This is a Hopf algebra, because

$$b_k^H(x)b_n^H(x) = \binom{k+n}{k} b_{k+n}(x). \qquad \square$$

(3.5) $K_*(CP_0)$. Since $\Delta = u^{-1}(e^{uD}-1)$ is the discrete derivative over K_*, the penumbral coalgebra is generated by the binomial coefficient polynomials $b_n^K(x) = \frac{1}{n!}[x]_{nu}$ of Example (1.7), where nu is the umbral integer n. As

$$b_n^K(mu) = u^n\binom{m}{n} \qquad \forall m,n,$$

each $b_n^K(x)$ is umbrally numerical. Moreover, $K_*(CP_0)$ is a Hopf algebra by virtue of the convolution identity

$$b_k^K(x)b_n^K(x) = \sum_\ell \binom{\ell}{k}\binom{k}{\ell-n} u^{k+n-\ell} b_\ell(x).$$

Since K_* is periodic, we have the version

$$K_*(CP_0) \cong K_* \otimes A$$

where A is the penumbral coalgebra for e^D-1 over $\mathbb{Z}$, and hence is generated by

$$\check{b}_n(x) = \binom{x}{n} = \frac{1}{n!}[x]_n.$$

In fact A is the well-known algebra of numerical polynomials whose infiltration into algebraic topology, for example documented in A. Baker et al. (1985), was a major stimulus to this work. $\square$

(3.6) $MU_*(CP_0)$. Being the universal case, this is the subtlest example; the umbral viewpoint seems to offer topologists a genuine alternative perspective. As hinted in Examples (1.7) & (1.12), the generators $b_n^{MU}(x)$ are best described in terms of the Bell polynomials $b_n(x;x_1,x_2,\ldots)$, described in Roman (1984). Recalling the conjugate umbra

$$\bar{\phi} = \,!CP \qquad\qquad (!CP)^n = n!CP^n,$$

we have

$$b_n^{MU}(x) = \frac{1}{n!}b_n(x;\bar{\phi}_1,\ldots,\bar{\phi}_{n-1}).$$

These may reasonably be labelled the (normalised) conjugate Bell polynomials.

These polynomials are extremely intricate, and those of us such as Ray (1969), who have laboured to compute a few cases (see Example (1.7)) would have been delighted to discover Riordan's tables before now! They have a deep combinatorial significance, e.g. see L. Comtet (1974), which is one of the main strands to our theory, and which surfaces in the next section. That they are umbrally numerical is again not directly evident.

The multiplicative structure of $MU_*(CP_0)$ is also complicated, since the $b_n^{MU}(x)$ are not multiplicatively closed over Φ_*. In a suitable sense, $MU_*(CP_0)$ is the smallest Hopf algebra containing the coalgebra $\Phi_*(CP_0) = \Pi(\Phi)_*$. □

A more suggestive and practical formula for the conjugate Bell polynomials is

$$\begin{aligned} b_n^{MU}(x) &= \frac{1}{n!}x(x+CP)(x+2CP)\ldots(x+(n-1)CP): \\ &= \frac{1}{n!}[x:]^{nCP} = \frac{1}{n!}[x:]_{-nCP}. \end{aligned} \tag{3.7}$$

Here the symbol : indicates an especially careful expansion of the brackets effected in such a way that k adjacent CP's multiply together to give CP^k, $1 \leq k \leq n-1$. Thus

$$\begin{aligned} b_3^{MU}(x) &= \frac{1}{6}x(x+CP)(x+2CP): \\ &= \frac{1}{6}(x^3 + 3CP^1x^2 + 2CP^2x), \end{aligned}$$

agreeing with Example (1.7).

The Todd genus now relates smoothly with (3.5), as $tdCP^n = (-u$ may be expressed

$$tdCP = -u \tag{3.8}$$

and then

$$\mathrm{td}b_n^{MU}(x) = \mathrm{td}[x:]_{-nCP} = [x:]_{nu} = [x]_{nu}$$
$$= b_n^K(x).$$

Our proof of (3.7) is related to the Redfield-Polya theory of cycle-type enumerators, outlined in C. Berge (1971). This should tantalise the cyclic cohomologists currently working with CP^∞.

<u>(3.9) $BP_*(CP_0)$.</u> We need only say here that this is the universal p-typical Leibniz example, and that the generators may be called the <u>p-typical</u> conjugate Bell polynomials, given by

$$b_n^{BP}(x) = \frac{1}{n!}[x:p]_{-nCP} \tag{3.10}$$

This elaboration of the notation of (3.7) indicates that we discard all but the p-typical clusters of CP's, i.e. those of $p-1,\ p^2-1,\ldots,\ p^i-1,\ldots$ adjacent terms. □

<u>(3.11) $G_*(CP_0)$,</u> another p-typical example. Here

$$b_n^G(x) = \frac{1}{n!}[x:p]_{nu},$$

and

$$G_*(CP_0) \cong G_* \otimes A(p),$$

where $A(p)$ has generators $\check{b}_n(x) = \frac{1}{n!}[x:p]_n$. So at $p = 3$, $b_4^G(x) = \frac{1}{24}(x^4 + 8u^2x^2)$, for example. □

It is a productive exercise to experiment with the concept of umbrally numerical in (3.9) and (3.11), as well as reformulations such as

$$b_n^G(x) = \sum \frac{u^{n-k}}{j_1!j_2!\ldots j_n!p^{j_2+2j_3+\ldots+(n-1)j_n}}x^k,$$

summing over $j_1+\ldots+j_n = k$ and $(p-1)j_2+\ldots+(p^{n-1}-1)j_n = n-k$.

In conclusion, we note that the fundamental fact of umbral calculus, (1.13), gives

$$\sum b_n^E(x)D^n = e^{x\bar{\Delta}} \quad \text{in} \quad E_*(CP_0) \hat{\otimes} E^*((D)).$$

This has recently been much used in computations with $E = MU$, and seems to have been rediscovered in this context by I. Hansen.

Our various formulae for $b_n^E(x)$ in fact determine generalised Stirling numbers for each E.

4. COMBINATORIAL ASPECTS

In this section we are sadly only able to offer a lightning sketch of a particularly rich and suggestive area which provides important insight. We are currently preparing a document, Ray (198?), purporting to explain the details and show that we may feed back the topology into the combinatorics to produce generalisations of the Mullin-Rota theory of binomial enumeration, as well as of the well-known chromatic and flow polynomials: e.g. see M. Aigner (1979).

The central theme, inspired by M. Content et al. (1980), is to enrich existing partition poset theory into the realm of category theory, and then to exploit the role played by the Bell polynomials in the universal example.

Following Aigner, we write $\mathcal{P}(n)$ for the poset of partitions of the set $\underline{n} = \{1,2,\ldots,n\}$. Thus we may display $\mathcal{P}(3)$ as

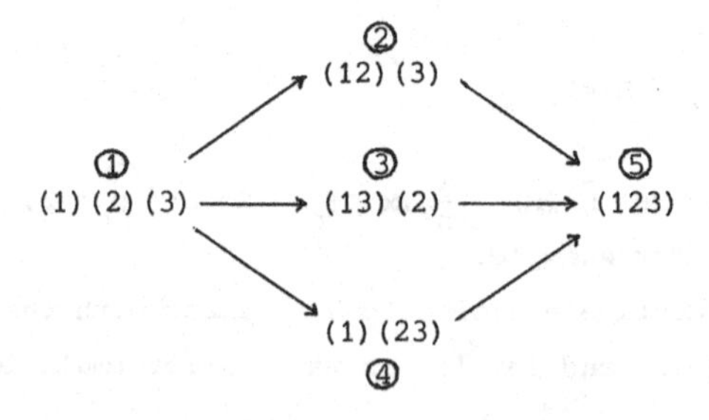

where each arrow represents an ordering, and the circled numbers give a compatible total ordering.

Each $\mathcal{P}(n)$ has a zeta-matrix Z_n, displaying incidence, which has an inverse Möbius matrix M_n, e.g. see E.K. Lloyd (1985). These requi the total ordering, to make them explicit.

(4.1) Example.

$$Z_3 = \begin{pmatrix} 1 & 1 & 1 & 1 & 1 \\ & 1 & 0 & 0 & 1 \\ & & 1 & 0 & 1 \\ & & & 1 & 1 \\ & 0 & & & 1 \end{pmatrix} \qquad M_3 = \begin{pmatrix} 1 & -1 & -1 & -1 & 2 \\ & 1 & 0 & 0 & -1 \\ & & 1 & 0 & -1 \\ & & & 1 & -1 \\ & 0 & & & 1 \end{pmatrix}$$ □

We now enrich $\mathcal{P}(n)$ into the category of partitions of n by labelling each partition according to type. We say that a partition π which brings together k_1 pairs of parts, k_2 triples,..., k_{n-1} n-tuples, is of type $\phi_1^{k_1}\phi_2^{k_2}\ldots\phi_{n-1}^{k_{n-1}}$. Thus the type of π in $\mathcal{P}(n)$ is always a monomial $\phi \in \phi_* = \mathbb{Z}[\phi_1, \phi_2, \ldots]$. We can now display the category $\mathcal{P}(3)$ as

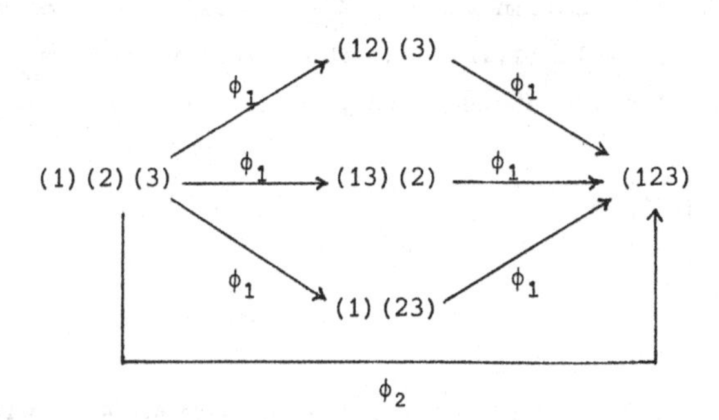

(4.2) Definitions. The zeta type-matrix of $\mathcal{P}(n)$ is the matrix Z_n^Φ which displays incidence-by-type in $\mathcal{P}(n)$. We call its inverse the Möbius type-matrix M_n^Φ. □

4.3 Example.

$$Z_3^\Phi = \begin{pmatrix} 1 & \phi_1 & \phi_1 & \phi_1 & \phi_2 \\ & 1 & 0 & 0 & \phi_1 \\ & & 1 & 0 & \phi_1 \\ & & & 1 & \phi_1 \\ & & & & 1 \end{pmatrix}, \quad M_3^\Phi = \begin{pmatrix} 1 & -\phi_1 & -\phi_1 & -\phi_1 & 3\phi_1^2 - \phi_2 \\ & 1 & 0 & 0 & -\phi_1 \\ & & 1 & 0 & -\phi_1 \\ & & & 1 & -\phi_1 \\ & & & & 1 \end{pmatrix}$$ □

(4.4) Note. We may determine the entries in M_n^Φ by the following procedure. Assign to each arrow of type ϕ_i in $\mathcal{P}(n)$ its Möbius type $-\phi_i$, and extend this assignation over composition in $\mathcal{P}(n)$ by using multiplication in Φ_*. Then the (r,s)th entry in M_n^Φ is the sum of the Möbius types of compositions between the elements r and s in $\mathcal{P}(n)$.

Thus the $(1,5)$th entry of M_3^Φ is the sum of the Möbius types

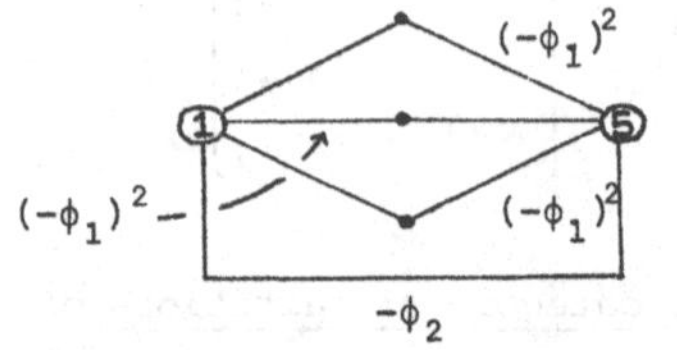

$(-\phi_1)^2$ $(-\phi_1)^2$ $(-\phi_1)^2$ $-\phi_2$ $= 3\phi_1^2 - \phi_2$.

This agrees with Example (4.3).

Now let $\underline{x}$ be the column vector $(x^n, x^{n-1}, x^{n-1}, \ldots, x^2, x^2, x)^t$, (where the number of entries x^i, $2 \leqslant i \leqslant n-1$ can be prescribed).

(4.5) Proposition. The column vector $Z_n^\Phi \underline{x}$ consists of the (unnormalised) Bell polynomials of Example (1.12), and the column vector $M_n^\Phi \underline{x}$ consists of the (unnormalised) conjugate Bell polynomials of (3.6). Thus

$$Z_n^\Phi \underline{x} = \begin{pmatrix} n!\bar{b}_n^\Phi(x) \\ \vdots \\ 1!\bar{b}_1^\Phi(x) \end{pmatrix} \quad \text{and} \quad M_n^\Phi \underline{x} = \begin{pmatrix} n!b_n^\Phi(x) \\ \vdots \\ 1!b_1^\Phi(x) \end{pmatrix}$$ □

In particular, tying together Proposition (4.5) with example (4.3) we see that

$$M_3^\Phi \begin{pmatrix} x^3 \\ x^2 \\ x^2 \\ x^2 \\ x \end{pmatrix} = \begin{pmatrix} x^3 - 3\phi_1 x^2 + (3\phi_1^2 - \phi_2)x \\ x^2 - \phi_1 x \\ x^2 - \phi_1 x \\ x^2 - \phi_1 x \\ x \end{pmatrix}, \tag{4.6}$$

thus confirming Example (1.7), and leading us to the underlying philosophical principle behind this section: the elements $\phi_i \in MU_{2i}$ constructed in 2.11, and the partition types ϕ_i of Definition 4.2 are the same!!

The heart of this correspondence is the proof of Proposition (4.5), which revolves around one of the many combinatorial descriptions of the Bell polynomials, first popularised by R. Frucht (1965).

One spin-off from the LIP and the fundamental fact (1.13) is a purely combinatorial method for evaluating CP^i in terms of the ϕ_i's; or, equivalently its Hurewicz image in $H_{2i}(MU)$. This tells us simply to sum all the Möbius types of paths in $\mathcal{P}(i+1)$ between the initial and final object, and divide by $i!$. Thus, using $\mathcal{P}(3)$ and Note (4.4),

$$CP^2 = \frac{1}{2!}(3\phi_1^2 - \phi_2) = 6b_1^2 - 3b_2 = \frac{1}{2!}\bar{\phi}_2, \tag{4.7}$$

as known!

We briefly digress to mention an alternative interpretation of Proposition (4.5).

The unnormalised Bell polynomials $n!\bar{b}_n^{\Phi}(x)$ enumerate forests of labelled complete graphs, or simplexes, or monotonic linear trees, on n vertices by type: i.e. they are typifiers. Similarly, their conjugates are typifiers for paths, or principal order ideals of such objects. Thus our equation (4.6)

$$3!b_3^{\Phi}(x) = x^3 - 3\phi_1 x^2 + (3\phi_1^2 - \phi_2)x$$

enumerates

1 path of type 1, from 3 to 3 parts
3 paths of type $-\phi_1$, from 3 to 2 parts
3 paths of type ϕ_1^2 ⎤ from 3 to 1 part
1 path of type $-\phi_2$ ⎦

We expand on this in Ray (198?).

We can now apply the combinatorial version of the Todd genus $td : \phi_i \longmapsto u^i$, and interpret u^i as the number-type of any partition amalgamating $i+1$ parts. We then obtain a realisation of our graded version of the exponential and falling factorial polynomials as enumerators.

The language of typifiers and enumerators offers an ideal framework for explaining the divided power properties (1.5) of the normalisation of all these polynomials, as well as the umbrally numerical phenomenon.

We have concentrated above on the universal case $E = \Phi$, only hinting at $E = K$. In fact each Δ-operator, and hence each complex oriented theory, has its accompanying partition theory. We conclude with a glimpse of a p-typical case, a viewpoint which seems to be rarely adopted by combinatorialists.

<u>(4.8) Example.</u> $E = G$ of (2.10), and $p = 3$.

As hinted in (3.10), the 3-typical Todd genus satisfies

$$\left.\begin{array}{ll} \bar{\phi}_1 = CP^1 \longmapsto 0 & \bar{\phi}_3 = 3!CP^3 \longmapsto 0 \\ \bar{\phi}_2 = 2!CP^2 \longmapsto 2!u^2 & \quad \cdots \\ \multicolumn{2}{c}{\bar{\phi}_{3^i-1} = (3^i-1)!CP^{3^i-1} \longmapsto (3^i-1)!(u^2)^{(3^i-1)/2}} \end{array}\right\} \quad (4.9)$$

So by our remarks preceding (4.7), we acquire a new type of arrow in $\mathcal{P}(n)$ only if $n = 3^i$, and its label is set by (4.9):

$\mathcal{P}(1)$	$\mathcal{P}(2)$	$\mathcal{P}(3)$
(1)	(1)(2)	$(1)(2)(3) \xrightarrow{-2u^2} (123)$
	$\phi_1 \longmapsto 0$	$\left.\begin{array}{l} -\phi_2 \\ \bar{\phi}_2 \end{array}\right] \longmapsto 2u^2$

$\mathcal{P}(4)$ (curtailed)

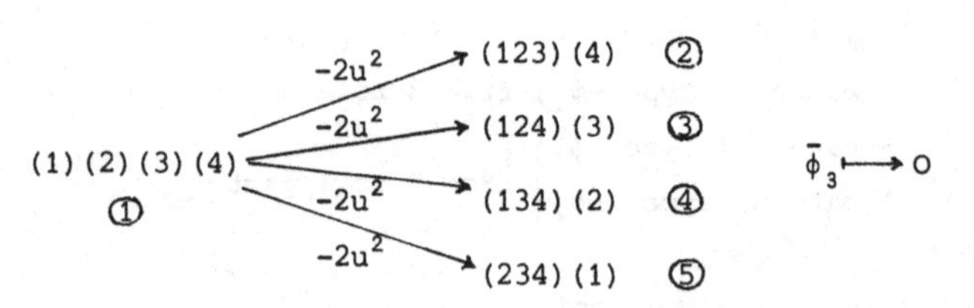

This gives the zeta and Möbius type-matrices

$$Z_4^G = \begin{pmatrix} 1 & -2u^2 & -2u^2 & -2u^2 & -2u^2 \\ & 1 & 0 & 0 & 0 \\ & & 1 & 0 & 0 \\ & 0 & & 1 & 0 \\ & & & & 1 \end{pmatrix}, \quad M_4^G = \begin{pmatrix} 1 & 2u^2 & 2u^2 & 2u^2 & 2u^2 \\ & 1 & 0 & 0 & 0 \\ & & 1 & 0 & 0 \\ & 0 & & 1 & 0 \\ & & & & 1 \end{pmatrix}$$

and so by Proposition (4.5), and $\wp(1)$, $\wp(2)$ and $\wp(3)$,

$$b_1^G(x) = x, \quad 2!b_2^G(x) = x^2, \quad 3!b_3^G(x) = x^3 + 2u^2x$$

$$4!b_4^G(x) = x^4 + 8u^2x^2.$$

This agrees with (3.10). □

A host of strategic calculations can be described in this fashion: the evident goal is BP, but G is a good warm-up!

5. STUNTED PROJECTIVE SPACE & DIRAC'S δ

In this section we introduce the operator calculus.

Given any $b \in \mathbb{Z}$, let CP_b denote the Thom spectrum, or stunted projective space, $T(b\eta)$ over CP^∞. For $b \geq 0$ this is a genuine complex of the form

$$S^{2b} \cup e^{2b+2} \cup \ldots ,$$

and for $b < 0$ we assume the spectrum chosen such that the same cell indexing holds. Note that $CP_0 = CP^\infty_+$.

Several results concerning CP_b follow at once from the traditional geometry of Thom complexes. They are sketched in H. Miller (1982), and we simply state them here.

There is a cofibration involving collapsing the bottom cell

$$S^{2b} \longrightarrow CP_b \xrightarrow{\ c\ } CP_{b+1} , \tag{5.0}$$

induced by $b\eta \subset (b+1)\eta$, with iterates

$$CP_b^{b+n-1} \longrightarrow CP_b \xrightarrow{\ c^n\ } CP_{b+n} \quad \text{for } n \geq 1. \tag{5.1}$$

This induces an inclusion

$$E^*(CP_b) \underset{(c^n)^*}{\hookleftarrow} E^*(CP_{b+n}). \tag{5.2}$$

There are compatible diagonal maps, such as

$$\begin{array}{ccc} CP_b & \longrightarrow & CP_0 \wedge CP_b \\ c\downarrow & & \downarrow 1\wedge c \\ CP_{b+1} & \longrightarrow & CP_0 \wedge CP_{b+1} \end{array} , \qquad (5.3)$$

for all b. These induce the Thom isomorphisms, which display $E^*(CP_b)$ as a module over $E^*(CP_0)$ (written by juxtaposition) after choice of Thom class ${}^bt \in E^{2b}(CP_b)$. For $b \geq 0$ we can select ${}^bt$ such that the inclusion $E^*(CP_b) \subset E^*(CP_0)$ of (5.1) identifies it with Δ^b. The module structure of (5.3) is then that of the ideal (Δ^b).

For $b < 0$, say $b = -d$, more care is needed since $E^*(CP_{-d})$ is not a ring. Taking the inclusion (5.1) in the form $E^*(CP_0) \subset E^*(CP_{-d})$, we may pick ${}^{-d}t$ so that $\Delta^{d+n} \cdot {}^{-d}t = \Delta^n$ in $E^*(CP_{-d})$, using the compatibility of (5.3). This suggests we write ${}^{-d}t = \Delta^{-d}$, so that the module structure is then given by

$$\Delta^{d+n}\Delta^{-d} = \Delta^n \quad \text{in} \quad E^{2n}(CP_{-b}), \qquad n \geq 1 .$$

By considering the diagonal

$$CP_{-d} \longrightarrow CP_{-c} \wedge CP_{c-d}$$

we may extend this to $\Delta^m\Delta^n = \Delta^{m+n}$ whenever $m,n,m+n \geq -d$. This useful, but hazardous notation was pioneered by Miller.

Here is a fundamental example.

(5.4) $H^*(CP_{-d})$. Following on from above,

$$\begin{aligned} H^*(CP_{-d}) &\cong H^*(CP_0)\{D^{-d}\} \\ &\cong \mathbb{Z}\{D^{-d},\ldots, D^{-1},1,D,D^2,\ldots\}. \end{aligned}$$

Dualising, there are elements $\delta^n \in H_{-2n}(CP_{-d})$ with

$$H_*(CP_{-d}) \cong \mathbb{Z}\{\delta^d,\ldots, \delta,1,x,\tfrac{x^2}{2},\ldots\}.$$

Further appeal to (5.3) shows that we may also define a partial $\cap$ produc operation, with $D = \frac{d}{dx}$, satisfying

$$D \cap 1 = \delta, \qquad D \cap \delta^n = \delta^{n+1},$$
$$D^{-1} \cap 1 = x, \qquad D^{-1} \cap \delta^n = \delta^{n-1}, \quad D^{-1} \cap x^n = \frac{x^{n+1}}{n+1} . \qquad \square$$

So to be consistent with §2, we would like to regard δ as the derivative of 1: which is where Heaviside's ideas, e.g. see Erdelyi (1962), have a role to play. We assume that $1, x, x^2, \ldots$ are as given for $x > 0$, and take the value 0 for $x \leq 0$. So 1 becomes Heaviside's unit function, and its derivative the Dirac δ, which is a generalised, or impulse function. We shall not go further into the Mikusinski-Raevski theory of generalised functions here. Suffice it to say that their methods of convolution quotients can be used to construct an algebra which subsumes both $H^*(CP_{-d})$ and $H_*(CP_{-d})$, in essence by identifying D with δ and D^{-1} with x. We shall retain only the residue of profit yielded by the language of Dirac's δ, and the notation δ^{n+1} for $D^n \cap \delta$.

We emphasise that D^{-1} actually exists only as a Thom isomorphism or an element of $H^*(CP_{-d})$, but not as an element of $H^*(CP_0)$. In disguise, this is the point of obscurity suffered by Heaviside's original approach.

As in §2, we now employ the Hurewicz and Boardman maps to extend the above philosophy to $E^*(\ )$.

(5.5) Proposition. As elements of $E\mathbb{Q}^*[[D]]$,

$$\Delta^{-d} = D^{-d}\left(\frac{D}{e^{\theta D} - 1}\right)^d \qquad\qquad \theta^{i+1} \equiv \theta_i .$$

Proof. This follows from the multiplicative properties of $\bar{e}$, and the diagrams (5.3). $\square$

(5.6) Definition. We call the elements ${}^dB^E_i \in E\mathbb{Q}_{2i}$, defined by

$$\left(\frac{D}{e^{\theta D}-1}\right)^d = e^{{}^dB^E D} \qquad\qquad ({}^dB^E)^i \equiv {}^dB^E_i ,$$

the d-th order E-Bernoulli numbers. If $d = 1$, we abbreviate to B^E_i, the E-Bernoulli numbers. $\square$

Thus

$$\Delta^{-d} = \sum_{i=-d}^{\infty} \frac{{}^{d}B^{E}_{d+i}}{(d+i)!} D^{i} \tag{5.7}$$

Our goal is now to characterise $E_*(CP_{-d})$ as a sub E_*-module of

$$H_*(E\mathbb{Q}) \otimes H_*(CP_{-d}) \cong E\mathbb{Q}_*\{\delta^{d},\ldots,\delta,1,x,x^2,\ldots\},$$

whose elements we may write as polynomials $r(\delta,x)$. From §3 we already have the notion of such a polynomial being umbrally numerical via the substitutions $r(m\theta)$, i.e.

$$\delta \longmapsto 0, \quad x \longmapsto \underbrace{\theta+\ldots+\theta}_{m} \quad \text{for} \quad m = 0,1,2,\ldots .$$

In addition, we say that $r(\delta,x)$ is <u>Bernoulli numerical</u> if the substitutions

$$\left.\begin{aligned} \delta^{j} &\longmapsto \frac{{}^{a}B^{E}_{a-j}}{(a-j)!} & 0 < j \leqslant a \leqslant d \\ x^{j} &\longmapsto \frac{j!\,{}^{a}B^{E}_{a+j}}{(a+j)!} & j \geqslant 0 \end{aligned}\right\} \tag{5.8}$$

yield an element $r({}^{a}B)$ of $E_* \subset E\mathbb{Q}_*$ for each $1 \leqslant a \leqslant d$.

Cumbersome though this formula may seem, it can work well in practice.

<u>(5.9) Theorem.</u> The submodule

$$E_*(CP_{-d}) \subset E\mathbb{Q}_*[x]\{\delta^{d},\ldots,\delta,1\}$$

consists of all polynomials that are both umbrally and Bernoulli numerical.

<u>Proof.</u> This parallels Theorem (3.3), after noting

$$E^*(CP_{-d}) \cong E^*\{\Delta^{-d},\ldots,\Delta^{-1},1,\Delta,\Delta^2,\ldots\},$$

and that $r({}^{a}B) = \langle \Delta^{-a} | r(\delta,x) \rangle$ in $E\mathbb{Q}_*$, by (5.7). □

We call the basis elements

$$b_n(\delta,x) \in E\mathbb{Q}_{2n}(CP_{-d}), \qquad n \geq -d,$$

dual to the powers of Δ, the <u>extended</u> associated sequence for Δ. Clearly $b_{-d}(\delta,x) = \delta^d$.

Our current investigations centre around the case $d = 1$. When $d = 1$, the extended associated sequence begins

$$\delta, \; 1 + \frac{1}{2}\theta_1\delta, \qquad x + (\frac{1}{6}\theta_2 - \frac{1}{4}\theta_1^2)\delta \; .$$

The combinatorial interpretations of §4 may be adapted so as to apply to extended associated sequences, and intriguing connections with zeta functions seem to emerge.

<u>(5.10) Corollary</u>. The submodule

$$E_*(CP_{-1}) \subset E\mathbb{Q}_*[x]\{\delta,1\}$$

contains all polynomials of the form

$$q(x) - q(B/)\delta \qquad\qquad (B/)^i \equiv \frac{B^E_{i+1}}{i+1}$$

where $q(x)$ is umbrally numerical. □

For example, the elements

$$x^n - \frac{B^E_{n+1}}{n+1}\delta \text{ in } E_{2n}(CP_{-1}), \qquad n \geq 0 \qquad (5.11)$$

play a major role in §6.

6. <u>TRANSFER & THE KUMMER CONGRUENCES</u>

In this section we employ §5 to bring together results of K-h. Knapp (1979) and Miller (1982) on the complex transfer, and of Baker (1982), (1984), on primitives and Kummer congruences. Throughout, we identify $H_*(MU)$ and $H_*(BU)$ via the canonical Thom isomorphism.

Our unifying theme is a study of two distinct Thom classes in $E^{-2}(CP_{-1})$. We begin with the universal Δ^{-1}, which we represent by a map

$$u : CP_{-1} \longrightarrow S^2MU \tag{6.1}$$

and we seek the effect of u_* on E-homology. This follows directly from the description $\Delta^{-1} = D^{-1}e^{B^{MU}D}$, but may be also and instructively obtained with the aid of the map $\perp : CP_0 \longrightarrow BU$ which classifies $-\eta$.

(6.2) Proposition. The map $\perp_* : H_*(CP_0) \longrightarrow H_*(BU)$ may be described by $x^n \longmapsto B_n^{MU}$, $n \geq 0$.

Proof. By definition,

$$\perp_*(1 + b_1^H T + \ldots + b_n^H T^n + \ldots) = (1 + b_1^H T + \ldots + b_n^H T^n + \ldots)^{-1}.$$

So $$\perp_* e^{xT} = (\tfrac{1}{T}(e^{\phi T} - 1))^{-1} = e^{B^{MU}T}, \qquad (B^{MU})^i \equiv B_i^{MU}$$

as sought. □

(6.3) Corollary. The map $u_* : E_*(CP_{-1}) \longrightarrow E_{*+2}(MU)$ may be described by the umbral substitution

$$r(\delta, x) \longmapsto r(1, B^{MU}/), \qquad (B^{MU}/)^i \equiv \frac{B_{i+1}^{MU}}{i+1}.$$

Proof. Since u is the Thom complexification of $\perp$, the result follows by observing that $u_*(\delta) = 1$ and $x^n \longmapsto \frac{x^{n+1}}{n+1}$ under the Thom isomorphism $H_*(CP_{-1}) \cong H_{*+2}(CP_0)$. □

We can now deduce a result of Baker and Miller.

(6.4) Proposition.

$$\frac{B_n^{MU}}{n} - \frac{B_n^E}{n} \quad \text{in } E\mathbb{Q}_{2n}(MU), \quad n \geq 1$$

is integral, in that it lies in $E_{2n}(MU)$.

Proof. Apply Corollary (6.3) to the element $x^{n-1} - \frac{B_n^E}{n}\delta$ of (5.11). □

Of course, we may deduce $\frac{B_n^F}{n} - \frac{B_n^E}{n} \in E_*(F)$ for any complex oriented F, and if $E = F$ we write $\frac{B_n^R}{n} - \frac{B_n^L}{n}$ to indicate which factor of

$E_\wedge E$ is involved. When $F = K$ we may iterate the construction to create deeper elements of $E_*(K)$, which may be interpreted as generalising the classical Kummer congruences. For details, see Baker et al. (1985).

The complex transfer t is defined by the cofibre diagram

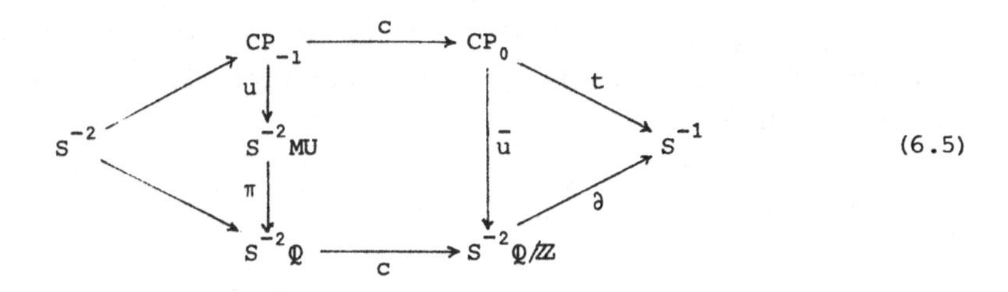

(6.5)

where π projects onto the bottom rational cell. We abbreviate $\pi \circ u$ to u.

(6.6) Proposition. The map $\bar{u}_*: E_*(CP_0) \longrightarrow E\mathbb{Q}/\mathbb{Z}_{*-2}$ may be described by the umbral substitution

$$p(x) \longmapsto -p(B/) \bmod E_*, \qquad (B/)^i \equiv \frac{B^E_{i+1}}{i+1} .$$

Proof. From Corollary (6.3) and (5.10), we note that

$$p(x) - p(B/)\delta \in E_*(CP_{-1})$$

maps via u_* to $-p(B/)$ in $E_{*+2}(S\mathbb{Q})$. Now apply c_* to both terms. □

This result, due to Knapp and Miller, allows us to carry out basic computations on the effect of t_* on stable homotopy.

Other of Knapp's and Miller's formulae can be efficiently obtained by symbolic methods. For example, the fundamental fact (1.13), in tandem with Proposition (6.6), gives

$$\begin{aligned}\sum_n \bar{u}_* b^E_n(x) D^n &= e^{-(B/)\bar{\Delta}} \\ &= -\frac{1}{\bar{\Delta}}(e^{B\bar{\Delta}}-1) \\ &= -\frac{1}{D} + \frac{1}{\bar{\Delta}} \bmod E_*\end{aligned}$$

So $$\bar{u}_* \sum_n b^E_n(x) D^{n+1} = \frac{D}{\bar{\Delta}} - 1 \quad \bmod E_* \tag{6.7}$$

Knapp's results on the transfers

$$CP_{-n} \xrightarrow{c} CP_{1-n} \xrightarrow{t} S^{1-2n} \qquad \text{(from (5.0))}$$

are also accessible by these means.

Our second Thom class arises from applying Knapp's work, and identifying S^2CP_{-1} with the Thom complex T of $(\eta-1) \otimes \eta$ over S^2CP_0 via the cofibration

$$S^1CP_0 \longrightarrow S^0 \longrightarrow S^2CP_{-1}.$$

This naturally describes a new Thom class $\Xi \in E^2(CP_{-1})$, and we represent the universal example by a map

$$w : S^2CP_{-1} \longrightarrow MU.$$

We mimic Corollary (6.3) by describing the effect of w_* on E-homology.

First remark that w is the Thom complexification of $\otimes : S^2CP_0 \longrightarrow BU$, representing $(\eta-1) \otimes \eta$.

<u>(6.7) Proposition.</u> The map $\otimes_* : H_*(CP_0) \longrightarrow H_{*+2}(BU)$ may be described by $x^n \longmapsto \Sigma^H_{n+1}$ $n \geq 0$, where Σ^H_{n+1} is the image of the Bott generator of $\pi_{2n+2}(BU)$.

<u>Proof.</u> See, for example, Switzer (1974). □

<u>(6.8) Corollary.</u> The map $w_* : E_*(CP_{-1}) \longrightarrow E_{*+2}(MU)$ may be described by the umbral substitution

$$r(\delta,x) \longmapsto r(1,\Sigma) \qquad \Sigma^i \equiv \Sigma^H_{i+1}$$

<u>Proof.</u> Combine Proposition (6.7) with the fact that $x^n \longmapsto x^n$ under the Thom isomorphism $H_*(S^2CP_{-1}) \cong H_*(T) \cong H_*(S^2CP_0+)$.

We deduce that in $MU^{-2}(CP_{-1})$,

$$\Xi = D^{-1} + \Sigma^H_1 + \Sigma^H_2 D + \ldots + \Sigma^H_{i+1} \frac{D^i}{i!} + \ldots \qquad (6.9)$$

So if we write d_n for the denominator of $\frac{B_n^{MU}}{n}$ in $MU\mathbb{Q}_{2n}$, we obtain a version of a result of Baker (1982).

(6.10) Proposition.

$$d_n \Sigma_n^H \in MU_{2n} \subset MU\mathbb{Q}_{2n}.$$

Proof. Since both are Thom classes, $\Delta^{-1} - \Xi$ lifts to $MU^{-2}(CP_0)$. Hence by the 1st divided power principle (2.6), along with (5.7) and (6.9),

$$\frac{B_n^{MU}}{n} - \Sigma_n^H \in MU_{2n}.$$

But $d_n \frac{B_n^{MU}}{n} \in MU_{2n}$, by definition. □

It is, of course well-known, and established directly in Ray (1986) (see §8 below), that $d_1 = 2$ and $d_{2m+1} = 1$ $\forall m > 0$.

The Thom class Ξ plays an important, albeit implicit role in Baker et al (1985).

7. VON STAUDT'S THEOREM FOR MORAVA K-THEORY

Having introduced the Bernoulli numbers in §5, we now indicate how symbolic methods enable computations to be made in currently fashionable cases. Throughout this section, we fix a prime p and let n be a natural number.

(7.1) Definition. Let K(n) be the Δ-operator with

$$K(n)_* = \mathbb{Z}[u^{p^n-1}, u^{1-p^n}] \subset K_*,$$

and Δ defined by applying the LIP to

$$D = \Delta + u^{p^n-1}\frac{\Delta^{p^n}}{p} + \ldots + u^{p^{ni}-1}\frac{\Delta^{p^{ni}}}{p^i} + \ldots .$$

We call K(n) the n-th (integral) Morava K-theory. Observe that $K(1)_{(p)} = G$ from (2.10). □

As a complex oriented cohomology theory, K(n) is usually defined after localisation at p. For a detailed algebraic discussion, see Morava (1985).

We propose to study the Bernoulli numbers $B_j^{K(n)} \in K(n)\mathbb{Q}_{2j}$, which can be non-zero only in the instances

$$B^{K(n)}_{d(p^n-1)} = B(n)_{d(p^n-1)} u^{d(p^n-1)} , \quad d \geq 1 . \tag{7.2}$$

Here the $B(n)$'s are rational numbers, which may be specified by

$$e^{B(n)D} = \frac{D}{\Delta} , \tag{7.3}$$

where Δ is defined over $\mathbb{Z}$ by

$$D = \sum_{i\geq 0} \frac{\Delta^{p^{ni}}}{p^i} .$$

We have merely set $u = 1$ in Definition (7.1), as we did in (3.10) for similar reasons. Purely for notational simplicity, from here on we work with this new Δ and the $B(n)$'s, thinking of them as the <u>Morava Bernoulli numbers.</u>

We now give a version of a classical theorem of von Staudt (1840).

<u>(7.4) Theorem.</u> In $\mathbb{Q}$,

$$B(1)_{d(p-1)} \equiv \frac{-1}{p} \bmod \mathbb{Z} \qquad \text{if } p \text{ odd;}$$

$$B(1)_d \equiv \begin{cases} \frac{1}{2} \bmod \mathbb{Z} & \text{if } d \text{ even or } 1 \\ 0 \bmod \mathbb{Z} & \text{otherwise.} \end{cases} \quad p = 2 . \qquad \square$$

Our aim is to prove an extension of this result for all n. We must beware that $B(n)_j$ is always integral for $n > 1$.

To set the scene, we combine definitions (7.1) and (7.3) into

$$e^{B(n)D} = 1 + \frac{\Delta^{p^n-1}}{p} + \ldots + \frac{\Delta^{p^{ni}-1}}{p^i} + \ldots , \tag{7.5}$$

and we define rationals $\mu(m,k)$ by

$$\left(\frac{D}{\Delta}\right)^{-m} = 1 + \mu(m,1)\Delta + \ldots + \mu(m,k)\frac{\Delta^k}{k!} + \ldots \tag{7.6}$$

Since (7.5) is the case $m = -1$, we deduce that $\mu(m,k) = 0$ unless $k = d(p^n-1)$, when $\mu(m,k) \in \mathbb{Z}$. Hereafter, we assume k to be of this form, and vary k by varying d.

(7.7) Lemma. For each $i \geq 0$,

$$\Delta^{p^{ni}-1} = D^{p^{ni}-1} + \ldots + (p^{ni}-1)[k-1]_{p^{ni}-2}\, \mu(k,k-p^{ni}+1)\frac{D^k}{k!} + \ldots$$

Proof. Apply the FLIP with $g(t) = t^{p^{ni}-1}$ and $\Delta = \bar{\Delta}$. □

We can now state our main result, which we label the von Staudt theorem for $K(n)$.

(7.8) Theorem. In $\mathbb{Q}$, for $n > 1$,

$$B(n)_{d(p^n-1)}/p^{d(p^{n-1}+\ldots+p^2+p)-(n-1)} \equiv \frac{\varepsilon}{p} \bmod \mathbb{Z}$$

where $\varepsilon \equiv (-1)^{d(n-1)+1} \bmod p$.

Proof. We describe the crucial steps.

Applying the binomial theorem to (7.3),(7.5),(7.6) yields

$$\mu(k,k-p^{ni}+1) = (k-p^{ni}+1)!\sum_{*}(-1)^{d_1+\ldots+d_s}\binom{k+d_1+\ldots+d_s-1}{k-1,d_1,\ldots,d_s}/p^{d_1+\ldots+sd_s},$$

where $*$ denotes summation over all $d_1,\ldots,d_s$ satisfying the equivalent conditions

$$d_1(p^n-1) + \ldots + d_s(p^{ns}-1) = d(p^n-1)-(p^{ni}-1)$$

or
$$d(p^n-1) + d_1 + \ldots + d_s - 1 = d_1p^n + \ldots + d_sp^{ns} + p^{ni}-2$$

or
$$ex = d - d_1 - \ldots - sd_s = \begin{cases} d_2(p^n-1) + \ldots + d_s(p^{n(s-1)} + \ldots + p^n-(s-1)) \\ + p^{n(i-1)} + \ldots + p^n+1. \end{cases}$$

Substituting into Lemma (7.7), and then (7.5), implies

$$B(n)_{d(p^n-1)} = \sum_{*,i} (-1)^{d_1+\dots+d_s} \begin{pmatrix} d_1p^n+\dots+d_sp^{ns}+p^{ni}-2 \\ d_1p^n,\dots,d_sp^{ns},p^{ni}-2 \end{pmatrix} \cdot$$

$$\cdot \frac{(d_1p^n)!}{d_1!p^{d_1}} \cdots \frac{(d_sp^{ns})!}{d_s!p^{sd_s}} \cdot \frac{(p^{ni}-1)!}{p^i} .$$

But $(p^n-1)! = \zeta p^{p^{n-1}+\dots+p-(n-1)}$ with $\zeta \equiv (-1)^n \bmod p$, so the p exponent ν_p of $B(n)_{d(p^n-1)}$ satisfies

$$\nu_p \geqslant d_1(p^{n-1}+\dots+p)+\dots+d_s(p^{ns-1}+\dots+p-(s-1))$$

$$+p^{ni-1}+\dots+p+1-i(n+1)$$

$$= d(p^{n-1}+\dots+p+1)-d_1-\dots-sd_s-i(n+1) \quad \text{by } *$$

$$= d(p^{n-1}+\dots+p)-(n-1)+ex-(i-1)(n+1)-2 \quad \text{by } * . \tag{7.9}$$

No other prime appears in the denominator.

To conclude, we note that in all cases (except $p = n = 2$), $ex-2 > (i-1)(n+1)$ unless

$$d_2 = \dots = d_s = 0 \quad \text{and} \quad i = 1 . \tag{7.10}$$

For the term given by (7.10), $ex = 1$ and

$$\nu_p = d(p^{n-1}+\dots+p^2+p) - n ,$$

whilst the other primes contribute $(-1)^{d(n-1)+1} \bmod p$. This provides the result, apart from the case $p = n = 2$, which requires a trifle more care. □

(7.11) Note. Precisely the same method yields Theorem (7.4) if $p \geqslant 5$. Again, if $p = 2$ or 3, extra care is needed. □

Other workers such as N. Katz (1977) have considered related questions from a number theoretic viewpoint.

8. UNIVERSAL BERNOULLI NUMBERS & FERMAT'S THEOREM

We conclude our survey of Bernoulli numbers by summarising some salient features of Ray (1986).

Above, we observed that the Bernoulli numbers for the Δ-operator K, or complex K-theory, satisfy

$$B_n^K = B_n u^n \quad \text{in} \quad K\mathbb{Q}_{2n},$$

where $B_n \in \mathbb{Q}$ is classical. This arises from setting $u = 1$ in K_* and considering the simplified Δ-operator $e^D - 1$ over $\mathbb{Z}$.

Of course, $B_1^K = \frac{1}{2}u$ and $B_{2m+1}^K = 0$.

So we may rewrite the results of von Staudt (1840), (1845) in terms of K. Here p is any odd prime, and ${}_{(p)}K_*$ is the localisation $K_* \otimes \mathbb{Z}_{(p)}$.

<u>(8.1) Theorem.</u> In $K\mathbb{Q}_*$

(i) $B_n^K/n \in {}_{(p)}K_{2n} \iff B_n^K \in {}_{(p)}K_{2n} \iff (p-1) \nmid n$

(ii) $B_{d(p-1)}^K \equiv -\frac{1}{p}u^{d(p-1)} \mod {}_{(p)}K_*$ □

Clearly we ought to investigate the Bernoulli numbers $B_n^{\Phi} = B_n^{MU} \in MU\mathbb{Q}_*$, which we label $\hat{B}_n$ to emphasise their universality. Their divisibility properties over Φ_* and MU_* are very different, and it transpires that MU_* is the correct ring of integers to choose for generalising Theorem (8.1). Thus there is some link between Leibniz and von Staudt type information.

By employing the Leibniz formula and the divided power principles we may establish the following.

<u>(8.2) Theorem.</u> In $MU\mathbb{Q}_*$, if p is odd then

(i) $\hat{B}_n/n \equiv {}_{(p)}MU_{2n} \iff \hat{B}_n \in {}_{(p)}MU_{2n} \iff (p-1) \nmid n$

(ii) $\hat{B}_{d(p-1)} \equiv -\frac{1}{p}\phi_{p-1}\phi_{(d-1)(p-1)} \mod {}_{(p)}MU_*$.

For $p = 2$,

(iii) $\hat{B}_{2m} \equiv \frac{1}{2}\phi_1\phi_{2m-1} \mod {}_{(2)}MU_*$, and

$\hat{B}_1 = -\frac{1}{2}\phi_1$, $\hat{B}_{2m+1}/2m+1 \in MU_{4m+2}$. □

The force of these formulae is, of course, provided by the fact that monomials in the ϕ's are indivisible in MU_* : thus we confirm results of Miller (1982). Theorem (8.2) is mapped to Theorem (8.1) by td.

Next, recall the work of Kummer (1850) on the Fermat problem, which is concerned with solving

$$x^n + y^n = z^n$$

by integers. It is sufficient to consider prime exponents, and for $n = 2$ many simple solutions exist. Kummer demonstrated that for exponent $p > 2$, the problem is insoluble for <u>regular</u> primes, i.e. those p satisfying

$$p \nmid \mathrm{num}(B_n/n) \qquad 1 \leq n < p-1,$$

where $\mathrm{num}(a/b)$ denotes the numerator of a rational a/b in its lowest terms.

So for each E we may call a prime <u>E-regular</u> if

$$p \nmid \mathrm{num}(B_n^E/n) \qquad 1 \leq n < p-1$$

in $E\mathbb{Q}_{2n}$. In particular, our calculations involved in proving Theorem (8.1) suggest that every prime, including 2, is MU-regular.

We are thus led naturally to construct the E-Fermat problem for any Δ-operator E.

<u>(8.3) The E-Fermat Problem.</u> Find those integers $n \geq 2$ for which there are non-zero umbral integers $a\theta$, $b\theta$, $c\theta$ satisfying

$$(a\theta)^n + (b\theta)^n = (c\theta)^n \qquad \theta^{i+1} \equiv \theta_i . \qquad \square$$

Thus the K-Fermat problem reduces to the classical case, with solutions known only for $n = 2$. However, the universal case is strikingly simple to resolve.

<u>(8.4) Proposition</u>. The MU-Fermat problem has no solutions. □

This is true even for $n = 2$, lending credence to our comments on MU-regularity. Hence any complex oriented cohomology theory E with a Todd genus $E \longrightarrow K$ acts as a go-between for the simple universal problem and the unsolved K-problem.

9. SEVERAL VARIABLES & SYMMETRISATION

In this section we offer a fleeting introduction to umbral calculus in several variables $x_1, \ldots, x_s$, $1 \leq s \leq \infty$, and a variant embodying group actions.

The starting point is the binomial coalgebra $E_*[x_1, \ldots, x_s]$, which admits the operators $D_r = \frac{\partial}{\partial x_r}$, $1 \leq r \leq s$. As before, its dual $E^*((D_1, \ldots, D_s))$ consists of formal divided power series in the D_r, and is called the umbral algebra. This contains partial Δ-operators

$$\Delta_r = e^{\theta D_r} - 1, \qquad 1 \leq r \leq s \qquad\qquad \theta^{i+1} \equiv \theta_i .$$

The penumbral coalgebra $\Pi_s(E)_*$ is the smallest E_*-module

$$E_*[x_1, \ldots, x_s] \hookrightarrow \Pi_s(E)_* \hookrightarrow E\mathbb{Q}_*[x_1, \ldots, x_s]$$

with the property that each $\Delta_r : (x_r) \cap \Pi_s(E)_* \longrightarrow \Pi_s(E)_*$ is an isomorphism.

(9.1) Theorem. We may characterise $\Pi_s(E)_*$ by the equivalent conditions

(i) $\Pi(E)_*^{\otimes s}$

(ii) $E_*\{b_{i_1}(x_1) \ldots b_{i_s}(x_s) \mid i_1, \ldots, i_s \geq 0\}$

(iii) the set of umbrally numerical polynomials $q(x_1, \ldots, x_s) \in E\mathbb{Q}_*[x_1, \ldots, x_s]$. □

The coalgebra structure is specified by (ii), whilst (iii) means that $q(m_1\theta, \ldots, m_s\theta) \in E_*$ for each integral vector $\underline{m} = (m_1, \ldots, m_s) \in \mathbb{Z}^s$. The continuous dual of $\Pi_s(E)_*$ is the formal power series algebra $E^*[[\Delta_1, \ldots, \Delta_s]]$. If Δ is Leibniz, then $\Pi_s(E)_*$ is an algebra.

Examples may be concocted from §1, as may combinatorial interpretations from §4. We can incorporate corresponding divided power principles, and LIPs in several variables as expounded in I. Goulden & D. Jackson (1983). The fundamental fact of umbral calculus is elegantly displayed as a Taylor's theorem

$$f(w_1+x_1, \ldots, w_s+x_s) = \sum_{i_1, \ldots, i_s \geq 0} b_{i_1}(x_1) \ldots b_{i_s}(x_s) \cdot \Delta_1^{i_1} \ldots \Delta_s^{i_s} f(w, \ldots, w_s) \tag{9.2}$$

for any $f(w_1,\ldots,w_s) \in E\mathbb{Q}_*[w_1,\ldots,w_s]$.

Now let the symmetric group Σ act by permuting $D_1,\ldots,D_s$. This may be alternatively described by $(D_1,\ldots,D_s)\tau$ through representing each permutation as a matrix $\tau \in GL(s,\mathbb{Z})$. We may then define the symmetric umbral algebra as the ring of invariants.

$$E^*((D_1,\ldots,D_s))^{\Sigma} \hookrightarrow E^*((D_1,\ldots,D_s)).$$

Dual is the projection σ of the binomial coalgebra onto the coalgebra $\mathrm{im}(\sigma) \subset E\mathbb{Q}_*[x_1,\ldots,x_s]$ of coinvariants, given by

$$\mathrm{im}(\sigma) \overset{\sigma}{\twoheadleftarrow} E_*[x_1,\ldots,x_s]$$

where

$$\sigma p(x_1,\ldots,x_s) = \frac{1}{s!}\sum_{\tau} \tau_* p(x_1,\ldots,x_s). \tag{9.3}$$

Here τ acts by

$$\tau_* p(x_1,\ldots,x_s) = p(\tau(x_1,\ldots,x_s)^t),$$

the dual action of Σ.

Given our Δ-operator E, the symmetric penumbral coalgebra $\Sigma_s(E)_*$ is the smallest E_*-module

$$\mathrm{im}(\sigma) \hookrightarrow \Sigma_s(E)_* \hookrightarrow E\mathbb{Q}_*[x_1,\ldots,x_s]^{\Sigma}$$

with the property that

$$\Delta_1\ldots\Delta_s : (x_1\ldots x_s) \longrightarrow \Sigma_s(E)_* \tag{9.4}$$

is an isomorphism.

(9.5) Theorem We may characterise $\Sigma_s(E)_*$ by the equivalent conditions

(i) $\sigma(\Pi_s(E)_*)$

(ii) $E_*\{\frac{1}{s!}\sum_{\tau} \tau_*(b_{i_1}(x_1)\ldots b_{i_s}(x_s)) \mid i_1 \geq \ldots \geq i_s \geq 0\}$

(iii) the set of symmetric umbrally numerical polynomials $q(x_1 \ldots, x_s) \in E\mathbb{Q}_*[x_1, \ldots, x_s]^{\Sigma}$. □

The coalgebra structure is specified by (ii), whilst (iii) means that $\tau_* q(x_1, \ldots, x_s) = q(x_1, \ldots, x_s)$ and

$$\frac{s!}{|S(\underline{m})|} q(m_1\theta, \ldots, m_s\theta) \in E_*,$$

where $S(\underline{m}) < \Sigma$ is the stabiliser of $\underline{m}$. Also, the continuous dual of the coalgebra epimorphism

$$\Sigma_s(E)_* \overset{\sigma}{\twoheadleftarrow} \Pi_s(E)_*$$

given by (i), is the inclusion of the algebra of invariants

$$E^*[[\Delta_1, \ldots, \Delta_s]]^{\Sigma} \hookrightarrow E^*[[\Delta_1, \ldots, \Delta_s]].$$

This fact can be utilised to give a simple proof of the theorem.

The corresponding form of the fundamental fact of umbral calculus is a symmetric version of Taylor's theorem for each $f(w_1, \ldots, w_s)$ as in (9.2):

$$\sum_{\tau} f((w_1, \ldots, w_s) + (x_1, \ldots, x_s)\tau) = \sum_{i_1 \geq \ldots \geq i_s} \{\sum_{\tau} \tau_*(b_{i_1}(x_1) \ldots b_{i_s}(x_s))\} \cdot \Delta_{(i_1, \ldots, i_s)} f(w_1, \ldots, w_s) \tag{9.6}$$

Here $\Delta_{(i_1, \ldots, i_s)}$ denotes the symmetric polynomial $\frac{1}{|S(\underline{i})|} \sum_{\tau} \Delta_1^{i_1} \ldots \Delta_s^{i_s}$.

Remark that $\Sigma_s(E)_*$ is not in general an algebra, even if E is Leibniz.

10. $E_*(BG)$ FOR CERTAIN LIE GROUPS G

To conclude, we outline a generalisation of part of a result due to S. Ochanine & L. Schwartz (1985).

We write T for the s-fold torus $\times_s S^1$, and remark that by combining our original methods with those of §9 we can readily describe $E_*(BT)$ for any complex oriented theory E.

(10.1) Proposition. There is an isomorphism

$$E_*(BT_+) \cong \Pi_s(E)_*,$$

whence we may identify $E_*(BT_+)$ with umbrally numerical polynomials in variables $x_1,\ldots,x_s$. □

The line bundle over BT classified by projection onto the r-th factor CP^∞ gives rise to a Thom isomorphism which is realised by Δ_r.

Let us now assume that G is a compact, connected Lie group of rank s and maximal torus T. Our aim is to characterise $E_*(BG_+)$ in the same spirit as Proposition (10.1). We shall conduct our analysis for those G which have $H_*(G)$ torsion free, and then explain how the case $G = U(s)$ corresponds to our results in §9.

We recall some properties of the Weyl group W of G. Each $\tau \in W$ gives rise to a homeomorphism $\tau : BT \longrightarrow BT$, specified by its action τ^* on

$$H^*(BT_+) \cong \mathbb{Z}[D_1,\ldots,D_s].$$

In turn, τ^* is determined on polynomials by an integral linear substitution

$$\tau^* f(D_1,\ldots,D_s) = f((D_1,\ldots,D_s)\tau), \tag{10.2}$$

where we now represent τ in $GL(s,\mathbb{Z})$. In homology, τ_* acts on $p(x_1,\ldots,x_s)$ by the transpose substitution.

Following F. Cohen (1976) and G. Cooke & L. Smith (1977), we act on the rationalised suspension $SBT_{\mathbb{Q}}$ by the idempotent

$$\sigma = \frac{1}{|W|}\sum_{\tau\in W}\tau \quad \text{in} \quad \mathbb{Q}W.$$

Consequently, we may split off a wedge summand

$$SBT_{\mathbb{Q}} \underset{\sigma^\infty}{\overset{i}{\rightleftarrows}} \text{Tel}, \tag{10.3}$$

where Tel is the iterated mapping telescope of σ. Furthermore, since by (10.2)

$$\sigma_* p(x_1, \ldots, x_s) = \frac{1}{|W|}\sum \tau_* p(x_1, \ldots, x_s),$$

we conclude that $H_{*+1}(\mathrm{Tel}_+)$ consists of the coalgebra of coinvariants in $\mathbb{Q}[x_1, \ldots, x_s]$.

By dual reasoning,

$$H\mathbb{Q}^{*+1}(\mathrm{Tel}_+) \cong \mathbb{Q}[D_1, \ldots, D_s]^W,$$

whence the composite

$$\mathrm{Tel} \xrightarrow{i} SBT_{\mathbb{Q}} \xrightarrow{j} SBG_{\mathbb{Q}},$$

where j includes the maximal torus, is an equivalence of rational spaces. We may choose an inverse h such that the diagram

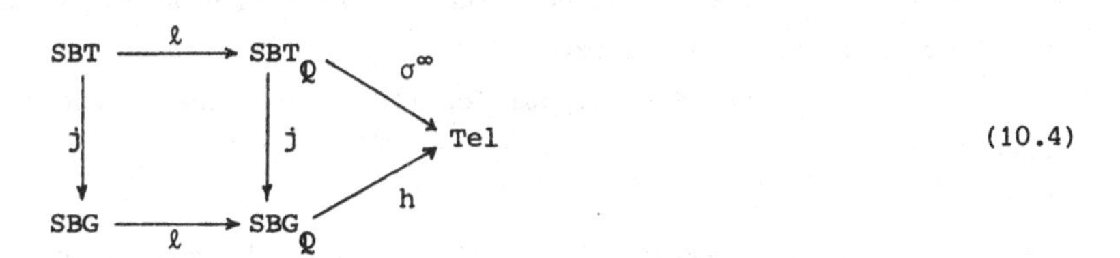

(10.4)

homotopy commutes, ℓ being rationalisation.

We might alternatively construct (10.4) by a transfer argument.

<u>(10.5) Theorem</u> The map j^* defines an isomorphism

$$E^*(BG_+) \cong E^*[[\Delta_1, \ldots, \Delta_s]]^W \qquad \square$$

This result is proved by the Atiyah-Hirzelbruch spectral sequence, and care is needed to induce the action of W by (10.2). Since $E_*(BG_+)$ is free over E_*, our task is to indentify the coinvariants of W.

<u>(10.6) Proposition</u>. We may characterise $E_*(BG_+)$ as a subcoalgebra of $E\mathbb{Q}_*[x_1, \ldots, x_s]^W$ by the equivalent conditions

(i) $\sigma_*(\Pi_s(E)_*)$

(ii) the free E_*-module generated by elements of the form

$\frac{1}{|W|}\sum_{\tau} b_{i_1}(x_1)\ldots b_{i_s}(x_s),\ i_1,\ldots,i_s \geq 0.$

Proof. Apply $E_*(\ )$ to (10.4). Since j_* is epic and ℓ_* is monic, $i_*h_*\ell_*$ describes $E_*(BG_+)$ as the subcoalgebra $\sigma_*\ell_*E_*(BT_+)$ of $E_*(BT_{\mathbb{Q}+})$. But ℓ_* coincides with the Hurewicz map of (2.4), so $E_*(BG_+)$ reduces to the image of $\pi_s(E)_*$ under σ_*. □

We could prove Proposition (10.6) without setting up (10.4), but it seems useful to make explicit the role of the variables x_r.

If $G = U(s)$, then $W = \Sigma$ and we have that $E_*(BU(s)_+) \cong \Sigma_s(E)_*$ from §9. Moreover, Theorem (9.5(iii)) furnishes a description more directly in keeping with Proposition (10.1). Such descriptions are also possible for more general G, but need extra care. Note further that for $G = U(s)$, we may identify (9.4) with the Thom isomorphism

$$t : E_*(MU(s)) \xrightarrow{\cong} E_*(BU(s)_+).$$

When $s = \infty$ this leads into new areas of double Δ-operators, and symbolic proofs of Hattori-Stong theorems.

We illustrate Proposition (10.6) with one of the simplest examples.

(10.7) $G = Sp(1)$. Here $s = 1$ and $W = \mathbb{Z}/2$, which acts on x_1 by $x_1 \longmapsto -x_1$. So

$$E_*(BSp(1)_+) \cong E_*\{\tfrac{1}{2}(b_{2i}(x) + b_{2i}(-x))\ |\ i \geq 0\}$$ □

This may easily be extended to cover $BSp(s)$ and $BSU(s)$; and also the classifying spaces $BS^{2p-3}_{(p)}$ due to D. Sullivan (1974) after adjoining a pth root of 1.

For each G, the corresponding form of Taylor's theorem is

$$\sum_{\tau} f((w_1,\ldots,w_s)+(x_1,\ldots,x_s)\tau) = \sum_{\underline{i}}\{\sum_{\tau} \tau_*(b_{i_1}(x_1)\ldots b_{i_s}(x_s))\}\cdot \Delta_{(i_1,\ldots,i_s)} f(w_1,\ldots,w_s) \tag{10.8}$$

where $\underline{i} \in \mathbb{Z}^s$ varies so as to yield a basis in Proposition (10.6(ii)).

Finally, we remark that the coaction of $E_*(E)$ on each of the coalgebra $E_*(X)$ under consideration may also be woven in symbolically.

REFERENCES

Adams, J.F. (1969). Lectures on generalised cohomology. Lecture Notes in Mathematics, 99, 1-138, Springer-Verlag.

Adams, J.F. (1974). Stable homotopy and generalised homology, Univ. of Chicago.

Aigner, M. (1979). Combinatorial theory, Springer-Verlag.

Baker, A. (1982). On weakly almost complex manifolds with vanishing decomposable Chern numbers. A.M.S. Contemp. Math., 19, 1-7.

Baker, A. (1984). Combinatorial and arithmetic identities based on formal group laws, preprint, Manchester Univ.

Baker, A., Clarke, F., Ray, N. & Schwartz, L. (1985). On the Kummer congruences and the stable homotopy of BU, preprint, Univ. Coll. Swansea.

Bell, E.T. (1927). Algebraic arithmetic, Colloq. Publications of the A.M.S., 7.

Bell, E.T. (1929). Certain invariant sequences of polynomials, Trans. Amer. Math. Soc., 31, 405-21.

Bell, E.T. (1938). The history of Blissard's symbolic method, with a sketch of its inventor's life, Amer. Math. Monthly, XLV, 414-21.

Berge, C. (1971). Principles of combinatorics, Academic Press.

Blissard, J. (1861). Theory of generic equations, Quart. Jour. of Pure and App. Math., 4, 279-305.

Cohen, F. (1976). Splitting certain suspensions via self-maps, Ill. Jour. Math., 20, 336-347.

Comtet, L. (1974), Advanced combinatorics, Reidel.

Content, M., Lemay, F. & Leroux, P. (1980). Catégories de Möbius et fonctorialités: un cadre général pour l'inversion de Möbius, Jour. Comb. Theory, A 28, 169-190.

Cooke, G. & Smith, L. (1977). Mod p decompositions of co H-spaces and applications, Math. Zeit, 157, 155-77.

Erdelyi, A. (1962). Operational calculus and generalized functions, Holt, Rinehart and Winston.

Frucht, R. (1965). A combinatorial approach to the Bell polynomials and their generalisations, in Recent Progress in Combinatorics, 69-74, Academic Press.

Goulden, I. & Jackson, D. (1983). Combinatorial enumeration, Wiley-Interscience.

Heaviside, O., (1893). On operators in physical mathematics, Proc. Roy. Soc., 52.

Jankowski, A. (1979). Splitting K-theory, and g_* characteristic numbers, Studies in algebraic topology, 189-212, Academic Press.

Joni, S.A. & Rota, G-C. (1979). Coalgebras and bialgebras in combinatorics, Studies in App. Math., 61, 93-139.

Katz, N. (1977). Formal groups and p-adic interpolation, Asterisque, 41-2, 55-65.

Knapp, K. (1979). Some applications of K-theory to framed bordism, Habilitationsschrift, Bonn.

Kummer, E.E. (1850). Allgemeiner Beweis des Fermatschen Satzes, Jour. Reine Angew. Math., 40, 130-8.

Lloyd, E.K. (1985). Enumeration, Handbook of applicable mathematics, 5, 531-621, Wiley.

Mikusinski, J.G. (1950). Une nouvelle justification du calcul de Heaviside, Att. Accad. Naz. Lincei. Mem. Cl. Sci. Fis. Mat. Nat., 8, Ser. 1,2, 113-21.

Mikusinski, J.G. (1959). Operational calculus, Pergamon Press.
Miller, H. (1982). Universal Bernoulli numbers and the S^1-transfer, Current trends in algebraic topology, 2 pt. 2, 437-49, CMS-AMS.
Morava, J. (1985). Noetherian localisations of categories of cobordism comodules, Ann. of Math., 121, 1-39.
Mullin, R. & Rota, G-C. (1970). On the foundations of combinatorial theory III: theory of binomial enumeration, in Graph theory and its applications, 168-213, Academic Press.
Ochanine, S. & Schwartz, L. (1985). Une remarque sur les générateurs du cobordisme complexe, Math. Zeit., 190, 543-57.
Ray, N. (1969). Thesis, Manchester Univ.
Ray, N. (1986). Extensions of umbral calculus I: penumbral coalgebras and generalised Bernoulli numbers. Adv. Math., to appear.
Ray, N. (198?). Umbral calculus, binomial enumeration and chromatic polynomials, in preparation.
Riordan, J. (1958). An introduction to combinatorial analysis, Wiley.
Roman, S. (1984). The umbral calculus, Academic Press.
Roman, S. & Rota, G-C. (1978). The umbral calculus, Adv. Math., 27, 95-188.
Shtokalo, I.Z. (1976). Operational calculus, Hindustan Pub. Corp.
Sullivan, D. (1974). Genetics of homotopy theory and the Adams conjecture, Ann. of Math., 100, 1-79.
Switzer, R.M. (1975). Algebraic topology-homotopy and homology, Springer-Verlag.
Sylvester, J.J. (1851). On the relation between the minor determinants of linearly equivalent quadratic functions, Phil. Mag., 1, 295-305.
Van der Pol, (1929). Simple proof and extension of Heaviside's operational calculus for invariable systems, Phil. Mag. 8, no.7, 1153.
Von Staudt, K.G.C. (1840). Beweis eines Lehrstatzes, die Bernoullischen Zahlen betreffend, Jour. Reine Angew. Math., 21, 373-4.
Von Staudt, K.G.C. (1845). De numeris Bernoullianis, commentationem alteram, Erlangen.
Whittaker, E.T. & Watson, G.N. (1927). A course of modern analysis, Cambridge Univ. Press.

AN OUTLINE
OF
HOW MANIFOLDS RELATE TO ALGEBRAIC K-THEORY

Friedhelm Waldhausen

Consider manifolds without boundary. Allow taking product with R^k. What is left? Certainly the homotopy type and the tangent bundle. But by an old theorem of Barry Mazur this is all that is left; and that is even true with parameters. In other words,

Thm. The forgetful map

$$\begin{array}{ccc} \text{space of (stable) manifolds} & \longrightarrow & \text{space of tangential spaces} \\ \text{(a suitable simplicial set)} & & \text{(another simplicial set)} \end{array}$$

is a homotopy equivalence.

(It goes without saying here, as they say, that we want our 'manifolds' and 'spaces' to have the homotopy types of finite CW complexes.)

We can reduce this to a simpler, but equivalent, statement by getting rid of the tangent bundles.

Equ. assertion. The map

$$\text{space of framed (stable) manifolds} \longrightarrow \text{space of spaces}$$

is a homotopy equivalence.

Mazur's theorem is strictly a result about non-compact manifolds. Namely define $C^{CAT}(X)$ as the homotopy fibre, at X, of the forgetful map

$$\text{space of } \textit{compact} \text{ framed stable manifolds} \longrightarrow \text{space of spaces}$$

where CAT is one of DIFF, PL, TOP and where stabilization is given by product with $[0,1]^k$ (plus rounding of corners in the DIFF case).

Then $C^{CAT}(X) \not\simeq *$ (in general). In fact, its loop space is given by,

Exercise. $\Omega\, C^{CAT}(X) \simeq$ (stable) CAT pseudo-isotopy space of X (if X is a CAT manifold).

(Hints. Use immersion theory and general position; cf. lemma 5.2 of [W2] for a related argument.)

One knows that $C^{PL}(X) \xrightarrow{\simeq} C^{TOP}(X)$ (cf. [H2]) whereas the same is not true with $C^{DIFF}(X)$ instead (in fact, as Hatcher has phrased it, the latter is so named because it is different). Specifically, $C^{PL}(*) \simeq *$ (by the Alexander trick), but already $C^{DIFF}(*)$ is highly non-trivial. For example, there is an infinite cyclic summand in each of its homotopy groups in dimensions 4, 8, 12, ... It is related to the summand which Borel has found, one dimension higher up, in the algebraic K-theory of the ring of natural integers.

This relation to algebraic K-theory, how does it come about? To relate a manifold gadget, such as $C^{CAT}(X)$, to a non-manifold gadget one must at some point get rid of the manifolds. For $C^{PL}(X)$ this can be achieved by the following result,

Thm. Let X be a finite polyhedron. Then $C^{PL}(X)$ is homotopy equivalent (naturally, in X) to the simplicial category with

- objects: (locally trivial families of) finite polyhedra containing X as a deformation retract,

- morphisms: maps whose point inverses are contractible.

Remarks. 1. The theorem is very typical of the game: After a lot of effort, the only thing one has learned in the end is that two rather complicated definitions lead to the same thing, up to homotopy.

2. The theorem has variants, technical and otherwise; for example the finite polyhedra in its statement may be replaced by finite simplicial sets.

This theorem is more or less the same as the *parametrized h-cobordism theorem* of Hatcher [H1]. The proof given by Hatcher is not quite satisfactory. There is another proof [W5], it goes roughly as follows. For each compact polyhedron Y one considers the pairs (M,p) where M is a compact PL manifold of some fixed dimension n, and $p: M \to Y$ is a map having contractible point inverses. Let $\mathcal{S}(Y,n)$ denote the 'space' of these pairs (a suitable simplicial set), and let $\mathcal{S}(Y) = \varinjlim \mathcal{S}(Y,n)$ (stabilize by allowing M to be replaced by $M\times[0,1]$). One shows,

(i) if $\mathcal{S}(Y)$ is contractible for all Y then the theorem follows. This is more or less formal: one applies a fibration criterion in the spirit of Quillen's theorem B [Q1].

(ii) $\mathcal{S}(Y)$ is contractible indeed. This is proved by a sort of induction on Y (the idea being to cut up Y into simpler pieces); the argument is based on a trick devised by Hatcher in the context of proving a theorem on 3-manifolds: the method of 'general position in patches' [H3].

As to algebraic K-theory now, let us begin by considering the *Euler characteristic*. It may be defined as an element of a certain universally defined group, namely the *class group* given in terms of generators and relations as follows,

- generators $[Y]$, where Y runs through the pointed spaces (of finite type)
- relations of two kinds
  (1) if $Y_1 \to Y_2$ is a cofibration, then $[Y_2] = [Y_1] + [Y_2/Y_1]$
  (2) if $Y \xrightarrow{\simeq} Y'$ is a weak homotopy equivalence, then $[Y] = [Y']$.

Obviously this definition of class group makes sense as soon as you have a category with notions of cofibration and weak equivalence. If moreover these notions have the usual familiar properties, and if you like to play with definitions, you will be able to write down a bisimplicial set in which these notions play a role, and which has the class group as its fundamental group, cf. [W1], [W3].

Def. The *algebraic K-theory* of that category (with cof. and w. eq.) is given by the loop space of (the geometric realization of) that bisimplicial set.

In particular, given a space X there is associated to it its K-theory $A(X)$ via a model category for an equivariant homotopy theory, attached to X, of spaces of finite type. In making this precise, one needs to make choices, viz.

space: topological space or simplicial set,

finite: finite on the nose or finite up to homotopy,

equivariant: spaces over X or spaces with an action of ΩX .

Thm. All these choices don't matter (if they are made right). In addition, $A(X)$ can also be expressed by the 'plus' construction of Quillen, using matrices over the 'ring up to homotopy' $Q(\Omega X_+)$ (cf. below).

The relation between K-theory and PL pseudo-isotopy theory is now given as follows. One introduces a connected de-loop $Wh^{PL}(X)$ of $C^{PL}(X)$ (this is to get rid of the dimension shift with respect to algebraic K-theory).

Thm. There is a map $A(X) \to Wh^{PL}(X)$ whose fibre is a homology theory (that is, as a functor of X it satisfies the excision axiom and, of course, the homotopy axiom).

The proof of this result, along with the aforementioned foundational material on K-theory, makes up the content of the rather long paper [W3]. The proof has nothing to do with manifolds: it uses the non-manifold translation of $C^{PL}(X)$, and hence $Wh^{PL}(X)$, described above.

There is nothing wrong with manifolds, however, and one can in fact write down manifold models for all the spaces involved, and natural maps between them, to represent the whole fibration of the theorem [W2]. (There is just one thing which is not clear from the manifold point of view, namely that one of the terms in the fibration happens to be a homology theory. In other words, the detour through non-manifolds is only required here to recognize a homology theory when one sees one!) The manifold models are essentially independent of the category of definition (their construction, that is, not necessarily of course the homotopy types that they represent). So one can use the natural forgetful map to compare categories. Smoothing theory now tells us that, in the situation at hand, the difference between DIFF and PL is itself only a homology theory. It thus results from the theorem that its analogue for smooth manifolds is also valid; that is [W2], there is a fibration, natural in X ,

$$h(X) \longrightarrow A(X) \longrightarrow Wh^{DIFF}(X)$$

where $X \mapsto h(X)$ is a homology theory. Now comes,

1st surprise. That fibration splits.

Reason. Functors can be *stabilized*, $F^S(X) = \varinjlim \Omega^n \text{fibre}(F(S^n \wedge X_+) \to F(*))$. Now it is known, as a consequence of Morlet's disjunction lemma [H2], that

$$(Wh^{DIFF})^S(X) \simeq * .$$

So one gets a diagram of fibrations

$$\begin{array}{ccccc} h(X) & \longrightarrow & A(X) & \longrightarrow & Wh^{DIFF}(X) \\ \simeq\downarrow & & \downarrow & & \downarrow \\ h^S(X) & \xrightarrow{\simeq} & A^S(X) & \longrightarrow & (Wh^{DIFF})^S(X) \end{array}$$

where the arrow $h(X) \to h^S(X)$ is a homotopy equivalence since stabilization doesn't change a homology theory and where, of course, $h^S(X) \to A^S(X)$ is a homotopy equivalence since $(Wh^{DIFF})^S(X)$ is contractible. Thus there is a splitting $A(X) \simeq A^S(X) \times Wh^{DIFF}(X)$. Next comes,

2nd surprise. $A^S(X)$ in turn splits as $Q(X_+) \times ?$.

Reason. By using an explicit description of $A^S(*)$ in terms of smooth manifolds one can check [W2] that the usual map $B\Sigma_\infty \to QS^0$ factors through $A^S(*)$; the argument in the general case is similar.

Putting the two surprises together one obtains a double splitting

$$A(X) \simeq Q(X_+) \times Wh^{DIFF}(X) \times \mu(X) .$$

The mysterious third factor has informally become known as the 'mystery homology theory'. Accordingly the next result might then be called the theorem of the vanishing mystery homology,

Thm. $\mu(X) \simeq *$.

An account of it may be found in [W4]. Rationally, that vanishing can be deduced [W1] from a vanishing in group homology, namely the rational vanishing of the kernel of the trace map

$$H_*(\, GL(Z), M(Z)\,) \longrightarrow H_*(\, GL(Z), Z\,) .$$

This was established by L^2-cohomology methods by Borel and by Farrell and Hsiang; more recently, Goodwillie has succeeded in giving an algebraic proof. Conversely it is now possible to turn the situation around and to use topology to compute the homology of the adjoint representation, including torsion. This goes as follows.

Given a ring R and a bimodule A over it, one can define a sort of particularly primitive K-theory, the so-called *stable K-theory* $K^S(R,A)$ (cf. [W6]). It is an elementary fact [W6], [K1] that this 'computes' the homology of the adjoint representation in the sense that there is a spectral sequence

$$H_p(\ GL(R),\ \pi_q K^S(R,A)\ ) \Longrightarrow H_{p+q}(\ GL(R),\ M(A)\ )$$

where the homology on the left is ordinary (i.e. untwisted). Moreover, as Bökstedt has pointed out, the spectral sequence will collapse in certain interesting cases, for example if $R = Z$ (this is due to the presence of a product structure). One is thus led to try and compute stable K-theory.

Now suppose that k is a ground ring over which R is an algebra. Then one can define the *Hochschild homology* $H_k(R,A)$ and one can construct a natural transformation

$$K^S(R,A) \longrightarrow H_k(R,A) \ .$$

It turns out that this natural transformation will be happy to become an equivalence as soon as one is prepared to give it a chance.

'Giving it a chance' means that one should not attempt to do something which, from a broader perspective perhaps, is openly unreasonable; for example, to try to take $k = Z$ here. Indeed, if the algebraic K-theory of spaces is looked at from the algebraic K-theory of rings point of view, one is forced to look at 'rings' which are a little unusual to the taste of many (e.g. $A(*) = K(R)$ where R is the 'ring up to homotopy' QS^o). Such 'rings' are not algebras over Z, in general, so that $H_Z(R,A)$ is not even defined.

One is thus led to try and take $k = QS^o$. Not surprisingly, this leads one into constructions which are technically rather involved (for example, it would be easy to mimick the usual definition of Hochschild homology to obtain a simplicial object in the homotopy category; but that is not enough, of course). Granting that all of this makes sense, unravelling of the definitions shows that, for this k , the statement " $\mu(*) \simeq *$ " is equivalent to the statement that the map

$$K^S(k,k) \longrightarrow H_k(k,k) \simeq k$$

is an equivalence. To proceed from this special case to the general case one has to use, among other things, a decomposition theorem for the K-theory of 'rings' which are free products, of sorts.

The 'topological Hochschild homology' $H_k(R,A)$, $k = QS^o$, has been computed by Marcel Bökstedt in the case $A = R = Z$ [B1]. The argument is difficult, but the result is easy enough to state. The homotopy groups of $H_k(Z,Z)$, $k = QS^o$, are the cyclic groups

$$Z , 0 , 0 , Z/2 , 0 , Z/3 , 0 , Z/4 , 0 , Z/5 , \ldots .$$

We see that we have a problem about ordinary rings here (for example the ring Z); namely the problem of how to compute the stable K-theory and hence the homology of the adjoint representation. But to solve the problem we must first consider it in the rather extended framework of 'rings up to homotopy'.

One wonders if perhaps a similar thing is true with regard to the problem of computing algebraic K-theory.

To conclude, here are a few remarks on the notion of *ring up to homotopy*. It is doubtful if there is one technical choice which is equally well suited for everything that one wants to do with it. There happens to exist a simple notion which suffices for many purposes [G1]. Namely let a quote-abelian group denote a functor from spaces to spaces which preserves connectivity and satisfies approximate excision. This may be regarded as another 'coordinate-free' way of specifying a (connected) homology theory, by stabilization, and hence a spectrum; that excision is asked only approximately here has the reason that (1) this is good enough, and (2) it allows one to keep functors such as the identity functor. By definition then a quote-ring is just a quote-abelian group together with a monad structure in the sense of category theory; and to define K-theory, say, is just a matter of writing it down [G1].

From a naive point of view, a 'ring up to homotopy' is nothing but a topological space R together with structure maps add: $R \times R \to R$, mult , and so on, so that the ring axioms are satisfied up to homotopy. In this situation one can, for example, define the 'space of homotopy invertible matrices' as the pullback in the diagram

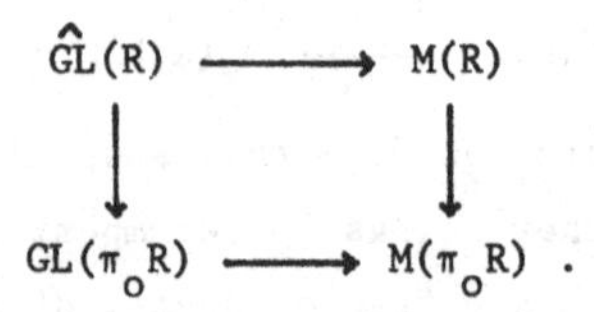

It is a multiplicative H-space by means of matrix multiplication. But it is plausible, on the other hand, that one needs more information to ensure the existence (or even functorial existence) of a classifying space for this H-space. Here is a neat example of a bad failure.

The determinant map $\widehat{GL}(QS^o) \longrightarrow \widehat{GL}_1(QS^o) = G$ is an H-map, and a retraction. Suppose it could be de-looped. Then the de-looped map would extend to the 'plus' construction (since BG has abelian fundamental group). That is, it would induce a map $A(*) \to BG$, and this map would be left inverse, more or less, to the (doubtlessly existing) map in the other direction, $BG \to A(*)$. It would therefore follow that the latter map is injective on homotopy groups. But this is not true as one sees by using the splitting $A(*) \simeq QS^o \times Wh^{DIFF}(*)$ together with the following facts,

(i) the composite map $BO \to BG \to A(*) \to Wh^{DIFF}(*)$ is trivial [W2],

(ii) the composite map $BG \to A(*) \to QS^o$ is non-trivial only at the prime 2 (and here just barely so, being 'multiplication by η') [B2].

References.

B1 M. Bökstedt, *Topological Hochschild homology*, preprint.

B2 M. Bökstedt and F. Waldhausen, *The map* $BG \to A(*) \to QS^0$, Proc. Conf. John Moore's birthday, Princeton Univ. Press.

G1 T. Gunnarson, *Algebraic K-theory of spaces as K-theory of monads*, Aarhus preprint.

H1 A. Hatcher, *Higher simple homotopy theory*, Ann. of Math. (2) 102 (1975), 101-137.

H2 ———, *Concordance spaces, higher simple homotopy theory, and applications*, Proc. Symp. Pure Math. vol. 32 (1978), 3-21.

H3 ———, *Homeomorphisms of sufficiently large* P^2*-irreducible 3-manifolds*, Topology 15 (1976), 343-347.

K1 C. Kassel, *La K-théorie stable*, Bull. Soc. Math. France 110 (1982), 381-416.

Q1 D. Quillen, *Higher algebraic K-theory. I*, Algebraic K-theory, Springer Lect. Notes Math. vol. 341 (1973), 85-147.

W1 F. Waldhausen, *Algebraic K-theory of topological spaces. I*, Proc. Symp. Pure Math. vol. 32 (1978), 35-60.

W2 ———, *Algebraic K-theory of spaces, a manifold approach*, Can. Math. Soc. Conf. Proc. vol. 2, part 1 (1982), 141-184.

W3 ———, *Algebraic K-theory of spaces*, Algebraic and geometric topology, Springer Lect. Notes Math. vol. 1126 (1984), 318-419.

W4 ———, *Algebraic K-theory of spaces, concordance, and stable homotopy*, Proc. Conf. John Moore's birthday, Princeton Univ. Press.

W5 ———, *Spaces of PL manifolds and categories of simple maps*, to appear (since '79).

W6 ———, *Algebraic K-theory of topological spaces. II*, Springer Lect. Notes Math. vol. 763 (1979), 356-394.

FAKULTÄT FÜR MATHEMATIK
UNIVERSITÄT BIELEFELD
4800 BIELEFELD, FRG.

Springer Theses

Recognizing Outstanding Ph.D. Research

Aims and Scope

The series "Springer Theses" brings together a selection of the very best Ph.D. theses from around the world and across the physical sciences. Nominated and endorsed by two recognized specialists, each published volume has been selected for its scientific excellence and the high impact of its contents for the pertinent field of research. For greater accessibility to non-specialists, the published versions include an extended introduction, as well as a foreword by the student's supervisor explaining the special relevance of the work for the field. As a whole, the series will provide a valuable resource both for newcomers to the research fields described, and for other scientists seeking detailed background information on special questions. Finally, it provides an accredited documentation of the valuable contributions made by today's younger generation of scientists.

Theses are accepted into the series by invited nomination only and must fulfill all of the following criteria

- They must be written in good English.
- The topic should fall within the confines of Chemistry, Physics, Earth Sciences, Engineering and related interdisciplinary fields such as Materials, Nanoscience, Chemical Engineering, Complex Systems and Biophysics.
- The work reported in the thesis must represent a significant scientific advance.
- If the thesis includes previously published material, permission to reproduce this must be gained from the respective copyright holder.
- They must have been examined and passed during the 12 months prior to nomination.
- Each thesis should include a foreword by the supervisor outlining the significance of its content.
- The theses should have a clearly defined structure including an introduction accessible to scientists not expert in that particular field.

More information about this series athttp://www.springer.com/series/8790

Stefan Thiele

Read-Out and Coherent Manipulation of an Isolated Nuclear Spin

Using a Single-Molecule Magnet Spin-Transistor

Doctoral Thesis accepted by
the University of Grenoble, France

Author
Dr. Stefan Thiele
Sensirion AG
Staefa
Switzerland

Supervisor
Dr. Wolfgang Wernsdorfer
Nano, Institut Neel
Grenoble
France

ISSN 2190-5053 ISSN 2190-5061 (electronic)
Springer Theses
ISBN 978-3-319-24056-5 ISBN 978-3-319-24058-9 (eBook)
DOI 10.1007/978-3-319-24058-9

Library of Congress Control Number: 2015953250

Springer Cham Heidelberg New York Dordrecht London

Printed on acid-free paper

Springer International Publishing AG Switzerland is part of Springer Science+Business Media
(www.springer.com)

Supervisor's Foreword

It is with great pleasure and the highest enthusiasm that I write this foreword to Stefan Thiele's thesis, which reports his most exciting results. Its endeavor is driven by one of the most ambitious, technological goals of today's scientists: the realization of an operational quantum computer. In this regard, the basic building block is generally composed of a two-level quantum system, namely a quantum bit (or qubit). Such a quantum system must be fully controllable and measurable, which requires a connection to the macroscopic world. In this context, solid-state devices, which establish electrical interconnections to the qubit, are of high interest, mainly due to the variety of methods available for fabrication of complex and scalable architectures. Moreover, outstanding improvements in the control of the qubit dynamics have been achieved in the last years. Among the different solid-state concepts, spin-based devices are very attractive because they already exhibit relatively long coherence times. For this reason, electrons possessing a spin 1/2 are conventionally thought as the natural carriers of quantum information. However, the strong coupling to the environment makes it extremely difficult to maintain a stable entanglement. Alternative concepts propose the use of nuclear spins as building blocks for quantum computing, as they benefit from longer coherence times compared to electronic spins, because of a better isolation from the environment. But weak coupling comes at a price: the detection and manipulation of individual nuclear spins remain difficult tasks. In this context, the main objective of the Ph.D. of Stefan Thiele was to lay the foundation of a new field called molecular quantum spintronics, which combines the disciplines of spintronics, molecular electronics, and quantum information processing. In particular, the objective was to fabricate, characterize, and study molecular devices (molecular spin-transistor, molecular spin-valve and spin filter, molecular double-dot devices, carbon nanotube, nano-SQUIDs, etc.) in order to read and manipulate the spin states of the molecule and to perform basic quantum operations. The visionary concept of the project is underpinned by worldwide research on molecular magnetism and supramolecular chemistry, and in particular within the European Institute of Molecular Magnetism (http://www.eimm.eu/), and collaboration with outstanding scientists in the close environment. For the project, we found the following funding: the main contributions came from the French Research Agency (ANR), the Contrat Plan Etat Région (CPER), the European Research Council (ERC), the Réseau Thématique de Recherche Avancé (RTRA), and the support from our institute. The Ph.D. of Stefan Thiele was funded by an ERC advanced grant. Among the most important results before Stefan Thiele's Ph.D., we showed the possibility of magnetic molecules to act as building blocks for the design of quantum spintronic devices and demonstrated the first important results in this new research area. For example, we have built a novel spin-valve device in which a nonmagnetic molecular quantum dot, consisting of a single-wall carbon nanotube contacted with nonmagnetic electrodes, is laterally coupled via supramolecular interactions to a $TbPc_2$ molecular magnet [Ph.D. of Matias Urdampilleta (2012)]. The localized magnetic moment of the SMM led to a magnetic field-dependent modulation of the conductance in the nanotube with magnetoresistance ratios of up to 300 % at low temperatures. We also provided the first experimental evidence for a strong spin–phonon coupling between a single-molecule spin and a carbon nanotube resonator [Ph.D. of Marc Ganzhorn (2013)]. Using a molecular spin-transistor, we achieved the electronic read-out of the nuclear spin of an individual metal atom embedded in a single-molecule magnet (SMM) [Ph.D. of Romain Vincent (2012)]. We could show very long spin lifetimes (several tens of seconds). Here, the Ph.D. of Stefan Thiele started with a completely new breakthrough. He proposed and demonstrated the possibility to perform quantum manipulation of a single nuclear spin by using an electrical field only. This has the advantage of reduced interferences with the device and less Joule heating of the sample. As an electric field is not able to interact with the spin directly, he used an intermediate quantum mechanical process, the so-called hyperfine Stark effect, to transform the electric field into an effective magnetic field. His project was designed to play a role of pathfinder in this—still largely unexplored—field. The main target concerned fundamental science, but applications in quantum electronics are expected in the long run.

Grenoble
April 2015

Dr. Wolfgang Wernsdorfer
Research Director

Abstract

The realization of a functional quantum computer is one of the most ambitious, technological goals of today's scientists. Its basic building block is composed of a two-level quantum system, namely a quantum bit (or qubit). Among the other existing concepts, spin-based devices are very attractive because they benefit from the steady progress in nanofabrication and allow for the electrical read-out of the qubit state. In this context, nuclear spin-based devices exhibit additional gain of coherence time with respect to electron spin-based devices due to their better isolation from the environment. But weak coupling comes at a price: the detection and manipulation of individual nuclear spins remain challenging tasks.

Very good experimental conditions were important for the success of this project. Besides innovative radio frequency filter systems and very low noise amplifiers, I developed new chip carriers and compact vector magnets with the support of the engineering departments at the institute. Each part was optimized in order to improve the overall performance of the setup and evaluated in a quantitative manner.

The device itself, a nuclear spin qubit transistor, consisted of a $TbPc_2$ single-molecule magnet coupled to source, drain, and gate electrodes and enabled us to read out electrically the state of a single nuclear spin. Moreover, the process of measuring the spin did not alter or demolish its quantum state. Therefore, by sampling the spin states faster than the characteristic relaxation time, we could record the quantum trajectory of an isolated nuclear qubit. This experiment shed light on the relaxation time T_1 of the nuclear spin and its dominating relaxation mechanism.

The coherent manipulation of the nuclear spin was performed by means of external electric fields instead of a magnetic field. This original idea has several advantages. Besides a tremendous reduction of Joule heating, electric fields allow for fast switching and spatially confined spin control. However, to couple the spin to an electric field, an intermediate quantum mechanical process is required. Such a process is the hyperfine interaction, which, if modified by an electric field, is also referred to as the hyperfine Stark effect. Using the effect, we performed coherent rotations of the nuclear spin and determined the dephasing time T_2^*. Moreover, exploiting the static hyperfine Stark effect we were able to tune the nuclear qubit in and out of resonance by means of the gate voltage. This could be used to establish the control of entanglement between different nuclear qubits.

In summary, we demonstrated the first single-molecule magnet based quantum bit and thus extended the potential of molecular spintronics beyond classical data storage. The great versatility of magnetic molecules holds a lot of promises for a variety of future applications and, maybe one day, culminates in a molecular quantum computer.

Acknowledgements

First of all, I want to say thank you to all the people I encountered during my Ph.D. and that helped me to make a lot of very enjoyable memories and the time in Grenoble a very special part of my life. In particular, I want to thank my supervisor Wolfgang Wernsdorfer, who guided me through the stormy waters of the Ph.D. Thank you for all your support and for being available any time I needed your advice. I am very grateful that you let me join your team and most of all that you encouraged and taught me to pursue many ideas which went beyond my project. This aspect was broadening my horizon unlike anything else and made this thesis one of the most enriching experiences in my life. Last but not least, I want to thank you for taking care of my experiment every time I was in need. Moreover, I want to thank Franck Balestro. Thank you for letting me measure on your setup, providing me with samples, and all the He transfers you did for me. But, most of all, thank you for always finding the words to encourage me, both in my work and personal life and of course for helping me finding the sample of my Ph.D. I am also very grateful to you for reading this manuscript and reshaping it into its almost perfect state;-).

I also want to thank Markus Holzmann, who helped me understanding and performing quantum Monte Carlo simulations. I appreciate very much your enthusiasm and the cheerful working atmosphere. I want to thank Rafik Ballou, who did not give up until he found a theory to properly describe the hyperfine Stark effect. It was always a pleasure to discuss with you. I am also very thankful to my referees Jakob Reichel and Wulf Wulfhekel as well as to my thesis comitee members Vincent Jacques, Patrice Bertet, and Mairbek Chshiev for the sincere work and fruitful comments on my thesis. Moreover, I want to thank all the researchers at the Néel institut I was interacting with, especially Tristan Meunier, Olivier Buisson, and Nicolas Roch for their help and advice.

Furthermore, I want to thank all the present and former members of the NanoSpin group. Thank you Jean-Pierre, Vitto, and Oksana for being very enjoyable office mates and never being tired to discuss personal and work affairs. Many thanks to Edgar and Christophe, who helped me solving a lot of informatics problems. Thank you Jarno, Viet, Antoine, Raoul, and Matias for all the fruitful discussions and coffee breaks. Thank you Romain for letting me measure on your sample and introducing me to the glorious world of Python. Thank you Marc for all your help and support especially during the hard time after the worldcup. Thank you Clément, I am very glad you joined our group and it was always a pleasure to talk to you.

Merci à Eric Eyraud, qui m'a aidé énormément avec mon cryostat et tout ce qui vient avec. Sans toi rien n'était possible. Merci égalment pour m'avoir montré comment on peux faire le "produit". Je voudrais remercier aussi Daniel Lepoittevin qui m'avait beaucoup aidé avec tout ce qui concerne l'électronique de mesure et pour m'avoir expliqué tout son fonctionnement. Un grand merci également á Christophe Hoarau pour m'avoir aidé avec les mesures et simulations á haute fréquence. Merci beaucoup aussi á Yves Deschanels qui m'avait aidé avec la construction des bobines supra et qui a passé une semaine dans la cave avec moi. Un grand merci aussi à Didier Dufeu, Richard Haettel, David Barral et Laurent Del-Rey pour toute leur aide.

I want to thank also all my friends who I have not mentioned yet. Thank you Carina for convincing me to come to Grenoble and for all your help and support. Thank you Angela for sharing the burden of writing this thesis, and of course the secret tiramisu recipe. Thank you Liza for motivating us all to go out for beers and for lending me your appartment to celebrate my birthday. Thank you Peter for your honesty and very funny way to see things. Thank you Claudio and Elena for unbreakable good mood, even if Beatrice stole you the last bit of sleep. Thank you Sven for all parties we had at your place especially in the evening of my defense. Thank you Cornelia and Roman for nice skiing trips. Thank you Ovidiu, Martin, and Marc for the nice evenings at D'Enfert. Thank you Simone and Francesca for the enjoyable evenings with the Scarpe Mobile. Thank you Clément, Dipankar, Angelo, Christoph, Hanno, Farida, and Tobias for all the nice moments in and outside the lab. Merci également à Gosia, Julian, Marine, Juan Pablo, Elenore, Aude et Clément pour m'avoir reçu très chaleureusement dans votre groupe. Je voudrais remercier mon équipe de volley avec qui j'ai passé beaucoup des bonnes moments, en particulier Malo et Olivier pour la meilleur saison de volley de toute ma vie. Vielen Dank auch an meine Freunde aus der Heimat Lars, Christian, Stefan, Daniel, Toni, Anja, Micha, Stefan, Rico für die schönen Momente mit euch.

Je voudrais remercier ma copine Sophie, qui m'a beaucoup soutenu et encouragé pendant ces deux dernières années. Ta constante bonne humeur était toujours une grande motivation pour moi. Un grand merci également à toute ta famille et en particulier à Vincent, Claudine et Céline. Zum Schluss möchte ich ganz besonders meiner Familie danken, die mich die ganze Zeit nach Kräften unterstützt haben und ohne die all dies nicht möglich gewesen wäre. Ein ganz besonderer Dank gilt vor allem meinen Eltern Andreas und Ines, meiner Schwester Sabrina, und meinen Großeltern Hans und Inge, weil ihr immer für mich da gewesen seid, wenn ich euch brauchte.

Stefan Thiele

Contents

Chapter 1
Introduction

1.1 Molecular Spintronics

The computer industry developed in the course of the last 60 years from its very infancy to one of the biggest global markets. This tremendous evolution was triggered by several historical milestones. In 1947, John Bardeen and Walter Brattain presented the world's first transistor [1] based on Walter Shockley's field-effect theory. Their discovery was soon after rewarded by the Nobel Prize in physics and led to the development of today's semiconductor industry.
Another groundbreaking discovery was made in 1977, when Alan Heeger, Hideki Shirakawa, and Alan MacDiarmid presented the first conducting polymer [2]. Their work opened the way for organic semiconductors, which stand for cheap and flexible electronics like organic LEDs, photovoltaic cells, and field-effect transistors. With still a lot of ongoing fundamental research, some fields already reached maturity. Especially, organic LEDs became an irreplaceable part of modern televisions in the last couple of years. The major impact of organic semiconductors was awarded by Royal Swedish Academy of Sciences with the Nobel Price in chemistry.
On decade later, in 1988, Peter Grünberg and Albert Fert reported an effect, which they called the giant magneto resistance (GMR) [3, 4]. In contrary to conventional electronic devices, which use charges as carriers of information, the GMR exploits the electronic spin degree of freedom. Their discovery led to the development of a completely new branch of research, which is these days referred to as spintronics. With the success of data-storage industry, in the last 25 years, devices using the GMR effect became a part of our everyday live.
The drive for steady innovation led researchers to think about new devices which unify these great ideas and would, therefore, be even more performing. The famous article of Datta and Das in 1990 [5] was the first step towards a new age of spintronic devices. Their proposal described a transistor, which could amplify signals using spins currents only. However, for this transistor to work, efficient spin-polarization, injection, and long relaxation times are necessary. Especially, the relaxation time is usually limited by spin-orbit coupling and the hyperfine interaction.

S. Thiele, *Read-Out and Coherent Manipulation of an Isolated Nuclear Spin*,
Springer Theses, DOI 10.1007/978-3-319-24058-9_1

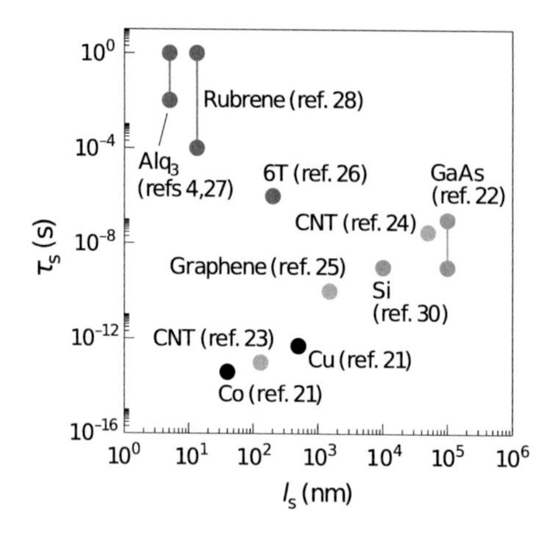

Fig. 1.1 Spin-relaxation time τ_S versus spin-diffusion length l_S. Organic semiconductors are situated in the upper left corner corresponding to long spin-lifetimes but short diffusion lengths. The figure was taken from [6], and the used references correspond to the ones from [6]

In this regard, organic spintronic devices might be a solution. They are known for their intrinsically small spin scattering, which allows for long spin relaxation times (see Fig. 1.1). This is because of the tiny spin-orbit interaction in organic materials. The latter is proportional to Z^4, with Z being the atomic number, which makes spin scattering very weak in carbon based devices.

In this context, single-molecule magnets (SMMs) are interesting candidates as building blocks for organic spintronic devices [7, 8]. Each molecule consists of a magnetic core, which is surrounded by organic ligands. The latter do not only protect the core from environmental influence but also tailor its magnetic properties. Replacing or modifying the ligands by means of organic chemistry alters the environmental coupling and makes selective bonding to specific surfaces possible [9]. Likewise, one can change the magnetic core, consisting of usually one or a few transition metal or rare earth ions, to alter the spin system, the spin-orbit coupling, or the hyperfine interaction of the molecule. Moreover, it is rather straight forward to synthesize billions of identical copies and embed them in virtually any matrix without changing their magnetic properties. It is this versatility, which makes them very attractive for spintronic devices.

The first, and most prominent, single-molecule magnet is the Mn_{12} acetate, which was discovered by Lis in 1980 [15]. It consists of 12 manganese atoms, which are surrounded by acetate ligands (see Fig. 1.2a). Another very famous single-molecule magnet is the Fe_8 [16], consisting of eight iron(III) ions surrounded by a macrocyclic ligand (see Fig. 1.2b). Both systems posses a total spin of S = 10 with an Ising type anisotropy resulting in an energy barrier separating the $m_S = \pm 10$ ground states by 63 K for Mn_{12} acetate [17] and by 25 K for Fe_8 [18]. In 1996, researches found the first evidence of quantum properties in SMM crystals. It was observed that the magnetization of the crystal is able to change its orientation via a tunnel process [19, 20]. A few years later, it was discovered that quantum inference during the tunnel process is possible [13]. And more recently, the coherent manipulation of

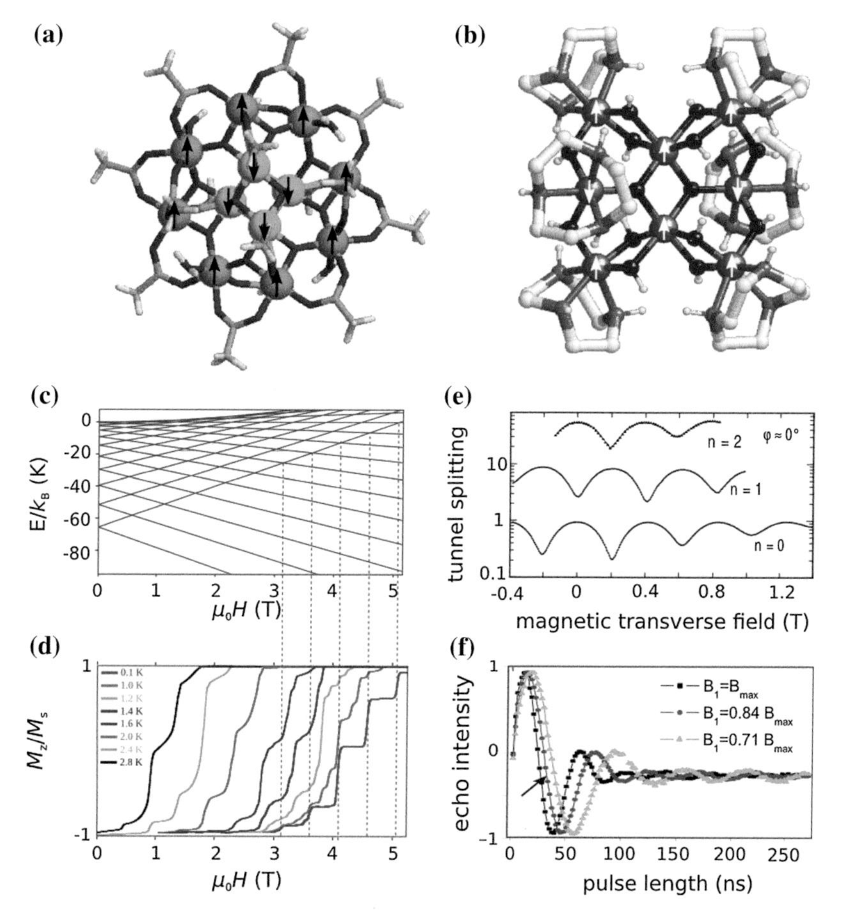

Fig. 1.2 **a** The Mn_{12} acetate SMM consists of 8 Mn(III) atoms with $S = 2$ (*orange*) and 4 Mn(IV) with $S = 3/2$ (*green*), which are connected via oxygen bonds. The spin of the twelve Mn atoms adds up to $S = 10$. Adapted from [10]. **b** The Fe_8 SMM, consists of eight Fe(III), which are interconnected by oxygen atoms (*red*). Each Fe(III) has spin of 5/2, which adds up to a total spin $S = 10$. Adapted from [11]. **c** Zeeman diagram of the Mn_{12} acetate obtained by exact numerical diagonalization. Important avoided level crossings are indicated by red dotted lines. **d** Magnetic hysteresis measurements obtained via Hall bar measurements of a microcrystal of Mn_{12}tBuAc. Adapted from [12]. **e** Quantum interference measurements obtained with a Fe_8 micro crystal. Adapted from [13]. **f** Rabi oscillations of a Fe_4 nano crystal. Adapted from [14]

the SMM's magnetic moment has been achieved for crystalline assemblies of SMMs [14, 21, 22].

The success of single-molecule magnets led to the discovery of a huge variety of new systems. A property which most of the experiments with SMMs have in common, is the use of a macroscopic amount of molecules in order to increase the detectable magnetic signal. However, a complete new type of experiments is possible when the molecules are measured isolated. Therefore, during the last couple of years, a lot

of effort was put into the construction of ultra sensitive detectors towards single-molecule sensitivity.
A promising concept to study isolated SMMs makes use of spin-polarized scanning-tunneling spectroscopy [23]. Therein, the molecule is deposited on a single crystalline metallic surface and studied via the tunnel current through a tiny movable tip. The advantage of this technique is the combination of transport measurements with atomic resolution imaging, which makes an explicit identification of the studied system possible. However, the electrical manipulation by means of a gate voltage is hard to implement, and consequently, this technique comes along with a tremendous reduction of the amount of information gained by transport measurements.
Therefore, our group followed two different strategies, both, allowing for the implementation of a back-gate, which adds an additional degree of freedom to the transport measurements.
In the first approach, two molecules were deposited onto a carbon nanotube [24, 25]. Due to a strong exchange coupling, the first molecule spin polarizes the current through the nanotube, whereas the second molecule acts as a detector. The conductance through the carbon nanotube is larger if the molecules were aligned parallel, with respect to an antiparallel alignment. This spin valve effect leads to a magneto resistance change of several hundred percent.
The second method, which was used in this thesis, traps the molecule in between to metallic electrodes, thus, creating a single-molecule magnet spin-transitor [26, 27]. The tunnel current through the transistor becomes again spin dependent due to the exchange coupling of the molecule's magnetic moment with the tunnel current, giving rise to an all electrical spin read-out.
However, in both techniques, a lack of imaging makes the unambiguous identification of the SMM very hard. That is why our group focused on terbium double-decker SMMs. They possess a large hyperfine splitting of molecule's electronic ground state levels, which can be used as a fingerprint and makes an unambiguous identification even without imaging possible. Moreover, the strong hyperfine interaction allows for the read-out of a single nuclear spin [26]. The latter is well protected from the environment and, therefore, a promising candidate for quantum information processing.

1.2 Quantum Information Processing

The construction of a quantum computer is one of the most ambitious goals of today's scientists. The idea was already born in 1982, when Richard Feynman stated that certain quantum mechanical effects cannot be simulated efficiently with classical computers [28]. Three years later, David Deutsch was the first who demonstrated that quantum computers are outperforming classical computers regarding certain problems [29], but concrete algorithms to program such a computer remained scarce. The beginning of a widespread interest in quantum computation was triggered by Peter Shor in the mid 90's (see Fig. 1.3). He presented a quantum prime factorization algorithm, which exponentially outperformed any classical algorithm [30]. Two

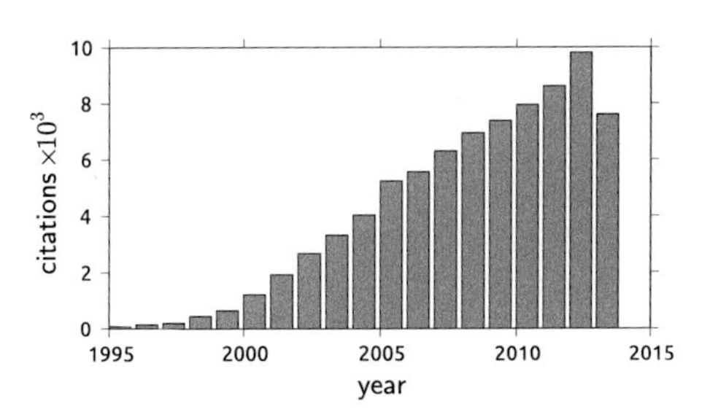

Fig. 1.3 Number of citations in Nature and Science whose topic contained quantum computing. Numbers were taken from Web of Science

years later, Grover demonstrated that using a quantum computer to find an element within an unsorted list would gain a polynomial speedup with respect to a classical computer [31].

In analogy to the classical bit, the smallest processing unit of a quantum computer is a quantum bit or qubit. It consists of a two level quantum system, whose states are usually denoted as $|0\rangle$ and $|1\rangle$. The difference to a classical bit, which can be either in 0 or 1, is that the qubit can be in the state $|0\rangle$, $|1\rangle$, or a superposition of both. This superposition state is mathematically described as $a|0\rangle + b|1\rangle$. In order to visualize a qubit, people often refer to the Bloch sphere (see Fig. 1.4). Therein, the $|0\rangle$ state corresponds to the north pole and the $|1\rangle$ state to the south pole of the sphere. In contrary to the classical bit, which is either at the north or the south pole, the qubit state can be at any point of the sphere, corresponding to a superposition state.

The real power of a quantum computer is believed to be in its exponential growth of the state space with increasing number of qubits. In contrary to a classical computer, which is able to address $2n$ different states with n bits, a quantum computer can address 2^n states with n bits.

Yet, to harness this power a real physical implementation of a quantum computer is necessary. In order to decide whether or not a quantum mechanical system is suited for constructing a quantum computer, DiVincenzo formulated the following five criteria [32].

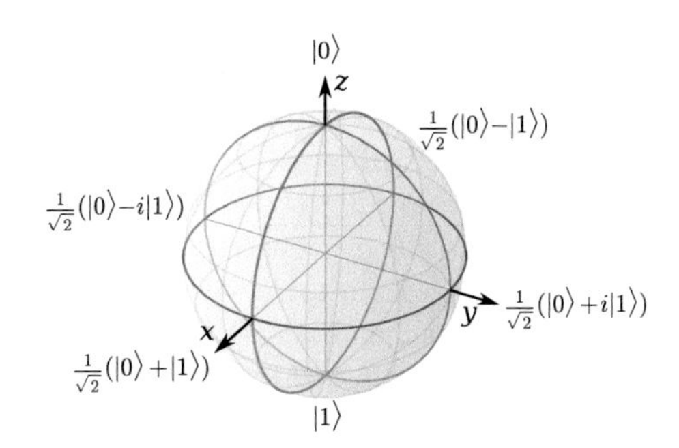

Fig. 1.4 Bloch sphere representation of a quantum bit. The two levels of a qubit $|0\rangle$ and $|1\rangle$ are represented by the north pole and the south pole of the sphere and any linear superposition can be visualized as a point on the surface of the sphere

- **Information storage on qubits**: the information is encoded on a quantum property of a scalable physical system which lives long enough to perform computations.
- **Initial state preparation**: the state of the qubit needs to be prepared before each computation.
- **Isolation**: the qubit must be protected from decoherence by isolation from the environment.
- **Gate implementation**: the manipulation of a quantum state must be performed with reasonable precision and much faster than the decoherence time T_2.
- **Read-out**: the final state of the qubit must be read-out with a sufficiently high precision.

One of the most delicate criteria for any quantum mechanical system is the isolation from the environment.
One of the earliest experiments fulfilling these criteria was performed in the group of David Wineland [33]. To create a qubit they were using electrically trapped ions, which were isolated from the environment using a ultra-high vacuum (see Fig. 1.5a). In another approach the group of Serge Haroche trapped light inside a cavity with an extremely high quality factor (see Fig. 1.5b). Using the light matter interaction they could read-out the quantum state of a photon. Both Wineland and Haroche were awarded the Nobel Prize in physics in 2012.
Yet, both techniques are experimentally very demanding. In order to get an easier access to a qubit system, researches were looking for solid state qubit systems which can be made using standard nano-fabrication techniques. A very promising candidate are Josephson junctions coupled to superconducting resonators [39, 40]. However, their size of several μm makes them extremely sensitive to external noise.
Another possibility to create qubits follows the proposal of Loss and DiVincenzo [41] (see Fig. 1.5d). Therein, the spin of electron inside a quantum dot is used as a two level quantum system. Since they are much smaller than superconducting circuits, they couple less strongly to the environment, but at the same time they are also harder to detect. The first single-shot read-out of an electron spin inside a quantum dot was reported in 2004 [42]. One year later, Stotz et al. demonstrated the coherent transport of an electron spin inside a semiconductor [43], and in 2006, the coherent manipulation of an electron spin in a GaAs quantum dot was presented by Koppens et al. [44].
Despite their big success, the coupling to the environment is still sufficiently strong to destroy coherence within several hundred nanoseconds. Alternative concepts propose the use of nuclear spins as building blocks for quantum computing since they benefit from inherently longer coherence times compared to electronic spins, because of a better isolation from the environment. But weak coupling comes at a price: the detection and manipulation of individual nuclear spins remain challenging tasks.
Despite the difficulties, scientists demonstrated operating nuclear spin qubits using optical detection of nitrogen vacancy centers [45] (Fig. 1.5e), or by performing single-shot electrical measurements in silicon based devices [46] (Fig.1.5f) and single-molecule magnet based devices [25, 26, 47] (Fig. 1.5g).

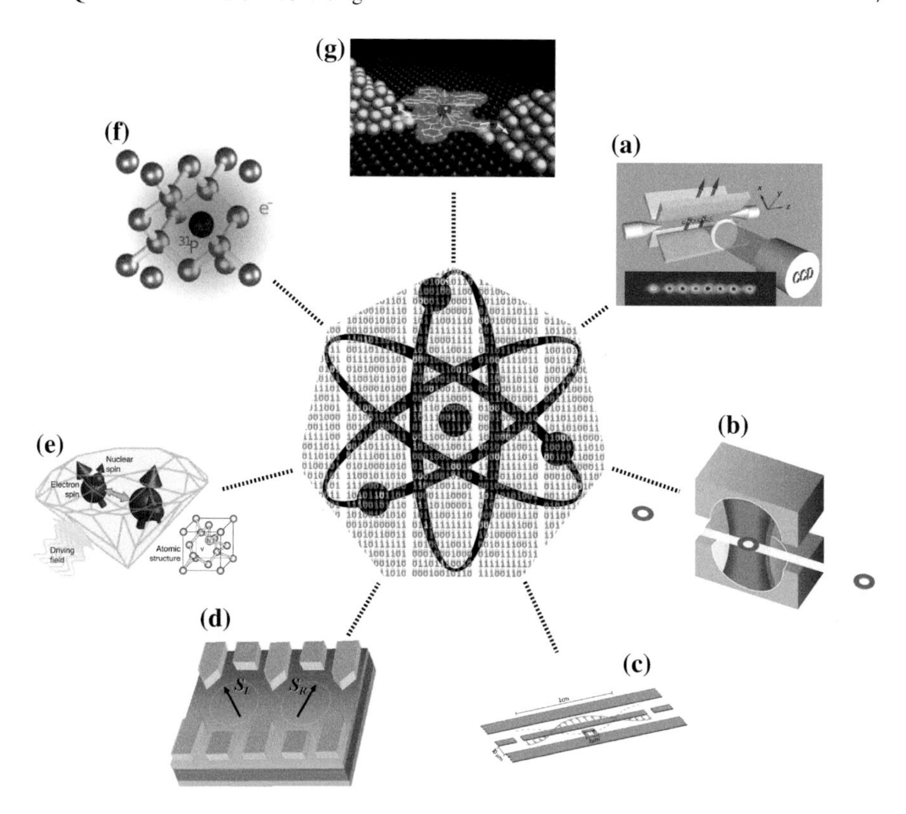

Fig. 1.5 Collection of different qubit types **a** ion traps taken from [34], **b** photons in a cavity source: Nobel Price Commite, **c** superconducting ciruits taken from [35], **d** quantum dots, source: Vitaly Golovach, **e** diamond color centers taken from [36], **f** ^{31}P impurities in silicon taken from [37] and **g** molecular magnets taken from [38]. Notice that selections focused on some important qubit families and is not a complete overview of all existing qubits

In order to solve the detection problem, the nuclear spin was measured indirectly through the hyperfine coupling to an electronic spin. Figure 1.6a explains this detection scheme exemplary using a NV defect, a color center in diamond [45]. The orbital ground state and the first excited state of the NV-center are $S = 1$ triplet states. Due to spin-spin interactions both states are split into a lower energy $m_S = 0$ ($|0_e\rangle$) state and two higher energy $m_S = \pm 1$ ($| \pm 1_e\rangle$) states. Their separation at zero magnetic field are 2.87 and 1.43 GHz for the ground state and excited state respectively. Optical transitions in NV-centers are spin preserving, leading to $\Delta m_S = 0$. If the spin is in the $m_S = 0$ ($m_S = \pm 1$) ground state, it can only be excited in the $m_S = 0$ ($m_S = \pm 1$) excited state. The average lifetime of the excited state is about 10 ns. After this time, a relaxation in the corresponding ground state takes place under the emission of a photon. If, however, the system was in the $m_S = \pm 1$ excited state, a relaxation via a non radiating metastable state into the $m_S = 0$ ground state is possible, causing a considerably smaller luminescence. The $|0_e\rangle$ and the $| \pm 1_e\rangle$ state are therefore

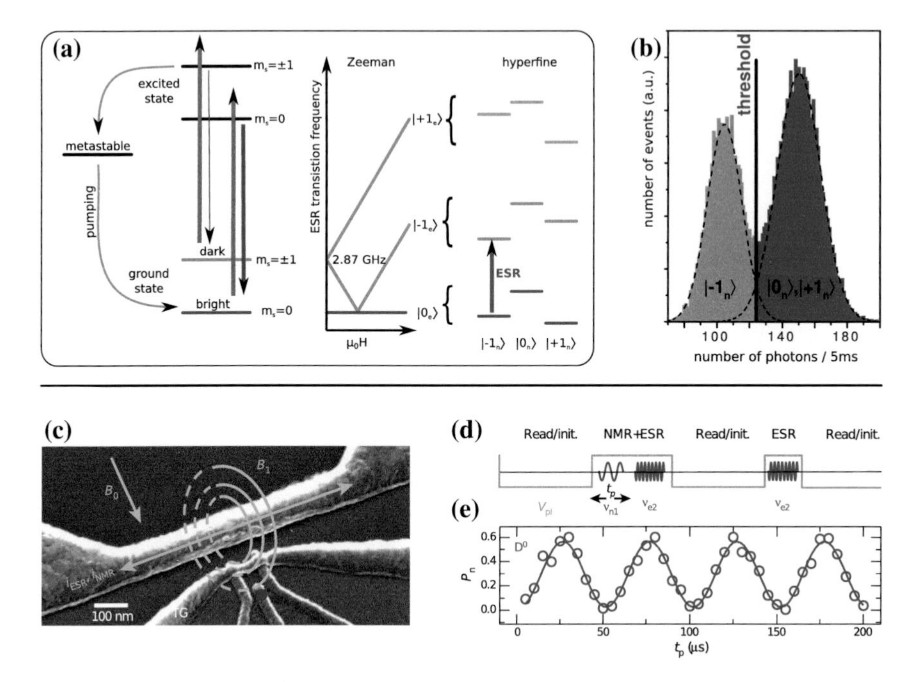

Fig. 1.6 **a** Energy diagrams of an NV-center. The left graph depicts radiative (*green and red arrows*) and non-radiative (*grey arrows*) transitions between the electronic ground state and the first excited state. In the center of the graph, the Zeeman diagram of the ground state triplet and its fine structure splitting were presented. The right graph shows the hyperfine splitting of each electronic state. Adapted from [45]. **b** Photon-counts histogram showing two Gausian like peaks. The left peak corresponds to the $|-1_\mathrm{n}\rangle$ state and the right peak to the $|0_\mathrm{n}\rangle$ and $|+1_\mathrm{n}\rangle$ states. Adapted from [45]. **c** Scanning electron micrograph of a Si qubit. **d** Pulse signal of a coherent nuclear spin rotation and the subsequent read-out. **e** Rabi oscillations of a single ^{31}P nuclear spin. **c**–**e** were taken from [46]

referred to as the bright and the dark state, respectively. This enables the optical detection of the magnetic resonance (ODMR). Furthermore, the relaxation process via the metastable state is pumping the system into the $|0_\mathrm{e}\rangle$ state, which is used to prepare the electronic spin in its initial state. The transition frequency between $|0_\mathrm{e}\rangle \rightarrow |-1_\mathrm{e}\rangle$ and $|0_\mathrm{e}\rangle \rightarrow |+1_\mathrm{e}\rangle$ can be changed by applying an external magnetic field along the quantization axis of the NV-center (see middle graph in Fig. 1.6a). Additionally, the hyperfine coupling to the nitrogen isotope ^{14}N, with a nuclear spin of $I = 1$, splits each electronic spin state into three, resulting in a nuclear spin dependent transition frequency under the influence of any external magnetic field. The three nuclear spin states will be referred to as $|-1_\mathrm{n}\rangle$, $|0_\mathrm{n}\rangle$, $|+1_\mathrm{n}\rangle$. To detect the nuclear spin, the system is first pumped into the $|0_\mathrm{e}\rangle$ state using a strong laser pulse. Afterward, a microwave pulse of precise duration and frequency is applied. If the frequency is matched to the $|0_\mathrm{e}\rangle|-1_\mathrm{n}\rangle \rightarrow |-1_\mathrm{e}\rangle|-1_\mathrm{n}\rangle$ level spacing, the electronic state will change from the bright into the dark state only if the nuclear spin was in the $|-1_\mathrm{n}\rangle$ state (see left graph in Fig. 1.6a). The read-out is done by repeating this procedure several times and recording the luminescence signal. If the nuclear spin was in the $|0_\mathrm{n}\rangle$ or $|+1_\mathrm{n}\rangle$ state, the luminescence signal is larger than for the $|-1_\mathrm{n}\rangle$ state (see Fig. 1.6b). Note that the detection of the nuclear spin state was realized by the read-out of the electronic spin state.

Quite similar to nitrogen color centers in diamond are ^{31}P impurities in silicon. However, owing to the small band gap of silicon the detection can be done electrically via a coupling to a close by quantum dot [46]. Notice that the nuclear spin read-out is again performed by exploiting the nuclear spin dependent electron spin resonance (ESR). Since the magnetic moment of the nuclear spin μ_N is about 2000 times smaller than the magnetic moment of the electronic spin μ_B, the manipulation of the former happens at times scales which are three orders of magnitude longer. In order to achieve a proper manipulation, large local AC magnetic fields are necessary. The group of Morello realized these fields by on-chip microwave strip lines (see Fig. 1.6c). The nuclear spin manipulation happened according to the following protocol (see Fig. 1.6d). First, the nuclear spin was prepared in its initial state. Afterward, a microwave pulse at the nuclear spin transition frequency of duration τ_p was applied. Depending on the pulse duration, the nuclear spin can be flipped with the probability P_n. Plotting P_n versus τ_p resulted in coherent Rabi oscillations (see Fig. 1.6e).

Nevertheless, the time scale of a manipulation remained in the order of 100 μs due to the tiny magnetic moment of the nuclear spin [46, 48]. Larger local alternating magnetic fields would increase this frequency, but they are difficult to generate using state of the art on-chip coils [49] due to the inevitable parasitic crosstalk to the detector and neighboring spin qubits.

To solve this problem, we propose and demonstrate in this thesis the single nuclear spin manipulation by means of an AC electric field. Indeed, it was already suggested by Kane [50] that the Stark effect of the hyperfine coupling could be used to tune different ^{31}P nuclear spins in and out of resonance using local DC gate voltages. He, therefore, established the individual addressability by applying only a global microwave field.

Our approach can be viewed as the extension of Kane's proposal to AC gate voltages. We will demonstrate coherent nuclear qubit manipulations using the hyperfine Stark effect to transform local electric fields into effective AC magnetic fields in the order of a few hundred mT and, hence, speeding up the clock speed of a single nuclear spin operation by two orders of magnitudes. In addition, we show that a local static gate voltage can shift the resonance frequency by several MHz, allowing for the individual addressability of several nuclear spin qubits.

1.3 Thesis Outline

My thesis was dedicated to study the read-out and manipulation of an isolated nuclear spin inside a single-molecule magnet. We made use of a three terminal transistor layout, in which the nuclear spin is electrically detected using a read-out quantum dot.

In order to give the reader a basic understanding of how the molecular spin-transistor works, we will recall in chapter two some fundamental transport properties of a quantum dot. In particular, we will focus on single electron tunneling, co-tunneling, and the Kondo effect since they are the most important transports characteristics observed in our devices.
In Chap. 3 we will concentrate on the magnetic properties of an isolated $TbPc_2$ single-molecule magnet. A lot of attention is directed to the electronic states of terbium ion, which are responsible for the observed magnetic properties of the device and, therefore, of paramount importance for this thesis.
A large part of my work was also devoted to the design and the construction of the experimental setup and is shown in chapter four. Starting from the dilution refrigerator I will explain each important part of the experiment which was added or modified in order to fabricate and measure a molecular spin-transistor.
Chapter 5 starts with explaining the mode of operation of the single-molecule magnet spin-transistor based on a simple model. The rest of the chapter details the conducted experiments in order to substantiate the aforementioned model.
In Chap. 6 we will use the spin-transistor to perform a time-resolved, quantum non-demolition read-out of the nuclear spin qubit state. We determined the relaxation time T_1 and the fidelity of the read-out. Furthermore, the experimental results are compared with quantum Monte Carlo simulations in order to deduce the dominating relaxation mechanism.
In Chap. 7 we propose and present the coherent manipulation of a single nuclear spin by means of the hyperfine Stark effect. Hence, using an AC electric field we generated and effective alternating magnetic field in the order of a few hundred mT. These results represent the first manipulation of a nuclear spin inside a single-molecule magnet and the first electrical manipulation of an isolated nuclear spin qubit.

References

1. J. Bardeen, W. Brattain, The transistor. A semi-conductor triode. Phys. Rev. **74**, 230–231 (1948)
2. C. Chiang, C. Fincher, Y. Park, A. Heeger, H. Shirakawa, E. Louis, S. Gau, A. MacDiarmid, Electrical conductivity in doped polyacetylene. Phys. Rev. Lett. **39**, 1098–1101 (1977)
3. J. Barnaś, A. Fuss, R. Camley, P. Grünberg, W. Zinn, Novel magnetoresistance effect in layered magnetic structures: theory and experiment. Phys. Rev. B **42**, 8110–8120 (1990)
4. M.N. Baibich, J.M. Broto, A. Fert, F.N. van Dau, F. Petroff, Giant magnetoresistance of (001)Fe/(001)Cr magnetic superlattices. Phys. Rev. Lett. **61**, 2472–2475 (1988)
5. S. Datta, B. Das, Electronic analog of the electro-optic modulator. Appl. Phys. Lett. **56**, 665 (1990)
6. G. Szulczewski, S. Sanvito, M. Coey, A spin of their own. Nat. Mater. **8**, 693–5 (2009)
7. D. Gatteschi, R. Sessoli, J. Villain, *Molecular Nanomagnets* (Oxford University Press, Oxford, 2006)
8. L. Bogani, W. Wernsdorfer, Molecular spintronics using single-molecule magnets. Nat. Mater. **7**, 179–86 (2008)

9. S. Klyatskaya, J.R.G. Mascarós, L. Bogani, F. Hennrich, M. Kappes, W. Wernsdorfer, M. Ruben, Anchoring of rare-earth-based single-molecule magnets on single-walled carbon nanotubes. J. Am. Chem. Soc. **131**, 15143–51 (2009)
10. R. Sessoli, H.L. Tsai, A.R. Schake, S. Wang, J.B. Vincent, K. Folting, D. Gatteschi, G. Christou, D.N. Hendrickson, High-spin molecules: [Mn12O12(O2CR)16(H2O)4]. J. Am. Chem. Soc. **115**, 1804–1816 (1993)
11. W. Wernsdorfer, Molecular nanomagnets: towards molecular spintronics. Int. J. Nanotechnol. **7**, 497 (2010)
12. W. Wernsdorfer, M. Murugesu, G. Christou, Resonant tunneling in truly axial symmetry Mn12 single-molecule magnets: sharp crossover between thermally assisted and pure quantum tunneling. Phy. Rev. Lett. **96**, 057208 (2006)
13. W. Wernsdorfer, Quantum phase interference and parity effects in magnetic molecular clusters. Science **284**, 133–135 (1999)
14. C. Schlegel, J. van Slageren, M. Manoli, E.K. Brechin, M. Dressel, Direct observation of quantum coherence in single-molecule magnets. Phys. Rev. Lett. **101**, 147203 (2008)
15. T. Lis, Preparation, structure, and magnetic properties of a dodecanuclear mixed-valence manganese carboxylate. Acta Crystallogr. B Struct. Crystallogr. Cryst. Chem. **36**, 2042–2046 (1980)
16. K. Weighardt, K. Pohl, I. Jibril, G. Huttner, Hydrolysis Products of the Monomeric Amine Complex $(C_6H_{15}N_3)FeCl_3$: The Structure of the Octameric Iron(III) Cation of $\{[(C_6H_{15}N_3)_6Fe_8(\mu_3\text{-}O)_2(\mu_2\text{-}OH)_{12}]Br_7(H_2O)\}Br \cdot 8H_2O$. Angew. Chem. Int. Ed. Engl. **23**, 77–78 (1984)
17. A. Caneschi, D. Gatteschi, R. Sessoli, A.L. Barra, L.C. Brunel, M. Guillot, Alternating current susceptibility, high field magnetization, and millimeter band EPR evidence for a ground S = 10 state in [Mn12O12(Ch3COO)16(H2O)4].2CH3COOH.4H2O. J. Am. Chem. Soc. **113**, 5873–5874 (1991)
18. A.-L. Barra, P. Debrunner, D. Gatteschi, C.E. Schulz, R. Sessoli, Superparamagnetic-like behavior in an octanuclear iron cluster. Europhys. Lett. (EPL) **35**, 133–138 (1996)
19. J.R. Friedman, M.P. Sarachik, R. Ziolo, Macroscopic measurement of resonant magnetization tunneling in high-spin molecules. Phys. Rev. Lett. **76**, 3830–3833 (1996)
20. L. Thomas, F. Lionti, R. Ballou, D. Gatteschi, R. Sessoli, B. Barbara, Macroscopic quantum tunnelling of magnetization in a single crystal of nanomagnets. Nature **383**, 145–147 (1996)
21. S. Bertaina, S. Gambarelli, A. Tkachuk, I.N. Kurkin, B. Malkin, A. Stepanov, B. Barbara, Rare-earth solid-state qubits. Nat. Nanotechnol. **2**, 39–42 (2007)
22. A. Ardavan, O. Rival, J. Morton, S. Blundell, A. Tyryshkin, G. Timco, R. Winpenny, Will spin-relaxation times in molecular magnets permit quantum information processing? Phys. Rev. Lett. **98**, 1–4 (2007)
23. J. Schwöbel, Y. Fu, J. Brede, A. Dilullo, G. Hoffmann, S. Klyatskaya, M. Ruben, R. Wiesendanger, Real-space observation of spin-split molecular orbitals of adsorbed single-molecule magnets. Nat. Commun. **3**, 953 (2012)
24. M. Urdampilleta, S. Klyatskaya, J.-P. Cleuziou, M. Ruben, W. Wernsdorfer, Supramolecular spin valves. Nat. Mater. **10**, 502–6 (2011)
25. M. Ganzhorn, S. Klyatskaya, M. Ruben, W. Wernsdorfer, Strong spin-phonon coupling between a single-molecule magnet and a carbon nanotube nanoelectromechanical system. Nat. Nanotechnol. **8**, 165–169 (2013)
26. R. Vincent, S. Klyatskaya, M. Ruben, W. Wernsdorfer, F. Balestro, Electronic read-out of a single nuclear spin using a molecular spin transistor. Nature **488**, 357–360 (2012)
27. E. Burzurí, A.S. Zyazin, A. Cornia, H.S.J. van der Zant, Direct observation of magnetic anisotropy in an individual Fe_4 single-molecule magnet. Phys. Rev. Lett. **109**, 147203 (2012)
28. R.P. Feynman, Simulating physics with computers. Int. J. Theoret. Phys. **21**, 467–488 (1982)
29. D. Deutsch, Quantum theory, the church-turing principle and the universal quantum computer. Proc. R. Soc. A Math. Phys. Eng. Sci. **400**, 97–117 (1985)
30. P. Shor, Algorithms for quantum computation: discrete logarithms and factoring, in *Proceedings 35th Annual Symposium on Foundations of Computer Science*, vol. 26 (IEEE Computer Society Press, 1994), pp. 124–134. ISBN 0-8186-6580-7

31. L.K. Grover, A fast quantum mechanical algorithm for database search, in *Proceedings of the Twenty-Eighth Annual ACM Symposium on Theory of Computing—STOC '96* (ACM Press, New York, New York, USA, 1996). ISBN 0897917855
32. D.P. DiVincenzo, *Topics in Quantum Computers*, vol. 345. NATO Advanced Study Institute. Ser. E Appl. Sci. vol. 345 (1996)
33. D. Wineland, R. Drullinger, F. Walls, Radiation-pressure cooling of bound resonant absorbers. Phys. Rev. Lett. **40**, 1639–1642 (1978)
34. R. Blatt, D. Wineland, Entangled states of trapped atomic ions. Nature **453**, 1008–15 (2008)
35. J. Clarke, F.K. Wilhelm, Superconducting quantum bits. Nature **453**, 1031–42 (2008)
36. S.C. Benjamin, J.M. Smith, Driving a hard bargain with diamond qubits. Physics **4**, 78 (2011)
37. A. Morello, Quantum information: atoms and circuits unite in silicon. Nat. Nanotechnol. **8**, 233–4 (2013)
38. S. Thiele, R. Vincent, M. Holzmann, S. Klyatskaya, M. Ruben, F. Balestro, W. Wernsdorfer, Electrical readout of individual nuclear spin trajectories in a single-molecule magnet spin transistor. Phys. Rev. Lett. **111**, 037203 (2013)
39. L. Dicarlo, M.D. Reed, L. Sun, B.R. Johnson, J.M. Chow, J.M. Gambetta, L. Frunzio, S.M. Girvin, M.H. Devoret, R.J. Schoelkopf, Preparation and measurement of three-qubit entanglement in a superconducting circuit. Nature **467**, 574–8 (2010)
40. M. Neeley, R.C. Bialczak, M. Lenander, E. Lucero, M. Mariantoni, A.D. O'Connell, D. Sank, H. Wang, M. Weides, J. Wenner, Y. Yin, T. Yamamoto, A.N. Cleland, J.M. Martinis, Generation of three-qubit entangled states using superconducting phase qubits. Nature **467**, 570–3 (2010)
41. D. Loss, D.P. DiVincenzo, Quantum computation with quantum dots. Phys. Rev. A **57**, 120–126 (1998)
42. J.M. Elzerman, R. Hanson, L.H. Willems Van Beveren, B. Witkamp, L.M.K. Vandersypen, L.P. Kouwenhoven, Single-shot read-out of an individual electron spin in a quantum dot. Nature **430**, 431–5 (2004)
43. J.A.H. Stotz, R. Hey, P.V. Santos, K.H. Ploog, Coherent spin transport through dynamic quantum dots. Nat. Mater. **4**, 585–8 (2005)
44. F.H.L. Koppens, C. Buizert, K.J. Tielrooij, I.T. Vink, K.C. Nowack, T. Meunier, L.P. Kouwenhoven, L.M.K. Vandersypen, Driven coherent oscillations of a single electron spin in a quantum dot. Nature **442**, 766–71 (2006)
45. P. Neumann, J. Beck, M. Steiner, F. Rempp, H. Fedder, P.R. Hemmer, J. Wrachtrup, F. Jelezko, Single-shot readout of a single nuclear spin. Science (New York, N.Y.) **329**, 542–4 (2010)
46. J.J. Pla, K.Y. Tan, J.P. Dehollain, W.H. Lim, J.J.L. Morton, F.A. Zwanenburg, D.N. Jamieson, A.S. Dzurak, A. Morello, High-fidelity readout and control of a nuclear spin qubit in silicon. Nature **496**, 334–338 (2013)
47. M. Urdampilleta, S. Klyatskaya, M. Ruben, W. Wernsdorfer, Landau-Zener tunneling of a single Tb^{3+} magnetic moment allowing the electronic read-out of a nuclear spin. Phys. Rev. B **87**, 195412 (2013)
48. W. Pfaff, T.H. Taminiau, L. Robledo, H. Bernien, M. Markham, D.J. Twitchen, R. Hanson, Demonstration of entanglement-by-measurement of solid-state qubits. Nat. Phys. **9**, 29–33 (2012)
49. T. Obata, M. Pioro-Ladrière, T. Kubo, K. Yoshida, Y. Tokura, S. Tarucha, Microwave band on-chip coil technique for single electron spin resonance in a quantum dot. Rev. Sci. Instrum. **78**, 104704 (2007)
50. B.E. Kane, A silicon-based nuclear spin quantum computer. Nature **393**, 133–137 (1998)

Chapter 2
Single Electron Transistor

The first single electron transistor (SET), made of small tunnel junctions, was realized in the Bell Laboratories in 1987 by Fulton and Dolan [1]. Since then, the fabrication of SETs became more and more sophisticated and allowed for operation at room temperature [2] or as sensors for electron spin detection [3]. In this thesis, a single electron transistor will be used to read-out the state of an isolated nuclear spin and, therefore, a basic knowledge of the transport properties in SETs and its associated effects such as Coulomb blockade, elastic and inelastic cotunneling, and the Kondo effect are necessary.

2.1 Equivalent Circuit

A single electron transistor consists of a conducting island or quantum dot, which is tunnel-coupled to the source and drain leads. Due to the small size of the dot the electronic energy levels E_n are discretized. In order to observe the characteristic single electron tunneling through the device, the resistance R_t of the tunnel barriers should be much higher than the quantum of resistance:

$$R_\mathrm{t} \gg \frac{h}{e^2} \tag{2.1.1}$$

where h is the Planck constant and e the elementary charge. This condition ensures that only one electron at the time is tunneling in or out of the quantum dot. A simple model to describe the electron transport through the dot was developed by Korotkov et al. [4], and reviewed by Kouwenhoven [5], and Hanson [6]. Therein, the quantum dot is coupled via constant source, drain, and gate capacitors (C_s, C_d, C_g) to the three

S. Thiele, *Read-Out and Coherent Manipulation of an Isolated Nuclear Spin*,
Springer Theses, DOI 10.1007/978-3-319-24058-9_2

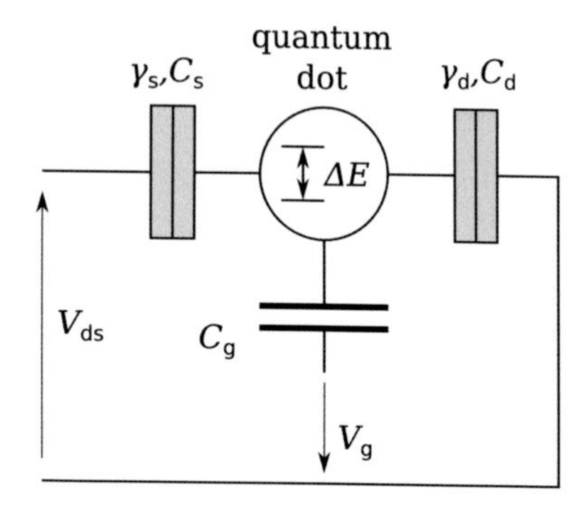

Fig. 2.1 Equivalent circuit of an SET. The electrostatic behavior of the dot is modeled by capacitors to the source, drain, and gate terminals

terminals as shown in Fig. 2.1. By applying a voltage to the three different terminals, the electrostatic potential U_{es} of the quantum dot is modified as:

$$U_{\mathrm{es}} = \frac{\left(C_{\mathrm{s}}V_{\mathrm{s}} + C_{\mathrm{d}}V_{\mathrm{d}} + C_{\mathrm{g}}V_{\mathrm{g}}\right)^2}{2C_\Sigma} \tag{2.1.2}$$

with $C_\Sigma = C_{\mathrm{s}}+C_{\mathrm{d}}+C_{\mathrm{g}}$ and V_{s}, V_{d}, and V_{g} being the source, drain, and gate voltages, respectively. Furthermore, due to the Coulomb repulsion, adding an electron to the quantum dot with N electrons ($N > 0$) will cost an additional energy:

$$U_{\mathrm{c}} = \frac{E_{\mathrm{c}}}{2} = \frac{e^2}{2C_\Sigma} \tag{2.1.3}$$

with E_{c} being the charging energy. Accordingly, to observe single electron tunneling, temperatures smaller than E_{c} are required since, otherwise, the tunnel process can be activate thermally.

$$E_{\mathrm{c}} \gg k_{\mathrm{B}}T \tag{2.1.4}$$

Putting all contributions together results in the total energy U of the quantum dot with N electrons:

$$U(N) = \frac{\left(-e(N-N_0) + C_{\mathrm{s}}V_{\mathrm{s}} + C_{\mathrm{d}}V_{\mathrm{d}} + C_{\mathrm{g}}V_{\mathrm{g}}\right)^2}{2C_\Sigma} + \sum_1^N E_n(B) \tag{2.1.5}$$

where N_0 is the offset charge and $E_n(B)$ the magnetic field dependent single electron energies. Experimentally, it is more convenient to work with the chemical potential, defined as the energy difference between two subsequent charge states $\mu_{\mathrm{dot}}(N) = U(N) - U(N-1)$. Inserting Eq. 2.1.5 into this expression gives:

$$\mu_{\mathrm{dot}}(N) = \left(N - \frac{1}{2}\right)E_{\mathrm{c}} - \frac{E_{\mathrm{c}}}{|e|}\left(C_{\mathrm{s}}V_{\mathrm{s}} + C_{\mathrm{d}}V_{\mathrm{d}} + C_{\mathrm{g}}V_{\mathrm{g}}\right) + E_N(B) \tag{2.1.6}$$

with E_N being the energy of the Nth electron in the quantum dot. Notice that the chemical potential depends linearly on the gate voltage, whereas the total energy shows a quadratic dependence. Therefore, the energy difference between the chemical potentials of different charge states remains constant for any applied voltages. The energy to add an electron to the quantum dot is called addition energy E_{add} and is defined as the difference between to subsequent chemical potentials.

$$E_{\mathrm{add}}(N) = \mu(N+1) - \mu(N) = E_{\mathrm{c}} + \Delta E \tag{2.1.7}$$

with ΔE being the energy spacing between two discrete energy levels.

2.2 Coulomb Blockade

The transport through the quantum dot is very sensitive to the alignment of the chemical potential μ inside the dot with respect to those of the source μ_{s} and drain μ_{d}. If we neglect the level broadening and any excited states of the quantum dot for a moment, then the transport through the SET can be explained with Fig. 2.2. Notice that V_{ds} and V_{g} are in arbitrary units and $V_{\mathrm{g}} = 0$ when $\mu_{\mathrm{dot}} = \mu_{\mathrm{s}} = \mu_{\mathrm{d}}$.

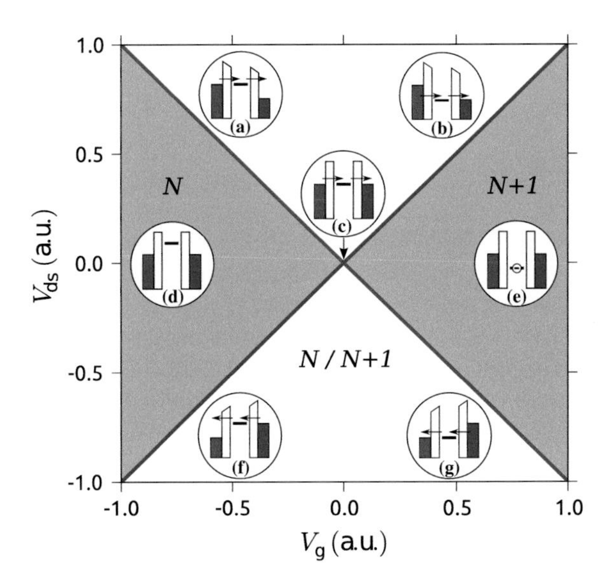

Fig. 2.2 Schematic of a stability diagram. Inside the *grey* regions the charge of the quantum dot is fixed to N (d) or $N+1$ (e), leading to the Coulomb blockade. Likewise, inside the white area electrons can tunnel in and out of the quantum dot. If the conductance dI/dV is measured instead of the current I, only the *red* and the *blue line* are visible, corresponding to a change in I. Along the *red line* the chemical potential of the quantum dot is aligned with the source chemical potential, whereas along the *blue line* it is aligned with the drain chemical potential

First we want to discuss what happens for zero bias $V_{ds} = V_d = V_s = 0$. If $V_g < 0$, the chemical potential of the dot is larger than the chemical potential of the leads, and the SET is in its off state (Fig. 2.2d). Increasing V_g to zero will align the three chemical potentials. Electrons can tunnel in and out of the dot from both sides leading to a finite conductance and a charge fluctuation between N and $N+1$. This particular working regime is called the charge degeneracy point (Fig. 2.2d). A further increase of V_g will push the chemical potential of the dot below the ones of source and drain, and the SET is again in its off state, but having $N+1$ electrons on the dot. Whenever the charge of the dot is fixed, the SET is in the Coulomb blockade regime since adding another electron would cost energy to overcome the electron-electron repulsion.
If we now increase the bias voltage to $V_{ds} \neq 0$, we shift the chemical potential between source and drain and open an energy or bias window of $\mu_s - \mu_d = eV_{ds}$, and a current is observed even for $V_g \neq 0$.
The red line in Fig. 2.2 corresponds to the situation where the chemical potential of the dot is aligned with μ_s (Fig. 2.2a, g). Crossing this line will turn the SET on or off, resulting in a conductance ridge along the line. The slope can be calculated from the equivalent circuit by setting the potential difference between dot and source to zero and is given by $-C_g/(C_g + C_s)$.
On the other hand, if μ_{dot} is aligned with the drain chemical potential, the SET turns also on or off, resulting in another conductance ridge (blue line in Fig. 2.2). Its slope is of opposite sign and calculated by setting the potential difference between drain and dot to zero, resulting in C_g/C_d. Therefore, inside the white region the transistor is turned on, whereas inside the grey region the SET is Coulomb blocked.

2.3 Cotunneling Effect

Up to now only transport through energetically allowed states was considered. This is usually sufficient if the tunnel barrier resistances are larger than 1 MΩ. However, for smaller tunnel barrier resistances, the time to exchange an electron between the dot and the leads becomes fast enough to allow for transport through energetically forbidden states. This is possible due to the Heisenberg uncertainty relation, which states that a system can violate energy conservation within a very short time $\tau = \hbar/E$, where $E \approx E_c$ for quantum dots. Therefore within the time τ an electron can enter the quantum dot whereas another is tunneling into the leads. Since this process involves two electrons it is called cotunneling. Note that the entire tunnel process is considered to be a single quantum event. We distinguish in the following two different cases of cotunneling events, namely, elastic and inelastic cotunneling [7]. If the electron entering the quantum dot occupies the same energy level as the outgoing one, the cotunneling is elastic and requires no additional energy (Fig. 2.3a). Experimentally, it can be observed as a conductance background inside the Coulomb blocked region. If, however, the electron entering the dot occupies an excited state, separated by ΔE from the chemical potential of the electron leaving the dot, the transport is inelastic (Fig. 2.3b). This process requires energy and happens only at finite bias voltages with

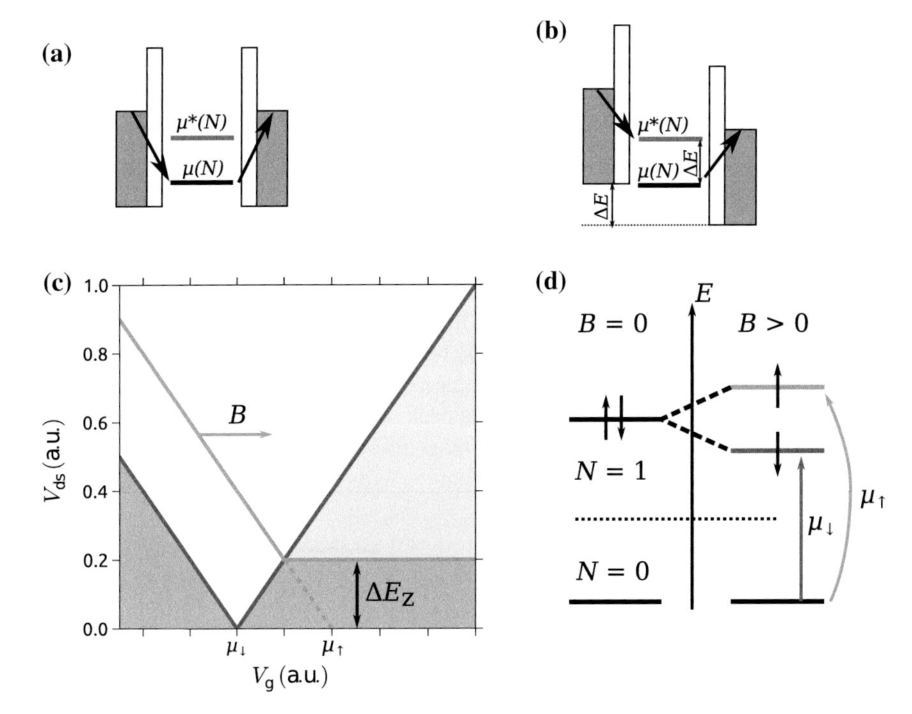

Fig. 2.3 Schematic showing the elastic **a** and inelastic **b** cotunneling process. **c** Stability diagram for a quantum dot whose ground and excited state are split by the Zeeman energy. The conductance step inside the Coulomb blocked region (*dark grey*) occurs at $e|V_{\text{ds}}| = \Delta E_Z$ and can be used to determine the Zeeman splitting. **d** Energy level diagram of a quantum dot with zero or one electron. Due to an external magnetic field, the degeneracy between *spin up* and *down* is lifted. The two lowest lying chemical potentials of the quantum dot correspond to the energy difference of the Zeeman split $N = 1$ doublet and the $N = 0$ singlet

$e|V_{\text{ds}}| > \Delta E$. The result is a conductance step inside the Coulomb blocked region. In the case of a very simple quantum dot as shown in Fig. 2.3c, d the conductance step can be used to determine the Zeeman splitting due to a magnetic field.

2.4 Kondo Effect

In the 1930s, de Haas et al. found out that while cooling down a long wire of gold, the resistance reaches a minimum at around 10 K and increases for further cooling [8]. Later, it was discovered that this effect was correlated to the presence of magnetic impurities, but a theoretical explanation of this phenomenon was only presented in the 1960s, by Jun Kondo [9]. In his model, an antiferromagnetic coupling between the conduction electrons and the residual magnetic impurities leads to the formation of a singlet state below a certain temperature T_K (Kondo temperature). This can be thought

of a cloud of conduction electrons, screening the magnetic impurity and therefore augmenting its effective cross section, which causes an increase in resistance.
The same effect can be found in quantum dots. If they are filled with an odd number of electrons, its total spin $S = 1/2$, which makes it an artificial magnetic impurity. If, furthermore, the coupling between the dot and the leads is large enough (tunnel resistances below 1 MΩ), electrons from the leads try to screen the artificial impurity by continuously flipping its spin via a tunnel process (Fig. 2.4a, b). This allows for a hybridization between the leads and the quantum dot, resulting in the appearance of two peaks in the quantum dot's DOS: one at Fermi level of the source and one at the Fermi level of the drain (Fig. 2.4c). The conductance through the quantum dot can be explained by the convolution of the two peaks. Since at zero V_{ds} the source and drain Fermi level coincide, the conductance will have a maximum and drops to zero for higher bias voltages, resulting in a peak, or Kondo ridge. If the temperature becomes comparable to T_K, the antiferromagnetic coupling between the magnetic

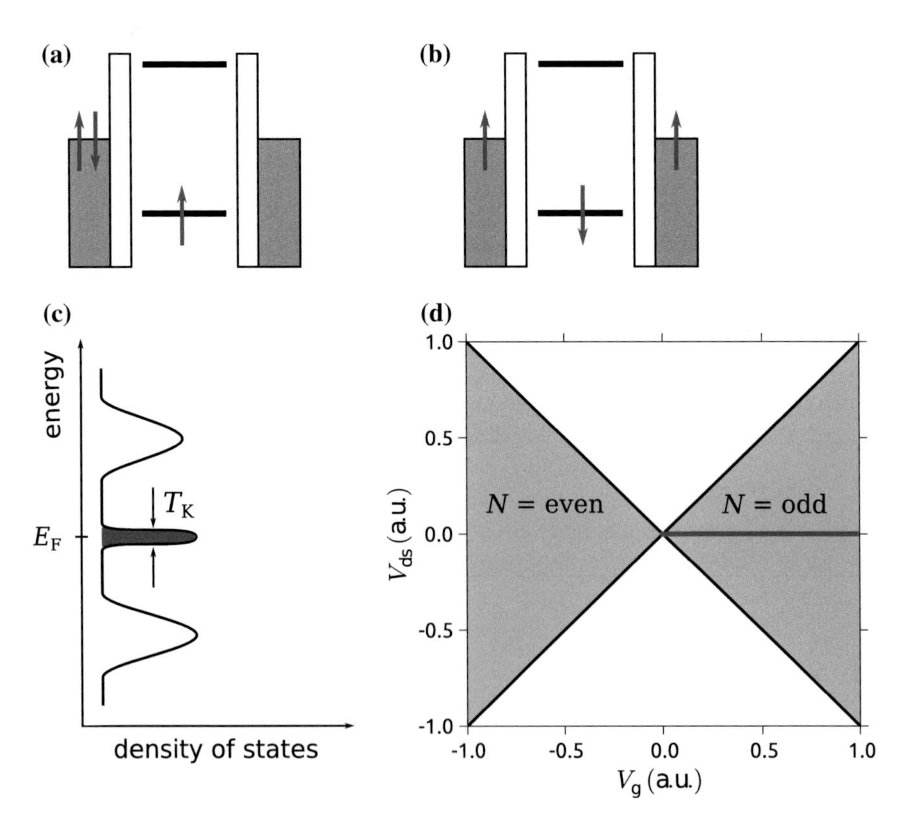

Fig. 2.4 Kondo transport mechanism with the initial state (**a**) and the final state (**b**). Note that the spin of the quantum dot flipped during the process. **c** The Kondo effect creates a peak in the density of states, whose width is given by the Kondo temperature. **d** Experimentally, the Kondo effect is observed as a conductance ridge or Kondo ridge (*red line*) inside the Coulomb blockade region of the stability diagram

impurity and the electrons in the leads is destroyed, resulting in the suppression of the conductance peak. The temperature dependance can be fitted by the empirical Goldhaber-Gordon equation [10]

$$G(T) = G_0 \left(\frac{T^2}{T_K^2} (2^{1/s} + 1) \right)^{-s} + G_c \tag{2.4.1}$$

and results in $G(T_K) = G_0/2$. The variable G_c accounts for a conductance offset caused by elastic cotunneling and $s = 0.22$.
Another possibility to study the Kondo effect is by applying a magnetic field. For a classical spin 1/2 in a magnetic field, the Zeeman effect will split the spin up and down levels by $g\mu_B B_c$. If this splitting becomes larger than the antiferromagnetic coupling given by $0.5k_B T_K$ [11], the Kondo ridge at zero bias is destroyed. However, applying a positive or negative bias voltage V_{ds} can compensate for the energy gap when $e|V_{ds}| = g\mu_B|B|$. This leads to the revival of the Kondo effect and is observed as two peaks, one at negative and one at positive bias. The separation of the Kondo peak as a function of the applied magnetic field is schematically displayed in Fig. 2.5a.

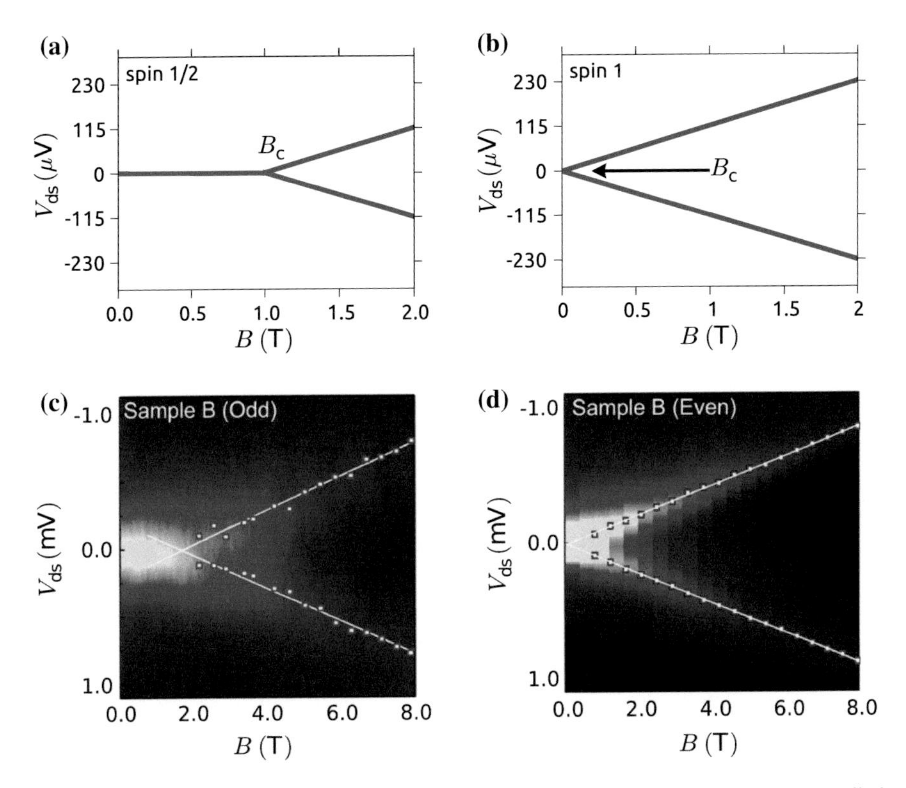

Fig. 2.5 Separation of the spin 1/2 (**a**) and spin 1 (**b**) Kondo peak as a function of the applied magnetic field. The critical field of 0.5 T was chosen arbitrarily. The slope corresponds corresponds to a g factor of 2. (**c**, **d**) Experimental data adapted from [12]

Using this model, the critical field can be used to estimate the Kondo temperature and the strength of the coupling.
If we add one electron to the quantum dot, the spin is either zero (singlet state) or one (triplet state). In case of zero spin, no Kondo effect will be observed. If, however, the triplet state becomes the ground state of the system, the situation changes. Similar to the spin 1/2 Kondo effect, electrons from the leads try to screen the artificial magnetic impurity, which now has a spin of 1. Therefore, the screening requires two conduction channels, one for each electron of the triplet.
In quantum dots, like the ones we used in our experiments, the coupling of different energy levels to the source and drain terminals is not symmetric in energy, resulting in two individual Kondo temperatures T_{K_1} and T_{K_2}. Hence, in the temperature window $T_{K_1} < T < T_{K_2}$ the screening of channel 1 is suppressed, whereas the screening of channel 2 is still working. This scenario is referred to as the underscreened Kondo effect. Its signature is such that the critical field needed to quench the conductance ridge is much smaller than $0.5k_B T_K$ [12]. This can be understood by a semi-classical consideration of the residual spin which was left unscreened. The ferromagnetic coupling between the two spins, which led to the formation of the triplet ground state, results in an effective magnetic field created by the unscreened spin at the site of the screen spin. This field weakens the antiferromagnetic coupling of the second spin to the electrons in the leads. Hence, already at very small external magnetic fields, the critical field is reached, leading to a shift of B_C towards zero (Fig. 2.5b). The magnitude of the shift is proportional to the ferromagnetic exchange coupling between the two spins but cannot be determined precisely due to the lack of knowledge of the fully screened B_C. However, it gives an estimate of its order of magnitude.

References

1. T. Fulton, G. Dolan, Observation of single-electron charging effects in small tunnel junctions. Phys. Rev. Lett. **59**, 109–112 (1987)
2. K. Matsumoto, M. Ishii, K. Segawa, Y. Oka, B.J. Vartanian, J.S. Harris, Room temperature operation of a single electron transistor made by the scanning tunneling microscope nanooxidation process for the TiOx/Ti system. Appl. Phys. Lett. **68**, 34 (1996)
3. J.M. Elzerman, R. Hanson, L.H. Willems Van Beveren, B. Witkamp, L.M.K. Vandersypen, L.P. Kouwenhoven, Single-shot read-out of an individual electron spin in a quantum dot. Nature **430**, 431–435 (2004)
4. A. Korotkov, D. Averin, K. Likharev, Single-electron charging of the quantum wells and dots. Phys. B Condens. Matter **165–166**, 927–928 (1990)
5. L.P. Kouwenhoven, C.M. Marcus, P.L. Mceuen, S. Tarucha, M. Robert, *Electron Transport in Quantum Dots* (Proceedings of the NATO Advanced Study Institute, Kluwer, 1997). ISBN 140207459X
6. R. Hanson, J.R. Petta, S. Tarucha, L.M.K. Vandersypen, Spins in few-electron quantum dots. Rev. Mod. Phys. **79**, 1217–1265 (2007)
7. M. Pustilnik, L. Glazman, Kondo effect in quantum dots. J. Phys. Condens. Matter **16**, R513–R537 (2004)

8. W. de Haas, J. de Boer, G. van dën Berg, The electrical resistance of gold, copper and lead at low temperatures. Physica **1**, 1115–1124 (1934)
9. J. Kondo, Resistance minimum in dilute magnetic alloys. Prog. Theor. Phys. **32**, 37–49 (1964)
10. D. Goldhaber-Gordon, J. Göres, M. Kastner, H. Shtrikman, D. Mahalu, U. Meirav, From the Kondo regime to the mixed-valence regime in a single-electron transistor. Phys. Rev. Lett. **81**, 5225–5228 (1998)
11. T.A. Costi, Kondo effect in a magnetic field and the magnetoresistivity of Kondo alloys. Phys. Rev. Lett. **85**, 1504–1507 (2000)
12. N. Roch, S. Florens, T.A. Costi, W. Wernsdorfer, F. Balestro, Observation of the underscreened Kondo effect in a molecular transistor. Phys. Rev. Lett. **103**, 197202 (2009)

Chapter 3
Magnetic Properties of $TbPc_2$

3.1 Structure of $TbPc_2$

The single molecule magnet which was investigated in this thesis is a metal-organic complex called bisphthalocyaninato terbium(III) ($[TbPc_2]^-$). The magnetic moment of the molecule arises from a single terbium ion (Tb^{3+}), situated in the center of the molecule. It is eightfold coordinate to the nitrogen atoms of the two phthalocyanine (Pc) ligands, which are stacked below and above the terbium ion resulting an approximate C_4 symmetry in the close environment of the Tb. The ligands are encapsulating the Tb^{3+} in order to preserve and tailor its magnetic properties. Its resemblance to the double-decker airplane of the 1920s is giving it its colloquial name—terbium double-decker (Fig. 3.1).

3.2 Electronic Configuration of Tb^{3+}

Naturally attained ^{159}Tb is one of the 22 elements with only one natural abundant isotope. With an atomic number of 65, it is situated within the lanthanide series in the periodic system of elements (see Fig. 3.2). Its name arises from the Swedish town Ytterby, where it was first discovered in 1843.
The electronic structure of Tb is $[Xe]4f^9 6s^2$. The 4f shell, which is not completely filled, is responsible for its paramagnetism. It is located inside the 6s, 5s, and 5p shell and therefore well protected from the environment. Like most of the lanthanides, Tb releases three electrons to form chemical bonds. These three electrons consist of two 6s electrons, which are on the outer most shell and therefore easy to remove, and one 4f electron. 4f electrons are most of the time inside the 5s and 5p shell, but they cannot come very close to the core neither, resulting in a smaller ionization energy than for 5s and 5p electrons. Thus, the electronic structure of the Tb^{3+} is $[Xe]4f^8$.
The energetic position of the different orbits and levels of the terbium ion is affected by several interactions, namely, the electron-electron interaction H_{ee}, the spin-

S. Thiele, *Read-Out and Coherent Manipulation of an Isolated Nuclear Spin*,
Springer Theses, DOI 10.1007/978-3-319-24058-9_3

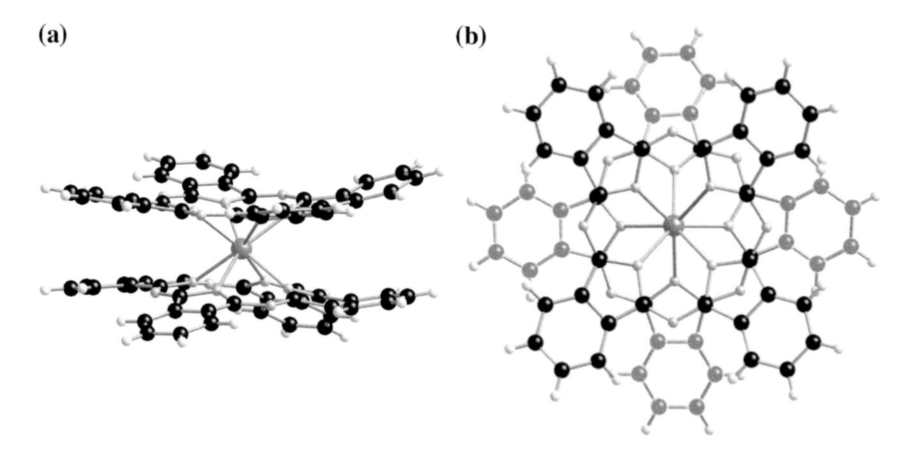

Fig. 3.1 *Side view* (**a**) and *top view* (**b**) of the $TbPc_2$. The *pink atom* in the center of the complex is the Tb^{3+} ion, which is eightfold coordinated to the nitrogen atoms (*blue*) of the two phthalocyanine ligands resulting in a local approximate C_4 symmetry.

1 H Hydrogen 1.00794																	2 He Helium 4.003
3 Li Lithium 6.941	4 Be Beryllium 9.012182											5 B Boron 10.811	6 C Carbon 12.0107	7 N Nitrogen 14.00674	8 O Oxygen 15.9994	9 F Fluorine 18.9984032	10 Ne Neon 20.1797
11 Na Sodium 22.989770	12 Mg Magnesium 24.3050											13 Al Aluminum 26.981538	14 Si Silicon 28.0855	15 P Phosphorus 30.973761	16 S Sulfur 32.066	17 Cl Chlorine 35.4527	18 Ar Argon 39.948
19 K Potassium 39.0983	20 Ca Calcium 40.078	21 Sc Scandium 44.955910	22 Ti Titanium 47.867	23 V Vanadium 50.9415	24 Cr Chromium 51.9961	25 Mn Manganese 54.938049	26 Fe Iron 55.845	27 Co Cobalt 58.933200	28 Ni Nickel 58.6934	29 Cu Copper 63.546	30 Zn Zinc 65.39	31 Ga Gallium 69.723	32 Ge Germanium 72.61	33 As Arsenic 74.92160	34 Se Selenium 78.96	35 Br Bromine 79.904	36 Kr Krypton 83.80
37 Rb Rubidium 85.4678	38 Sr Strontium 87.62	39 Y Yttrium 88.90585	40 Zr Zirconium 91.224	41 Nb Niobium 92.90638	42 Mo Molybdenum 95.94	43 Tc Technetium (98)	44 Ru Ruthenium 101.07	45 Rh Rhodium 102.90550	46 Pd Palladium 106.42	47 Ag Silver 107.8682	48 Cd Cadmium 112.411	49 In Indium 114.818	50 Sn Tin 118.710	51 Sb Antimony 121.760	52 Te Tellurium 127.60	53 I Iodine 126.90447	54 Xe Xenon 131.29
55 Cs Cesium 132.90545	56 Ba Barium 137.327	57 La Lanthanum 138.9055	72 Hf Hafnium 178.49	73 Ta Tantalum 180.9479	74 W Tungsten 183.84	75 Re Rhenium 186.207	76 Os Osmium 190.23	77 Ir Iridium 192.217	78 Pt Platinum 195.078	79 Au Gold 196.96655	80 Hg Mercury 200.59	81 Tl Thallium 204.3833	82 Pb Lead 207.2	83 Bi Bismuth 208.98038	84 Po Polonium (209)	85 At Astatine (210)	86 Rn Radon (222)
87 Fr Francium (223)	88 Ra Radium (226)	89 Ac Actinium (227)	104 Rf Rutherfordium (261)	105 Db Dubnium (262)	106 Sg Seaborgium (263)	107 Bh Bohrium (262)	108 Hs Hassium (265)	109 Mt Meitnerium (266)	110 (269)	111 (272)	112 (277)	113	114				
				58 Ce Cerium 140.116	59 Pr Praseodymium 140.90765	60 Nd Neodymium 144.24	61 Pm Promethium (145)	62 Sm Samarium 150.36	63 Eu Europium 151.964	64 Gd Gadolinium 157.25	65 Tb Terbium 158.92534	66 Dy Dysprosium 162.50	67 Ho Holmium 164.93032	68 Er Erbium 167.26	69 Tm Thulium 168.93421	70 Yb Ytterbium 173.04	71 Lu Lutetium 174.967
				90 Th Thorium 232.0381	91 Pa Protactinium 231.03588	92 U Uranium 238.0289	93 Np Neptunium (237)	94 Pu Plutonium (244)	95 Am Americium (243)	96 Cm Curium (247)	97 Bk Berkelium (247)	98 Cf Californium (251)	99 Es Einsteinium (252)	100 Fm Fermium (257)	101 Md Mendelevium (258)	102 No Nobelium (259)	103 Lr Lawrencium (262)

Fig. 3.2 Periodic table of elements. The element ^{159}Tb belongs to the lanthanide series and possesses only one stable isotope.

orbit coupling H_{SO}, the ligand field potential H_{lf}, the exchange interaction H_{ex}, the hyperfine-coupling H_{hf}, and the magnetic field H_{Z}. An overview of the magnitude of these energetic effects on the 4f electrons is given in Table 3.1.

In the following we want to briefly discuss the different interactions starting with the Zeeman effect.

Table 3.1 Energy scale of different effects acting on 4f electrons.

Interaction	Energy equivalent (cm^{-1})
Electron-electron interaction H_{ee}	$\approx 10^4$
Spin-orbit coupling H_{so}	$\approx 10^3$
Ligand-field potential H_{lf}	$\approx 10^2$
Exchange interaction H_{ex}	< 1
Hyperfine interaction H_{hf}	$\approx 10^{-1}$
Magnetic field H_Z at 1T	≈ 0.5

Taken from [1, 2]

3.3 Zeeman Effect

From classical mechanics it is known that a magnetic moment $\boldsymbol{\mu}$ exposed to an external magnetic field $\boldsymbol{B}$ will change its potential energy by $E_{pot} = -\boldsymbol{\mu}\boldsymbol{B}$. The quantum mechanical equivalent is called the Zeeman effect. To calculate the Zeeman energy we write down the Zeeman Hamiltonian:

$$H_Z = g\mu_B \boldsymbol{J}\boldsymbol{B} \tag{3.1}$$

where g is the Landée factor, $\mu_B = e\hbar/2m_e$ the Bohr magneton, and $\boldsymbol{J} = \boldsymbol{L} + \boldsymbol{S}$ the total angular momentum of the system. For a more general derivation of this formula see Appendix A.1 and A.2. In the case of a free electron with $\boldsymbol{J} = \boldsymbol{S}$ and $B = (0, 0, B_Z)$, the Zeeman Hamiltonian becomes:

$$H_Z = g\mu_B S_Z B_Z \tag{3.2}$$

with S_Z being the Pauli matrix. Diagonalizing this Hamiltonian at different magnetic fields results in Fig. 3.3, which is referred to as the Zeeman diagram. It shows that the spin degeneracy if lifted at $B \neq 0$.

3.4 Electron-Electron Interaction

As we have seen in Table. 3.1, the electron-electron interaction is the strongest of all interactions and is mainly responsible for the orbital energies and the shell filling. The latter is well explained by the famous Hund's rules:

1. Hund's rule The electrons within a shell are arranged such that their total spin S is maximized. $\sum s_i \rightarrow max$. This can be understood in terms of Coulomb repulsion. Electrons with the same spin have to be in different orbitals due to the Pauli principle. Since they are in different orbitals, they are in average further apart from each other, resulting in a reduced Coulomb repulsion.

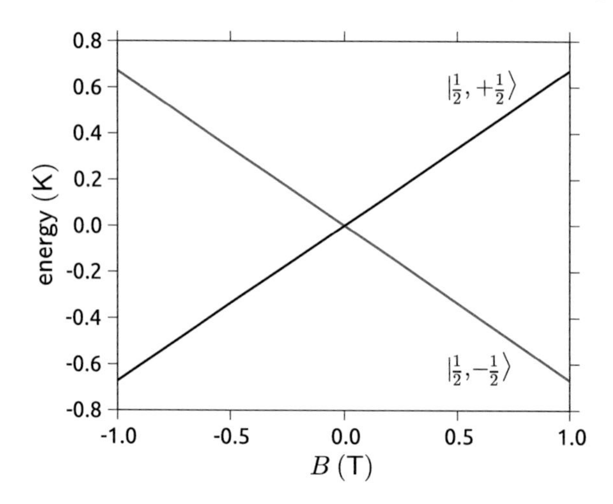

Fig. 3.3 Zeeman diagram of a free electron.

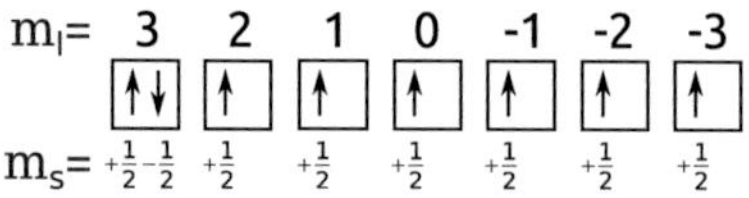

m_l= 3 2 1 0 -1 -2 -3

m_s= $+\frac{1}{2}-\frac{1}{2}$ $+\frac{1}{2}$ $+\frac{1}{2}$ $+\frac{1}{2}$ $+\frac{1}{2}$ $+\frac{1}{2}$ $+\frac{1}{2}$

Fig. 3.4 Electronic structure of the Tb^{3+} 4f shell. $L = \sum_i m_l^i = 3$ and $S = \sum_i m_s^i = 3$

2. Hund's rule For a given spin, the electrons are arranged within the shell such that their total angular momentum L is maximized. $\sum l_i \rightarrow max$. This Hund's rule also origins from the Coulomb repulsion. Electrons with similar angular momentum are revolving more synchronous and avoiding each other therefore more effectively.
3. Hund's rule For less than half-filled sub-shells the total angular momentum $J = |L-S|$, whereas for more than half filled sub-shells the total angular momentum $J = |L + S|$. This rules arises from minimizing the spin-orbit coupling energy and cannot be explained easily with hand-waving arguments.

In order to fill up the 4f shell of Tb^{3+} we start with rule number one by putting seven electrons with spin up in the seven different orbitals and therefore maximize the spin S. The last electron is put in the $m_l = 3$ state according to the second rule. This already results in the final shell filling with a total spin $S = 7 \times \frac{1}{2} - \frac{1}{2} = 3$ and an angular momentum $L = 3 + 2 + 1 + 0 - 1 - 2 - 3 + 3 = 3$ as shown in Fig. 3.4.

3.5 Spin-Orbit Interaction

The spin-orbit interaction is the coupling of the electron's spin $\boldsymbol{s}$ with its orbital momentum $\boldsymbol{l}$. In the semi-classical picture the electron's orbital motion creates a magnetic moment $\boldsymbol{\mu}_l = -\frac{\mu_B}{\hbar}\boldsymbol{l}$. Furthermore, since the famous Stern-Gerlach experi-

ment from 1921, it is known that electrons possess a magnetic moment $\boldsymbol{\mu}_\mathrm{S}$ generated by its inherent spin $\boldsymbol{s}$. Due to dipole interactions, the formerly independent momenta are connected, resulting in the total momentum $\boldsymbol{j} = \boldsymbol{l} + \boldsymbol{s}$. The energy change resulting from this interaction is $\Delta E = -\boldsymbol{\mu}_\mathrm{S}\boldsymbol{B}_\mathrm{l} \propto \boldsymbol{l}\boldsymbol{s}$. Applying the correspondence principle leads to the spin-orbit Hamiltonian $H_\mathrm{SO} = \xi\, \boldsymbol{l}\boldsymbol{s}$, where ξ is the one-electron spin-orbit coupling parameter. A more exact derivation of the spin-orbit interaction is given in Appendix A.4 for the interested reader.
Since Tb^{3+} has eight electrons in the 4f shell we have to consider more than just one spin and orbital momentum. If, however, the coupling between different orbital momenta $H_{\mathrm{l_i-l_j}} = a_{ij}\boldsymbol{l}_i\boldsymbol{l}_j$ and the different spins $H_{\mathrm{S_i-S_j}} = b_{ij}\boldsymbol{s}_i\boldsymbol{s}_j$ is large compared to the spin-orbit coupling $H_{\mathrm{l_iS_i}}$, the momenta itself couple first to a total spin $\boldsymbol{S} = \sum_i \boldsymbol{s}_i$ and a total orbital momentum $\boldsymbol{L} = \sum_i \boldsymbol{l}_i$, before coupling to the total momentum $\boldsymbol{J} = \boldsymbol{L} + \boldsymbol{S}$, and the spin-orbit Hamiltonian modifies to:

$$H_\mathrm{SO} = \lambda(\boldsymbol{r})\, \boldsymbol{L}\boldsymbol{S} \tag{3.3}$$

Minimizing this energy for the Tb^{3+} leads to the third Hund's rule with a ground state of $J = L+S = 6$, which is $2J+1 = 13$ times degenerate. All possible combinations are displayed in Fig. 3.5. In the following paragraph we will see how the spin-orbit interaction can be computed within the framework of first order perturbation theory. Without spin-orbit coupling all spins would couple to the total spin $\boldsymbol{S}$ and all orbital momenta would combine to $\boldsymbol{L}$, leading to $(2L+1)\times(2S+1)$ degenerate states. Since the spin-orbit contribution to the electron energy is small with respect to the electron-

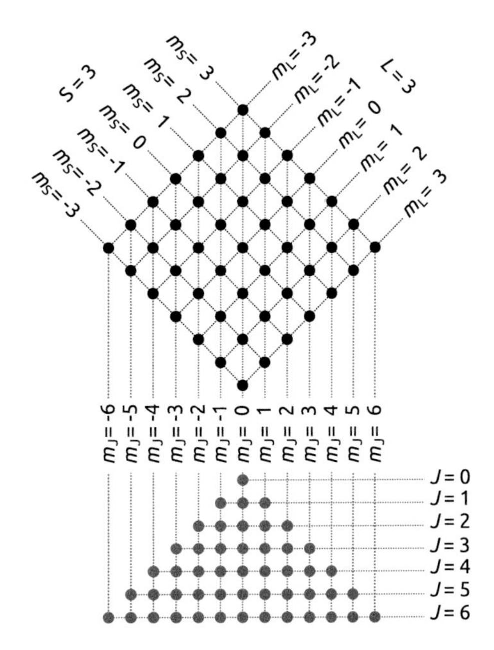

Fig. 3.5 Due to the spin-orbit coupling, the total spin $\boldsymbol{S}$, with its $2S+1$ states, is coupling to the total orbital momentum $\boldsymbol{L}$, with its $2L+1$ states, resulting in a total momentum $\boldsymbol{J} = \boldsymbol{L} + \boldsymbol{S}$ with $(2S+1)(2L+1)$ states.

electron interaction, first-order degenerate perturbation theory can be applied. We perform our calculations using the L and S along with their projections m_L and m_S as good quantum numbers to describe our unperturbed states. To calculate the energy correction to first order for N degenerate states, we have to write the spin-orbit Hamiltonian in this basis as a $N \times N$ matrix and perform an exact numerical diagonalization. We assume that $\Psi^{(0)}_{u,v}$ is the unperturbed electron wave function and $u, v = [0..(2L+1) \times (2S+1)]$. Applying the product ansatz splits the wave function into a radial, angular and spin-dependent part: $\Psi^{(0)}_{u,v} = |R\rangle|m_\mathrm{L}\rangle|m_\mathrm{S}\rangle$. Since the operators $\boldsymbol{L}$ and $\boldsymbol{S}$ are not acting on the radial part we can write the Hamiltonian as:

$$H_\mathrm{SO} = \zeta\, \boldsymbol{LS} \tag{3.4}$$

$$\boldsymbol{LS} = L_\mathrm{x}S_\mathrm{x} + L_\mathrm{y}S_\mathrm{y} + L_\mathrm{z}S_\mathrm{z} \tag{3.5}$$

where $\zeta = \langle R|\lambda(\boldsymbol{r})|R'\rangle$ is the one-electron spin-orbit coupling constant. In order to expand this equation into a matrix, we make use of the following transformation:

$$L_\mathrm{x} = \frac{1}{2}(L_+ + L_-); \qquad L_\mathrm{y} = \frac{1}{2i}(L_+ - L_-) \tag{3.6}$$

$$S_\mathrm{x} = \frac{1}{2}(S_+ + S_-); \qquad S_\mathrm{y} = \frac{1}{2i}(S_+ - S_-) \tag{3.7}$$

with

$$L_\pm|m'_\mathrm{L}\rangle = \sqrt{L(L+1) - m_\mathrm{L}(m_\mathrm{L} \pm 1)}|m'_\mathrm{L} \pm 1\rangle \tag{3.8}$$

$$S_\pm|m'_\mathrm{S}\rangle = \sqrt{S(S+1) - m_\mathrm{S}(m_\mathrm{S} \pm 1)}|m'_\mathrm{S} \pm 1\rangle \tag{3.9}$$

Inserting Eq. 3.6 and 3.8 into Eq. 3.4 results in the final spin-orbit Hamiltonian:

$$H_\mathrm{SO} = \zeta \left[L_\mathrm{z}S_\mathrm{z} + \frac{1}{2}(L_+S_- + L_-S_+) \right] \tag{3.10}$$

What is left is the definition of the operators L_i and S_i, with i being z, $+$, or $-$. Each of them is defined as a generalized Pauli matrix σ^N of order N, with N being $(2L+1)$ or $(2S+1)$ respectively (see Appendix A.3). To expand the dimension of these operators to a $(2L+1) \times (2S+1)$ Hilbert space we apply the Kronecker product $\otimes$. It is not commutative, and the order of the multiplication needs to be preserved. The operators L_i and S_i are therefore:

$$L_i = \sigma_i^{2L+1} \otimes \mathbb{I}^{2S+1}$$

$$S_i = \mathbb{I}^{2L+1} \otimes \sigma_i^{2S+1}$$

with $\mathbb{I}^M$ being the identity matrix of order M. Setting $\zeta = -336$ K and diagonalizing the Hamiltonian results in the eigenvalues as shown in Fig. 3.6. The calculated eigen-

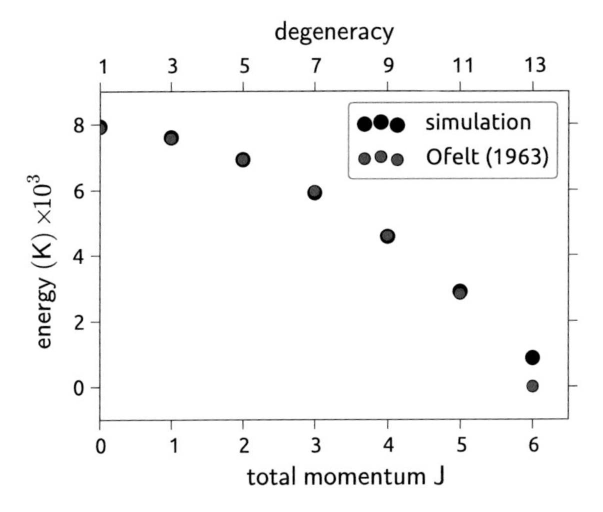

Fig. 3.6 This graph is obtained by calculating the eigenvalues of Eq. 3.10 and $\zeta = -336K$. The simulated values were shifted vertically to coincident with the values taken from [3]. As depicted the spin-orbit coupling lifts the degeneracy of the 49 states resulting in seven different multiplets with $J = 6$ as a new ground state.

Table 3.2 Energy splitting between the ground state (GS) $|J\rangle$ and excited state (ES) $|J - 1\rangle$ multiplets for pure LS coupling [2, 5]

Ion	Tb^{3+}	Dy^{3+}	Ho^{3+}	Er^{3+}	Tm^{3+}	Yb^{3+}
Elec. Conf.	$4f^8$	$4f^9$	$4f^{10}$	$4f^{11}$	$4f^{12}$	$4f^{13}$
GS	$^7F_6$	$^6H_{\frac{15}{2}}$	$^5I_8$	$^4I_{\frac{15}{2}}$	$^3H_6$	$^2F_{\frac{7}{2}}$
ES	$^7F_5$	$^6H_{\frac{13}{5}}$	$^5I_7$	$^4I_{\frac{13}{2}}$	$^3H_5$ (^{1})	$^2F_{\frac{5}{2}}$
ΔE (K)	2900	4300	7300	9400	11900(^{1})	14400

(^{1}) For Tm^{3+} $^3H_4$ lies below $^3H_5$ [2]

values and the experimentally obtained ones fit very well except for $J = 6$, where higher order perturbation theory is necessary. Nevertheless, a large energy splitting between the new ground state $J = 6$ and the new first excited state $J = 5$ of 2900 K [3, 4] is observed, making it possible to simplify the calculation of the magnetic properties by considering the 13 ground states only.
As shown in Table. 3.2 the large splitting between the ground state and the first excited state is a general property of rare earth ions and increases with the atomic number.

3.6 Ligand-Field Interaction

The ligand-field theory describes the electrostatic interaction between the coordination center of a complex and its ligands, leading to a modification of the electronic states of the former. Since the 4f shell of the lanthanides is situated inside the 5s and 5p shell, it is to a large part protected from its surrounding environment. However, the effect on the energy levels is still in the order of a few hundred Kelvin and acts as a perturbation on the spin-orbit coupling. We will start our considerations with a brief introduction into the ligand-field theory and apply this formalism to the terbium double-decker in the following. The electrostatic potential V_{lf} created by the ligand can be expressed in a very general way:

$$V_{\mathrm{lf}}(\boldsymbol{r}) = \int \frac{\rho(\boldsymbol{r}')}{4\pi\epsilon_0|\boldsymbol{r}-\boldsymbol{r}'|} d^3R \tag{3.11}$$

where $\boldsymbol{r}$ is the position of the electron and $\rho(\boldsymbol{r}')$ the charge density of the ligands. Since symmetry plays a very important role in this theory, we will express $1/|\boldsymbol{r}-\boldsymbol{r}'|$ in terms of spherical harmonics:

$$\frac{1}{|\boldsymbol{r}-\boldsymbol{r}'|} = \sum_{k=0}^{\infty} \frac{4\pi}{2k+1} \frac{r^k}{R^{k+1}} \sum_{q=-k}^{k} Y_k^q(\Theta, \Phi) Y_k^q(\theta, \phi) \tag{3.12}$$

where $Y_k^q(\Theta, \Phi)$ describes the position of the ligands, and $Y_k^q(\theta, \phi)$ describes the position of the electron. Therefore the ligand-field potential becomes:

$$V_{\mathrm{lf}}(\boldsymbol{r}) = \sum_{k=0}^{\infty} \sum_{q=-k}^{k} r^k \underbrace{\frac{4\pi}{2k+1} Y_k^q(\theta, \phi)}_{C_k^q} \underbrace{\int \frac{\rho(\boldsymbol{r}')}{4\pi\epsilon_0 R^{k+1}} Y_k^q(\Theta, \Phi) d^3R}_{A_k^q} \tag{3.13}$$

The term C_k^q is the so-called Racah tensor and depends only on the ligand position. The last term A_k^q is the geometrical coordination factor, which is a constant that can be determined experimentally. As V_{lf} can be treated as a perturbation to the spin-orbit ground-state multiplet, J remains a good quantum number, and the wave function Ψ can be written as $\Psi = |J, m_{\mathrm{J}}\rangle$. It is very convenient to replace the operator C_k^q by the Stevens operators O_k^q, which are linear combinations of the total angular momentum operators and simplify the calculation in this basis [6]. Additional factors u_k (Stevens factors) account for the proper transformation [7]. The symmetry of the O_k^q is identical to the spherical harmonics Y_k^q, where $k-q$ is the number of nodes in the polar direction and q the number of nodes in the azimuthal direction with $-k \leq q \leq k$. The matrices for $q = 0$ have only diagonal elements, whereas for $q \neq 0$ off-diagonal elements occur, introducing a coupling between different states. The term O_0^0 has spherical symmetry and gives rise to a constant potential, which can be omitted. Furthermore, due to time reversal symmetry, all odd values of k vanish since they involve J_{z} to odd powers. It is sufficient to carry out the summation up to $k \leq 2J$ [8], with higher order terms being usually smaller than lower order terms. The ligand-field Hamiltonian H_{lf} is therefore:

Table 3.3 (**a**) The Stevens factors [7] and (**b**) the ligand-field parameters [10] for $TbPc_2$

(a)	u_2	u_4	u_6		
	$-\frac{1}{99}$	$\frac{2}{16335}$	$-\frac{1}{891891}$		
(b)	$A_2^0\langle r^2\rangle$	$A_4^0\langle r^4\rangle$	$A_4^4\langle r^4\rangle$	$A_6^0\langle r^6\rangle$	$A_6^4\langle r^6\rangle$
	595.7 K	−328.1 K	14.4 K	47.5 K	0 K

$$H_{\mathrm{lf}} = \sum_{k=0}^{\infty} \sum_{q=-k}^{k} A_k^q \langle r^k \rangle u_k O_k^q (J_{\mathrm{x}}, J_{\mathrm{y}}.J_{\mathrm{z}}) \tag{3.14}$$

The matrix elements O_k^q of the Stevens operators are tabulated in [2] and Appendix B, and the terms $A_k^q\langle$ and $r^k\rangle$ can be determined experimentally using absorption spectra (Table 3.3).

Now we turn to the Hamiltonian for $TbPc_2$. Time reversal symmetry tells us that at zero magnetic field m_{J} and $-m_{\mathrm{J}}$ are degenerate. Therefore, only even k values are allowed. Due to the decreasing weight of terms with higher order, we can limit the allowed k values to 2, 4, and 6. Furthermore, due to the local approximate C_4 symmetry of $TbPc_2$ the only remaining q values are $q = 0, 4$. With these considerations the final Hamiltonian of the $TbPc_2$ becomes [9]

$$H_{\mathrm{lf}} = \langle r^2\rangle u_2\, A_2^0 O_2^0 + \langle r^4\rangle u_4\, \left(A_4^0 O_4^0 + A_4^4 O_4^4\right) + \langle r^6\rangle u_6\, \left(A_6^0 O_6^0 + A_6^4 O_6^4\right) \tag{3.15}$$

with

The terms O_k^0, contain the operator J_{z} up to the power of k and are introducing a strong uni-axial anisotropy in z-direction. As a result, the degeneracy between $|J, m_j\rangle$ and $|J, m_j \pm 1\rangle$ is lifted, whereas due to the even powers of J_{z} the $|J, m_j\rangle$ and $|J, -m_j\rangle$ states remains degenerate. An exact numerical diagonalization of $H_{\mathrm{lf}} + g_{\mathrm{J}}\mu_{\mathrm{B}} J_{\mathrm{z}} B_{\mathrm{z}}$ at different magnetic fields results in Fig. 3.7a. The ligand field induces an energy gap of a few hundred Kelvin between the ground state $|6, \pm 6\rangle$ and the first excited state $|6, \pm 5\rangle$. Therefore, already at liquid nitrogen temperatures, the magnetic properties of this complex are almost exclusively determined by the new ground state doublet $m_{\mathrm{J}} = \pm 6$. At room temperature the ground state population is still at 69 %. If we would replace the terbium ion by another rare earth ion like Dy^{3+}, Ho^{3+}, Er^{3+}, Tm^{3+}, or Yb^{3+}, this splitting would decrease as shown in Fig. 3.8 [11]. The terms O_4^4 and O_6^4 occur due to the slight misalignment between the two phthalocyanine ligands, which are not exactly rotated by 45°. Note that for an angle of 45 degrees the system would have a higher symmetry namely D_{4d}, resulting in the suppression of these two terms. Since the misalignment is only a few degrees, the geometrical coordination factor A_6^4 is still too tiny to be measured and can be omitted. The term O_4^4 contains the operators J_+^4 and J_-^4, which are mixing the ground state doublet and lift their degeneracy by $\Delta \simeq 1\ \mu$K (see Fig. 3.7b). This so-called avoided level crossing gives rise to zero field tunneling of the magnetization, which will be explained in Sect. 3.8.1.

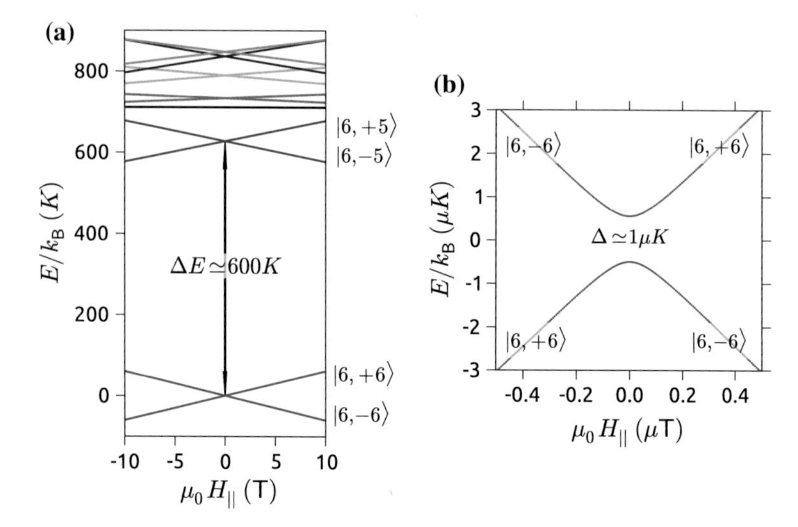

Fig. 3.7 Zeeman diagram of the $TbPc_2$. **a** The ligand field splits the ground state (*red*) and first excited state (*blue*) by around 600 K, leaving only two spin degrees of freedom at low temperature, which makes the molecule an ideal two level quantum system. Higher order excited states are $|6, 0\rangle$ (*black*), $|6, \pm1\rangle$ (*green*), $|6, \pm2\rangle$ (*orange*), $|6, \pm3\rangle$ (*grey*) and $|6, \pm4\rangle$ (*purple*). **b** Additional terms in the ligand field Hamiltonian (A_4^4, A_6^4) lift the degeneracy of the ground state doublet by $\Delta \simeq 1\ \mu$K and introduce an avoided level crossing in the Zeeman diagram.

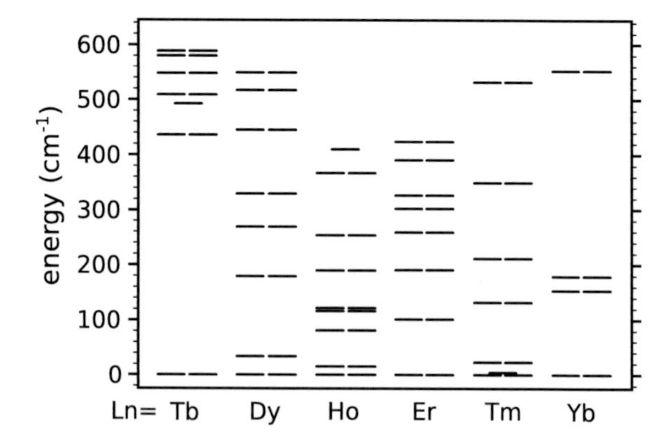

Fig. 3.8 Crystal field splitting of the bis-phthalocyaninato complex with different rare earth atoms as coordination centers (adapted from [11]).

3.7 Hyperfine Interaction

The nucleus of the terbium ion has, besides its electrical charge, also an inherent angular moment $I = 3/2$, resulting in an additional magnetic dipole moment:

$$\boldsymbol{\mu}_\mathrm{I} = g_\mathrm{I}\mu_\mathrm{N}\boldsymbol{I} \qquad (3.16)$$

with $g_{\mathrm{I}} = 1.354$ [12] and μ_{N} the nuclear magneton. Similar to the spin-orbit interaction, this magnetic moment interacts via dipole coupling with the magnetic moment $\boldsymbol{\mu}_{\mathrm{J}}$ created by total angular momentum $\boldsymbol{J}$. The Hamiltonian accounting for this interaction is formulated as:

$$H_{\mathrm{dip}} = A\, \boldsymbol{I}\boldsymbol{J} \tag{3.17}$$

$$\boldsymbol{I}\boldsymbol{J} = I_{\mathrm{z}} J_{\mathrm{z}} + \frac{1}{2}(I_{+}J_{-} + I_{-}J_{+}) \tag{3.18}$$

with A being the hyperfine constant. To obtain Eq. 3.18 we use the same transformation as in Eq. 3.6.
In addition, the nuclear spin possesses an electric quadrupole moment which makes it sensitive to electric field inhomogeneities, such as produced by the electrons in the 4f orbitals. The Hamiltonian which accounts for this interaction can be written as:

$$H_{\mathrm{quad}} = P\, (\boldsymbol{I}\boldsymbol{J})^2 \tag{3.19}$$

$$(\boldsymbol{I}\boldsymbol{J})^2 = (I_{\mathrm{z}} J_{\mathrm{z}} + \frac{1}{2}(I_{+}J_{-} + I_{-}J_{+}))^2 \tag{3.20}$$

with P being the hyperfine quadrupole constant. The hyperfine Hamiltonian is now simply the sum of the magnetic dipole interaction and the electric quadrupole contribution.

$$H_{\mathrm{hf}} = A\, \boldsymbol{I}\boldsymbol{J} + P\, (\boldsymbol{I}\boldsymbol{J})^2 \tag{3.21}$$

For the terbium ion the two parameters A and P are given in Table. 3.4.
By diagonalizing the full Hamiltonian

$$H = H_{\mathrm{lf}} + H_{\mathrm{hf}} + H_{\mathrm{Z}} \tag{3.22}$$

at different magnetic fields and plotting the eight lowest lying eigenvalues, we obtain Fig. 3.9a. Due to the hyperfine interaction each electronic ground state is split in to four. The lines with a positive (negative) slope correspond to the electronic spin $|+6\rangle$ ($|-6\rangle$) and lines with the same color (blue, green, red, black) to the same nuclear spin state ($|+3/2\rangle$, $|+1/2\rangle$, $|-1/2\rangle$, $|-3/2\rangle$). The splittings of the electronic levels are unequal due to the quadrupole contribution of the hyperfine interaction and calculated as 2.(5) GHz, 3.(1) GHz and 3.(7) GHz as depicted in Fig. 3.9b. Moreover, the anticrossing, which was formerly at $B = 0$ T, is now split into four anticrossings,

Table 3.4 Hyperfine constant A and the quadrupole parameter P for the terbium ion according to Ishikawa et al. [10].

A	P
24.9 mK	0.4 mK

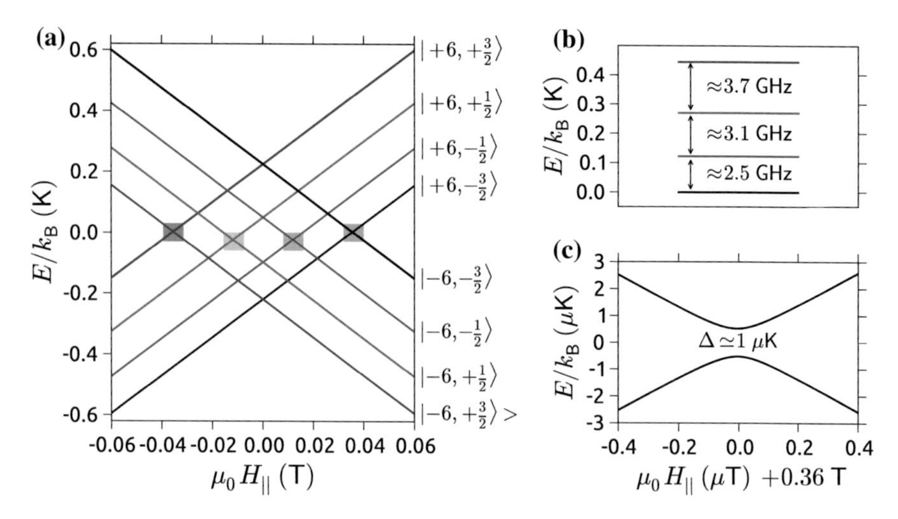

Fig. 3.9 **a** Zeeman diagram of the $TbPc_2$. The *colored rectangles* indicate avoided level crossing between two states of opposite electronic spin and identical nuclear spin. **b** Energy spacing between the different nuclear spin states. **c** Magnification of the avoided level crossing between $|6, \frac{3}{2}\rangle$ and $|-6, \frac{3}{2}\rangle$.

one for each nuclear spin state (colored rectangles in Fig. 3.9a). The energy gap at each anticrossing remains about 1 μK (Fig. 3.9c).

3.8 Magnetization Reversal

Changing the external magnetic field parallel to the easy axis of the $TbPc_2$ allows for the reversal of the molecule's magnetic moment. Hence, when sweeping the magnetic field periodically between positive and negative values we can measure a hysteresis loop as depicted in Fig. 3.10a. Is shows that the magnetization reverses in a step-like shape at small magnetic fields, followed by a continuous reversal at larger magnetic fields. The hysteresis shape can be understood by considering two completely different reversal mechanisms: a direct relaxation, dominating at larger magnetic fields; and the quantum tunneling of magnetization, dominating at small magnetic fields.

3.8.1 Quantum Tunneling of Magnetization

The quantum tunneling of magnetization (QTM) is a tunnel transition between two different spin states $|S, m_s\rangle$ and $|S, m_s'\rangle$. It requires a finite overlap of the two wave-functions, which is caused by off-diagonal terms in the Hamiltonian. Since these

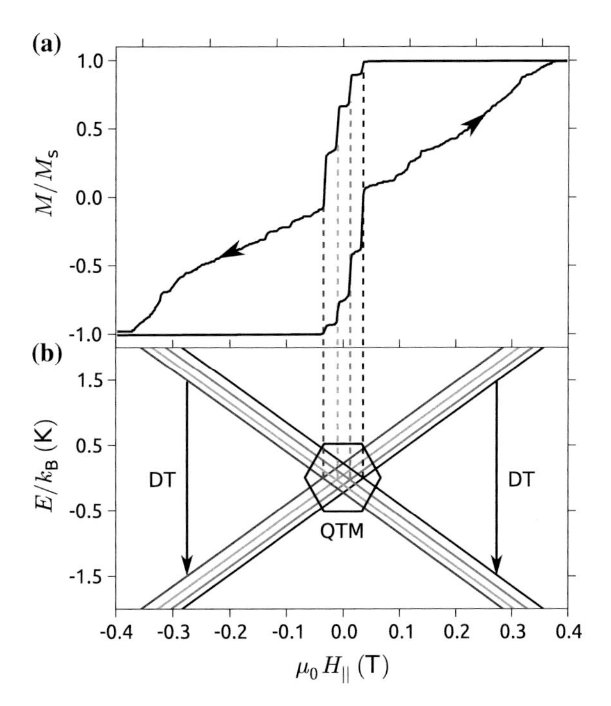

Fig. 3.10 **a** Normalized hysteresis loop of a single $TbPc_2$ single molecule magnet obtained by integration of 1000 field sweeps. Adapted from [13]. **b** Zeeman diagram calculated by diagonalizing Eq. 3.22. The steps in the hysteresis loop of (**a**) coincide with the avoided level crossings and are caused by the quantum tunneling of magnetization (QTM). The remaining magnetization reversal of (**a**) can be explained by direct transitions (DT) from the excited state into the ground state involving the creation of a phonon.

terms are usually small compared to the diagonal terms, the overlap is negligible except for those longitudinal magnetic fields, where the diagonal terms in the Hamiltonian start to vanish. The consequence is the formation of an avoided energy level crossing at those magnetic fields (see Fig. 3.11). When sweeping the longitudinal magnetic field over this anticrossing (see Fig. 3.11) the spin can tunnel from the $|S, m_S\rangle$ into $|S, m'_S\rangle$ state with the probability P given by the Landau-Zener formula [14, 15]:

$$P_{m,m'} = 1 - exp\left(-\frac{\pi \Delta_{m,m'}}{2\hbar g \mu_B |m - m'| \mu_0 dH_{||}/dt}\right) \tag{3.23}$$

Formula 3.23 states that the transition probability increases exponentially with the level splitting Δ and decreases exponentially with the sweep-rate $\mu_0 dH_{||}/dt$ of the longitudinal magnetic field.

As described in Sect. 3.7 the $TbPc_2$ possesses four of these avoided level crossings. This results in four distinct steps at small magnetic field in Fig. 3.10a, which can be used as a fingerprint to identify the single-molecule magnet, as it was shown by Vincent et al. [16].

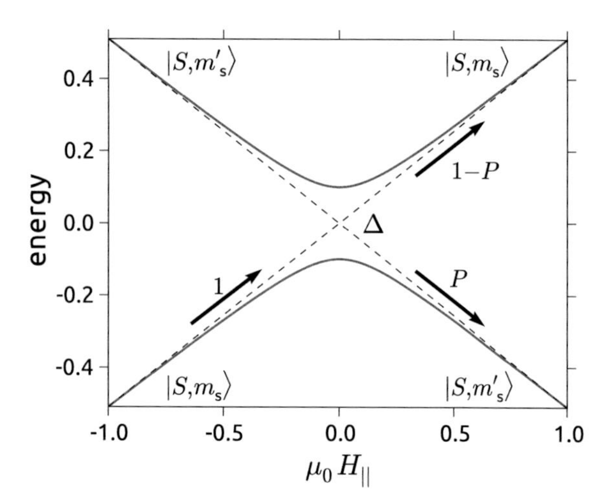

Fig. 3.11 Avoided level crossing between the two states $|S, m_S\rangle$ and $|S, m'_S\rangle$, which leads to the quantum tunneling of magnetization. While sweeping the parallel field over the anticrossing the probability P to tunnel from one state into the other is given by the Landau-Zener formula (Eq. 3.23).

3.8.2 Direct Transtions

In addition to the QTM, the magnetic moment of the molecule can reverse in a direct transition. This is an inelastic process and involves the creation and/or annihilation of phonons to account for the energy and momentum conservation. Therefore, this process is often referred to as phonon assisted or spin-lattice relaxation.

Depending on the temperature we can distinguish between three types of relaxation processes. At low temperature the spin of an SMM is most likely reversed in a direct relaxation process under the emission of one phonon to the thermal bath (see Fig. 3.12a). This process becomes more likely at higher magnetic fields and scales with $(\mu_0 H)^3$. Increasing the temperature allows for a two phonon relaxation process. Therein, the molecule is excited into the state $|e\rangle$ while absorbing a phonon of energy $\hbar\omega_1$ and subsequently relaxes into the ground state via the emission of a phonon of energy $\hbar\omega_2$. Depending on whether the excited state is a real or virtual state, we distinguish between the Orbach process (see Fig. 3.12b) or the Raman process (see

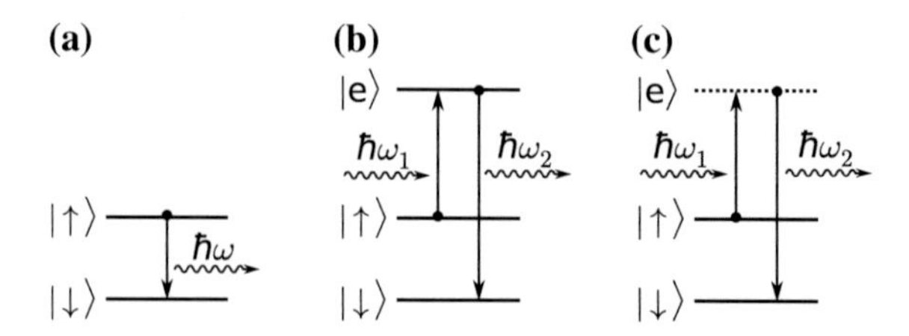

Fig. 3.12 Phonon assisted relaxation. **a** Direct relaxation into the ground state under the emission of a phonon with energy $\hbar\omega$. **b** The two phonon Orbach process involves the absorption of a phonon of energy $\hbar\omega_1$, exciting the molecule into the state $|e\rangle$, and a subsequent emission of another phonon of energy $\hbar\omega_2$, relaxing the molecule into its ground state. **c** The two phonon Raman process is similar to the Orbach process, however, the excited state $|e\rangle$ is a virtual state.

Fig. 3.12a). The Orbach process shows an exponential temperature dependence $\frac{1}{\tau} \propto exp(\Delta / k_B T)$, whereas the Raman process has a polynomial temperature dependence $\frac{1}{\tau} \propto (k_B T)^7$, with τ being the relaxation time, $\Delta = \hbar\omega_2 - \hbar\omega_1$, and T the temperature.

References

1. H. Lueken, *Course of lectures on magnetism of lanthanide ions under varying ligand and magnetic fields* (Institute of Inorganic Chemistry, RWTH Aachen, 2008)
2. A. Abragam, B. Bleaney, *Electron Paramagnetic Resonance of Transition Ions (Oxford Classic Texts in the Physical Sciences)* (Oxford University Press, USA, 2012). ISBN 0199651523
3. G.S. Ofelt, Structure of the f6 Configuration with Application to Rare-Earth Ions. J. Chem. Phys. **38**, 2171 (1963)
4. K.S. Thomas, S. Singh, G.H. Dieke, Energy Levels of Tb3+ in LaCl3 and Other Chlorides. J. Chem. Phys. **38**, 2180 (1963)
5. K. Binnemans, R. Van Deun, C. Görller-Walrand, J. Adam, Spectroscopic properties of trivalent lanthanide ions in fluorophosphate glasses. J. Non-Cryst. Solids **238**, 11–29 (1998)
6. D. Smith, J.H.M. Thornley, The use of operator equivalents. Proc. Phys. Soc. **89**, 779–781 (1966)
7. K.W.H. Stevens, Matrix elements and operator equivalents connected with the magnetic properties of rare earth ions. Proc. Phys. Soc. Sect. A **65**, 209–215 (1952)
8. D. Gatteschi, R. Sessoli, J. Villain, *Molecular Nanomagnets* (Oxford University Press, USA, 2006). ISBN 0198567537
9. C. Gröller-Walrand, K. Binnemans, *Handbook on the Physics and Chemistry of Rare Earths*. Elsevier Amsterdam, 23 edition (1996). (ISBN 9780444825070)
10. N. Ishikawa, M. Sugita, W. Wernsdorfer, Quantum tunneling of magnetization in lanthanide single-molecule magnets: bis(phthalocyaninato)terbium and bis(phthalocyaninato)dysprosium anions. Angewandte Chemie (International ed. in English) **44**, 2931–2935 (2005)
11. N. Ishikawa, M. Sugita, T. Okubo, N. Tanaka, T. Iino, Y. Kaizu, Determination of ligand-field parameters and f-electronic structures of double-decker bis(phthalocyaninato)lanthanide complexes. Inorg. Chem. **42**, 2440–2446 (2003)
12. J.M. Baker, J.R. Chadwick, G. Garton, J.P. Hurrell, E.p.r. and Endor of Tb Formula in Thoria. Proc. R. Soc. A: Math. Phys. Eng. Sci. **286**, 352–365 (1965)
13. R. Vincent, S. Klyatskaya, M. Ruben, W. Wernsdorfer, F. Balestro, Electronic read-out of a single nuclear spin using a molecular spin transistor. Nature **488**, 357–360 (2012)
14. L. Landau, Zur Theorie der Energieubertragung II. Phys. Sov. Union **2**, 46–51 (1932)
15. C. Zener, Non-adiabatic crossing of energy levels. Proc. Roy. Soc. A: Math. Phys. Eng. Sci. **137**, 696–702 (1932)
16. R. Vincent, S. Klyatskaya, M. Ruben, W. Wernsdorfer, F. Balestro, Electronic read-out of a single nuclear spin using a molecular spin transistor. Nature **488**, 357–360 (2012)

Chapter 4
Experimental Details

4.1 Overview Setup

During my thesis, my work was dedicated to the study of single-molecule magnet based transistors in order to perform a coherent quantum manipulation and a non-destructive read-out of a single nuclear spin. Towards this goal, I designed an experimental setup to perform ultra low noise electrical measurements at very low temperature (40 mK), under the influence of fast sweeping 3D magnetic fields and RF electromagnetic fields. An overview of the entire setup is presented in Fig. 4.1. In interaction with the different technical supports of the Néel Institut, I fabricated and tested the different parts of the setup, which were designed to fulfill the diverse experimental constraints of this experiment.
The molecular spin transistor is a three terminal device, consisting of a single-molecule magnet ($TbPc_2$), which is electrically coupled to source, drain, and gate electrodes. In order cool down the device to very low temperatures, it was mounted onto the cold finger of a dilution refrigerator whose base temperature is about 40 mK.
The transistor was microbonded on a specially designed chip carrier consisting of a 50 Ω broadband waveguide and 24 DC strip lines. To avoid 4 K radiation, this chip carrier was encapsulated in a fixed radiation shield anchorage to the mixing chamber. A large sweep-rate, three-dimensional vector magnet, surrounding the chip carrier, was developed to control and read-out the anisotropic electronic moment carried by the single molecular magnet. Electrical connections of the spin transistor to the outside world were established via low temperature pi-filters (1 MHz–1 GHz) and home-made Eccosorb filters (from 1 GHz). Subsequently, at room temperature, the signal can be amplified by two different current-voltage converters. One was dedicated to the electromigration procedure, while the other one was designed for very sensitive low current measurements. They were directly connected to the cryostat to minimize the electro-magnetic and electro-mechanic noise pick up. Additionally, we used room temperature low pass filters and voltage dividers on the bias and gate voltages wires to reduce the noise which was send to the sample. All this room

S. Thiele, *Read-Out and Coherent Manipulation of an Isolated Nuclear Spin*,
Springer Theses, DOI 10.1007/978-3-319-24058-9_4

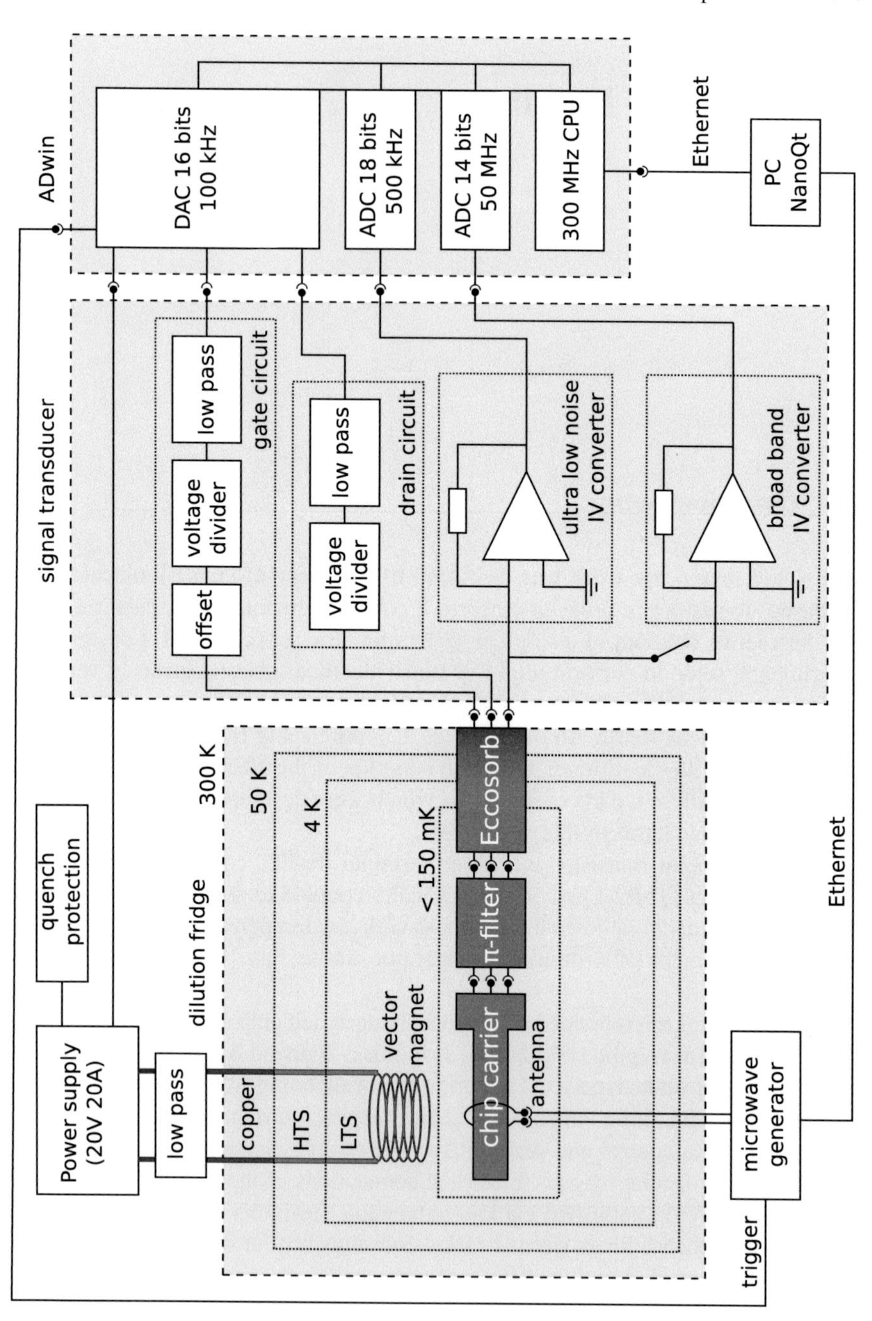

Fig. 4.1 Schematic of the experimental setup

temperature electronics is integrated into a doubly shielded box, which we refer to as the signal transducer.
Finally, the signal transducer is controlled via an independent real time digital-analog converter (ADwin). The latter drives also the 3D magnetic field and triggers the microwave pulse generator, which guarantees a synchronized operation of all devices.

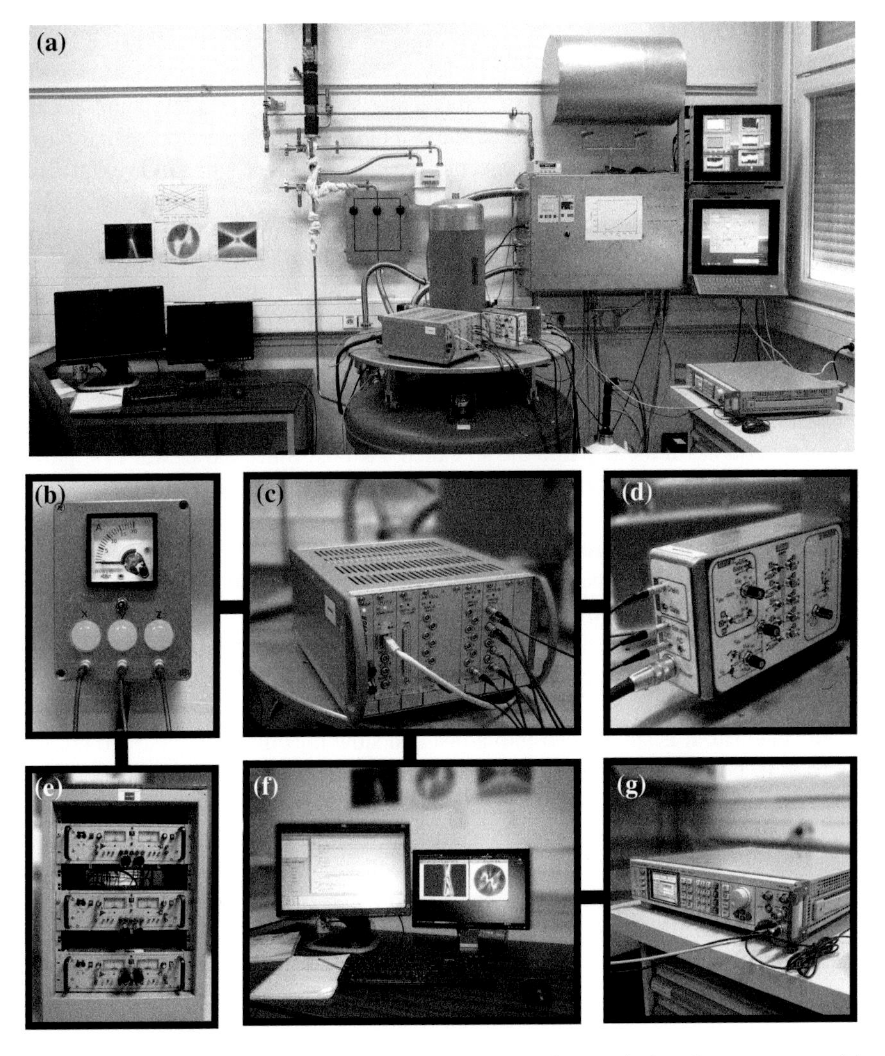

Fig. 4.2 **a** Picture of the experimental setup. The computer (**f**) controls the microwave source (**g**) and the ADwin real time data acquisition unit (**c**). The latter, in turn, operates the signal transducer (**d**) and the power supplies of the vector magnet, which are situated in the basement (**e**) via a remote terminal (**b**). The three LEDs in (**b**) indicate if the coils are operating. In case of a quench, the corresponding power supply is shut down automatically, and the LEDs will turn off. Further details of the individual parts of the experiment are explained in the text

The ADwin is connected over Ethernet with a standard PC, which interfaces the unit using a home-made software called NanoQt.
A picture of the entire setup is shown in Fig. 4.2, whereas a more detailed description of all experimental parts is given in the following sections.

4.2 Dilution Refrigerator

To explore of the quantum world of a molecular spin transistor a low temperature environment is required, which makes the use of dilution refrigerators (DR) indispensable. Among many different concepts, we chose to work with an inverted DR, which combines a fast cool down (about 3 h) with a spacious low temperature stage. The basic working principle of this DR will be explained in the following paragraphs. The schematic in Fig. 4.3 shows that cryostat consists of six different thermal stages, each encapsulated by another with higher temperature. Vacuum isolates one level from another so that each stage functions as a radiation shields for the next inner lying. To cool down the cryostat, two independent cooling circuits operate simultaneously. The secondary open cycle cooling circuit replaces the liquid ^{4}He bath of conventional cryostats (green circuit in Fig. 4.3). It operate with liquid ^{4}He, which is injected from a Dewar underneath the DR into the so called 4 K box. Since the Dewar is slightly over pressured, a sufficiently large ^{4}He circulation is established to guarantee a steady state operation. An additional pump inside the circuit is only needed during the cool down from room temperature, since high cooling power and hence high flow rates are necessary. The liquid helium inside the 4 K box is used to cool the 4 K stage, whereas the vapor created by the boiling liquid ^{4}He is ejected into a spiral counter-flow heat-exchanger. While leaving the cryostat, it gradually cools down the primary cooling circuit as well as the 20 K and the 100 K stages.
The primary cooling circuit is a closed cycle cooling circuit, containing a mixture of ^{3}He and ^{4}He. It is subdivided into a fast and slow injection (blue and red circuit in Fig. 4.3), both entering the DR via the counter-flow heat exchanger. Due to the cooling power extracted from the secondary circuit, the gas is gradually cooled down to 4.2 K. Afterward ,the fast injection is directly thermalized onto the 1 K stage and leaves the cryostat via the mixing chamber, the discrete exchangers, and the still. It has a larger cross section than the slow injection and is used to precool the colder parts of the cryostat to 4.2 K during the cool down from room temperature.
The slow injection on the other hand is responsible for the condensation of the mixture followed by the steady state operation. In order to condense the mixture, an external compressor pressurizes the gas to 4 bar before injecting it into the spiral heat exchanger of the cryostat. Leaving the latter at a temperature of 4.2 K, it passes a second heat exchanger, which is terminated by a flow impedance. The resulting pressure gradient leads to a Joule-Thomson expansion and lowers the temperature of the gas by $\approx$2 K before entering the still. After having passed the latter, the mixture traverses a set of continuous and discrete heat exchangers before being injected into the mixing chamber (Fig. 4.4).

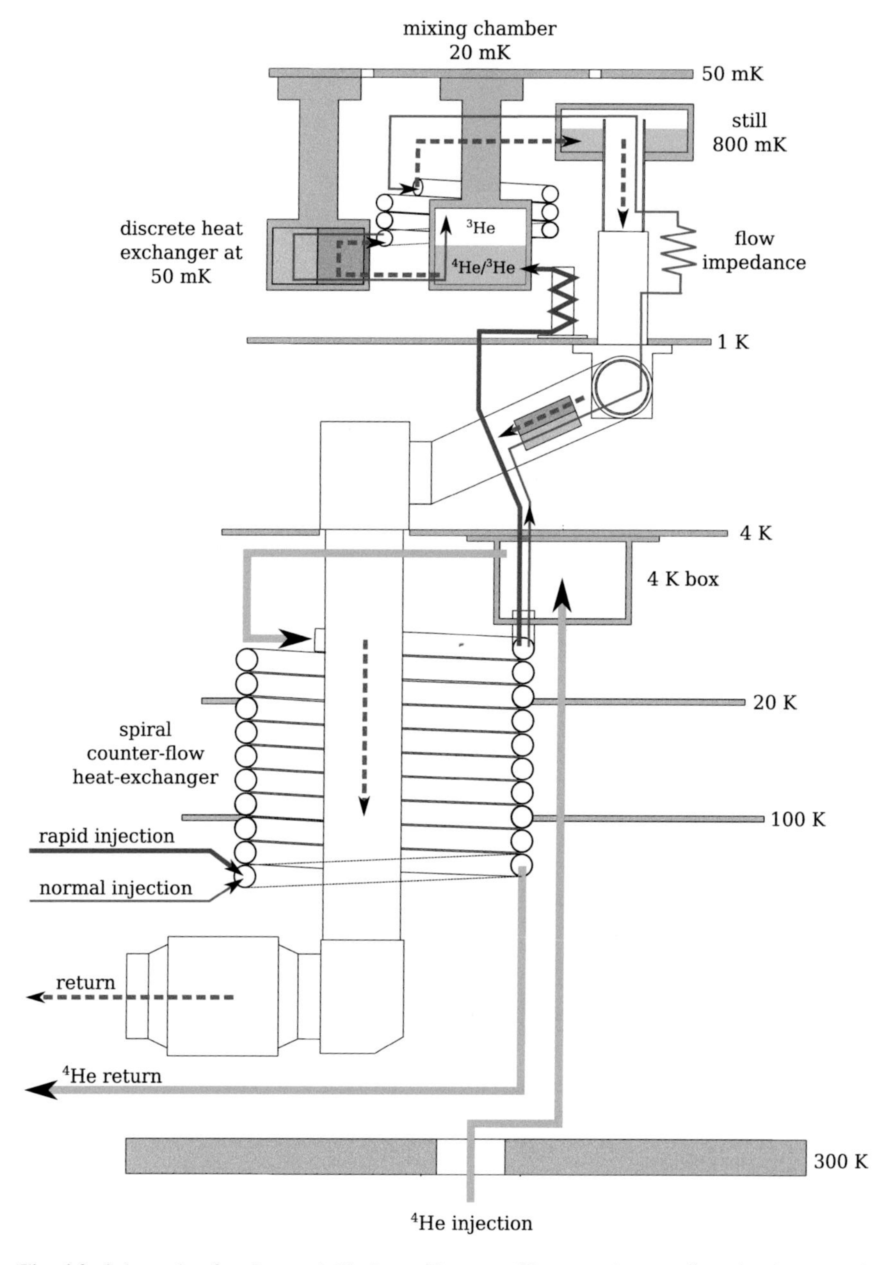

Fig. 4.3 Schematic of an inverted dilution refrigerator. The secondary cooling circuit (*green*) is precooling the primary circuit consisting of the normal injection (*red*) and rapid injection (*blue*). The latter is only used during the cool down from room temperature. During the steady state operation, ^{3}He is injected via the normal injection into the ^{3}He rich phase of mixing chamber and extracted from the diluted phase (*violet*)

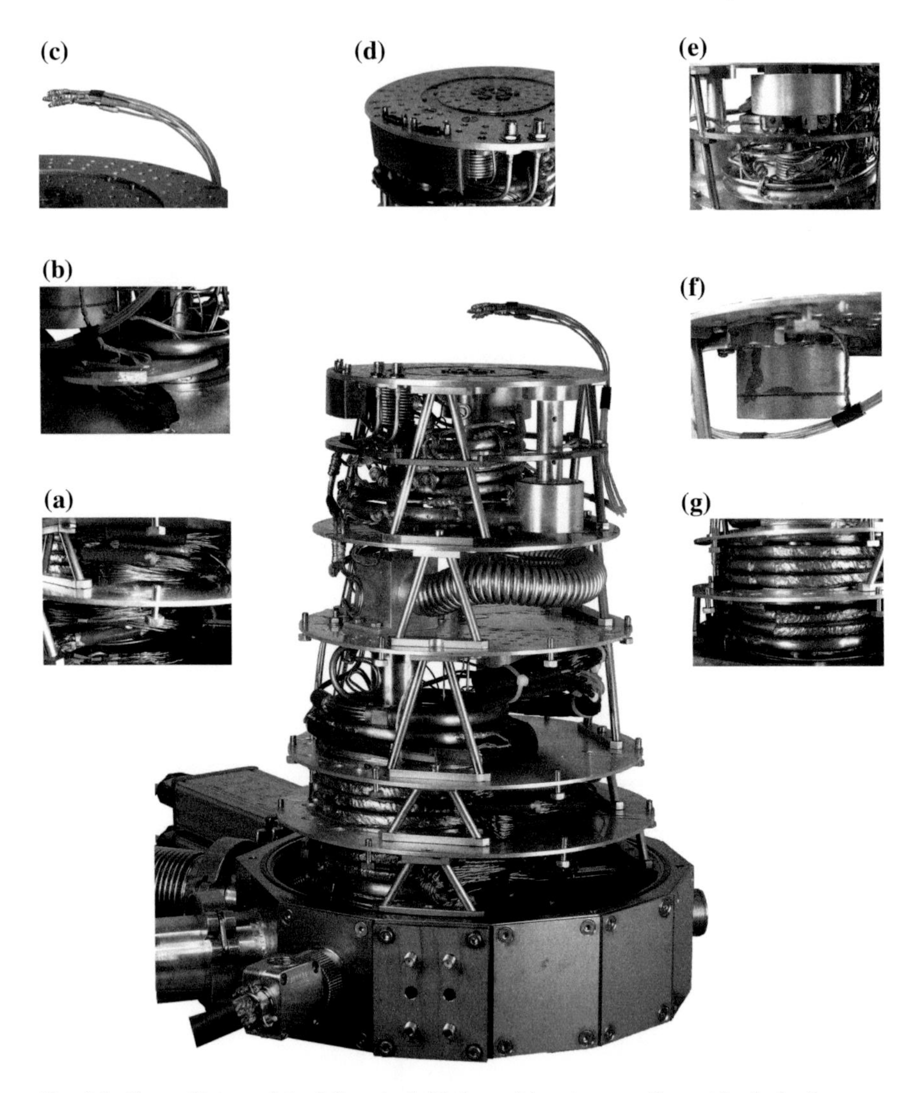

Fig. 4.4 *Center* Picture of the fully wired dilution refrigerator. **a–c** Current leads for the superconducting vector magnet consisting of copper (**a**), high temperature superconducting (**b**), and low temperature superconducting cables (**c**). **d** Cold stage showing the DC and microwave connectors. The sample holder (not shown) is situated in the *center* of the cold stage. **e–g** Important parts of the primary and secondary cooling circuit showing the still (**e**), the 4 K box (**f**), and the spiral counter flow heat exchanger (**g**)

External pumps are decreasing the pressure inside the mixing chamber below 0.1 mbar, allowing for another adiabatic expansion, which results in the condensation of the mixture. The cold gas evaporating from the liquid is being pumped out through the numerous heat-exchangers cooling down the incoming mixture. Hence,

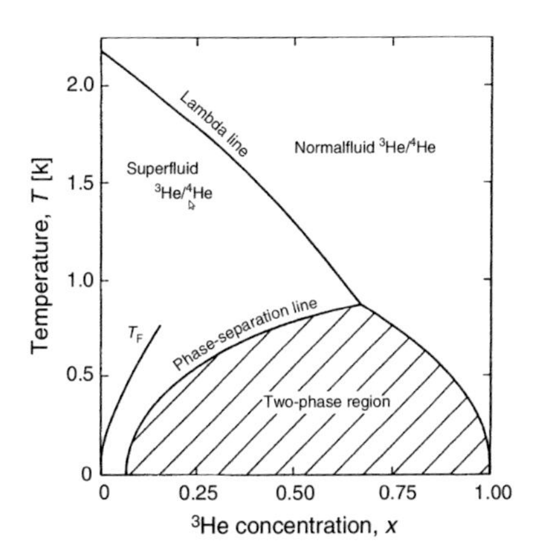

Fig. 4.5 Phase diagram of liquid ^{3}He and ^{4}He mixtures at saturated vapour pressure taken from [1]. Below the critical temperature of 867 mK, the mixture separates into two phases, a ^{3}He rich and a ^{3}He diluted phase

more and more gas condenses, gradually filling up every part from the mixing chamber to the still with liquid. At a temperature of around 800 mK, a phase separation into a lighter ^{3}He rich phase and heavier ^{3}He dilute phase is taking place inside the mixing chamber. The exact ration of ^{3}He/^{4}He in each phase depends on temperature and is shown in Fig. 4.5.

The diluted phase expands from the bottom of the mixing chamber to the still. It contains mainly super-fluid ^{4}He, which can be viewed as inert and noninteracting with the ^{3}He. Nevertheless, the vapor inside the still contains, despite the high concentration of ^{4}He, 97 % of ^{3}He due to its low boiling point. By pumping on the still and re-injecting the gas in the ^{3}He rich phase, a ^{3}He circulation is established. In order to maintain the equilibrium concentration, ^{3}He from the rich phase is pushed into the diluted phase. This is an endothermic process, providing the cooling power to cool down to mK temperatures. This process can also be viewed as an evaporation of liquid ^{3}He from the rich into the diluted phase since the ^{4}He, which requires heat and continues even to the lowest temperatures since the concentration of ^{3}He in the diluted phase remains finite. The base temperature of the cryostat is only determined by residual heat leaks and remains usually above 10 mK for most of the DRs. A picture of the fully mounted dilution refrigerator is shown in Fig. 4.4.

4.3 3D Vector Magnet

The observation and manipulation of a single-molecule magnet (SMM), which is the centerpiece of a molecular spin transistor, demands external magnetic fields in arbitrary directions. A way to create such three dimensional fields comprises three coils mounted perpendicular to each other like the axes of a coordinate system. The orientation and magnitude of the magnetic field is controlled by adjusting the current through each coil so that the resulting field is simply the vector sum of three respective fields.

Conventional state of the art 3D vector magnets consist of a cylindrical coil surrounded by two Helmholtz coils. They are capable of creating a magnetic field of 1 T with the Helmholtz coils and around 2 T with the cylindrical coils. However, their size is typically in the order of $200 \times 200 \times 200$ mm. Despite the fact that they are not fitting inside our cryostat, they have a very high inductance making it impossible to reach high sweep rates. Furthermore, their huge heat capacity would be severely retarding every cool down. Therefore, we aimed to build very small 3D vector magnets with approximately the same magnetic field specifications. The fabrication process was supported by Yves Deschanels from the Institute Néel.
The cryogenic environment of the DR allows for the use of superconducting wires, which are creating much higher fields than conventional copper wires. Among the several available types, we chose a multifilament NbTi superconducting wire embedded in a CuNi matrix. The multifilament layout diminishes flux jumps and reduces the total amount of vortices, leading to higher stability and smaller remanence. The NbTi superconducting core is known to be less fragile than the Nb_3Sn core, which was important during the fabrication process. The CuNi matrix was chosen because of smaller Eddy-currents compared to a pure Cu matrix, hence, allowing for higher sweep rates.
Since the DR operates in vacuum, the maximum current per coil was fixed to 20 A for field pulses and 10 A for steady state operation as safety precautions. Looking up the different specifications of available SC wires, we found the low current SC wire from SUPERCON Inc. with an outer diameter of 152 μm and 18 NbTi filaments as most suited for our purpose.
A first design study was then carried out to optimize the central-cylindrical coil, also referred to as the z-coil, using the COMSOL Multiphysics software. The field at the sample, situated inside the z-coil, should be around 1 T at 10 A in order to be comparable with commercial state of the art electromagnets. The inner diameter of the coil was set to 6 mm, thus, being still large enough to insert the chip carrier with the sample later on. Given the current, the wire diameter, and an ideality factor of coil of 98 %, the parameters remaining for optimization were the coil length L and the width W of the accumulated layers. The calculated magnetic field in the parameter space of L and W is displayed in Fig. 4.6. In order to maximize the field of the other two coils, W must be as small as possible. As shown in Fig. 4.6 the optimal dimensions were found to be $W = 2$ mm and $L = 25$ mm.
With the above mentioned dimensions of the z-coil ($W = 2$ mm and $L = 25$ mm), we calculated the spacial magnetic field distribution. The result of this simulation (Fig. 4.7) shows an almost uniform field distribution within a radius of 3 mm around the center, which is about the size of our sample.
Having set the dimensions of the z-coil, we started the design-study of the x and y split-pair magnets. Their separation of 10 mm is given by the outer diameter of the z-coil. In order to reach fields of around 1 T at 10 A, with a coil separation of 10 mm, we developed a new design concept. In a first approach, we replaced the standard cylindrical Helmholtz coils by conically shaped coils, thus, increasing the volume share at equal dimensions. Fixing the smaller diameter of the cone to 10 mm, we end up with two variable parameters, namely, the inner diameter D and the coil thickness

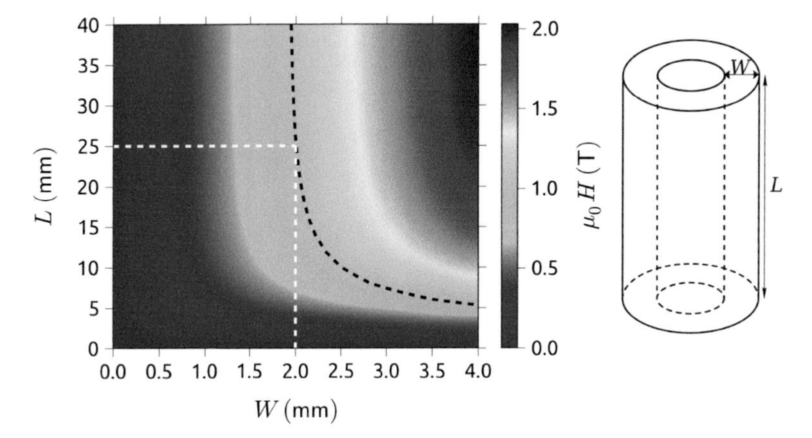

Fig. 4.6 Magnitude of the magnetic field in the center of the z-coil as a function of the coil length L and the accumulated width W of the wire layers. The *black dashed line* is the isofield line of 1 T. In order to reach 1 T at 10 A the minimum width needs to be 2 mm resulting in a length of 25 mm

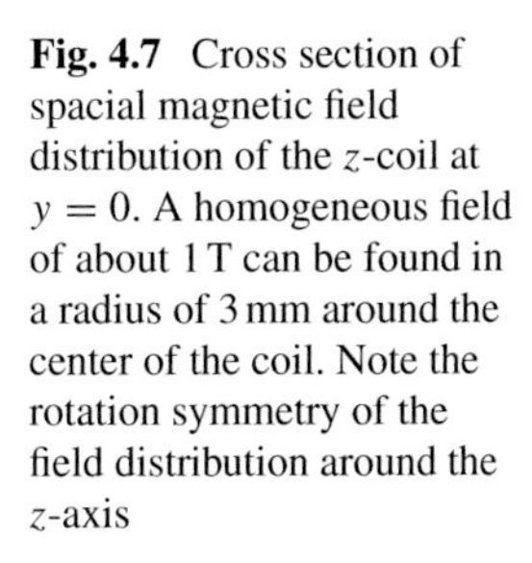
Fig. 4.7 Cross section of spacial magnetic field distribution of the z-coil at $y = 0$. A homogeneous field of about 1 T can be found in a radius of 3 mm around the center of the coil. Note the rotation symmetry of the field distribution around the z-axis

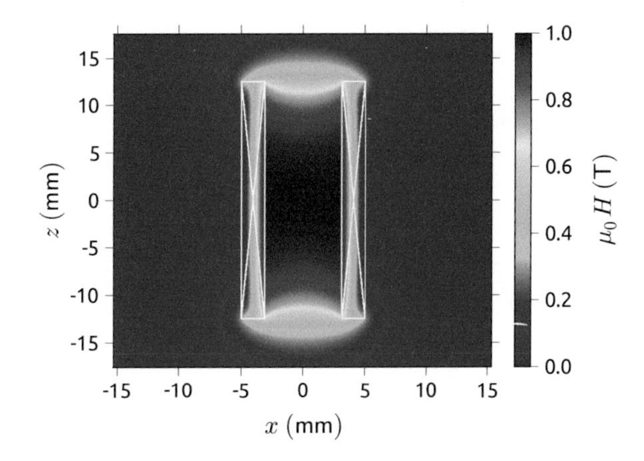

L. In order to find the optimal parameters, a second design study was carried out. The calculated magnetic field in the parameter space of D and L is shown in Fig. 4.8. For engineering reasons the conical shape needed to be approximated by a step like shape. In the first iteration, we introduced only one step in each coil. The position of this step was subsequently optimized, by keeping the above mentioned thickness and inner diameter, in order to obtain 1 T at 10 A in the center of the vector magnet. The final result of the shape and magnetic field magnitude for the x and y coil is shown in Fig. 4.9a, b respectively. Notice that the highest field of the split coils is in the center of the inner wall and is much higher than the field at the center of the vector magnet.

A picture of the fully mounted vector magnet is shown in Fig. 4.10. The first tests were carried out in liquid helium. We measured the maximum magnetic field as well

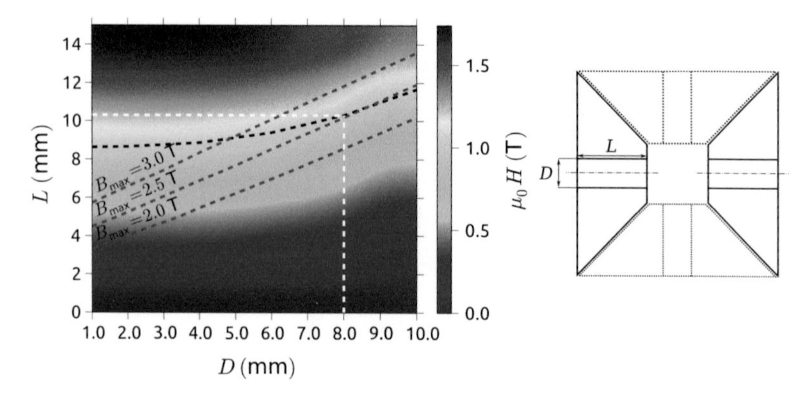

Fig. 4.8 Magnitude of the magnetic field in the center of the two conically shaped split coils as a function of the inner diameter D and the thickness L. The *black dashed line* is the isofield line of 1 T. The *red line* correspond to the maximum field at the inner wall of the coils. In order to reduce that maximum field, we set the dimensions to $D = 8$ mm and $L = 10$ mm

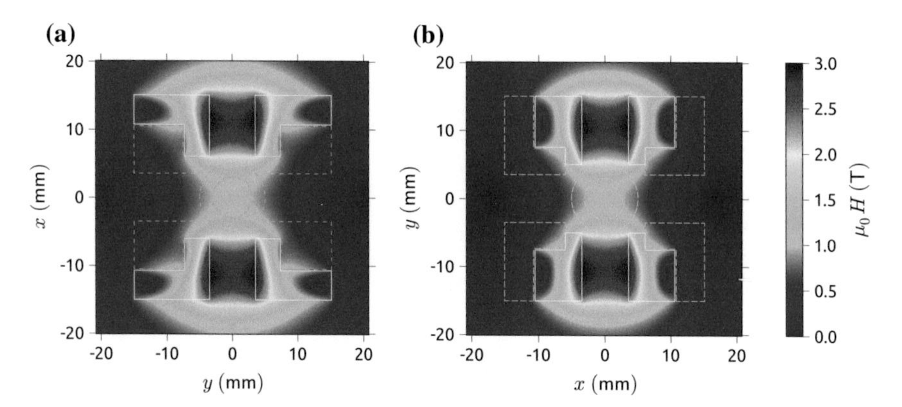

Fig. 4.9 Cross section of the magnetic field distribution of the x-coil (**a**) and y-coil (**b**) at $z = 0$. The contour of the respective coil is shown in *white*, whereas the contours of the other two coils are drawn as a *grey dotted line*. The generated field in the center of the vector magnet is about 1 T at 10 A for each coil and the maximal field is around 2.7 T. Note that the field distribution of the x- and y-coil has an axial symmetry around the x- and y-axis respectively

as the maximum sweep rate, the coils could resist before quenching. The results are shown in Table 4.1.

Each coil was able to produce a field of 1 T when operated alone, however, the x and y coil generated this field only at 11 and 12.5 A respectively. When operating all three coils simultaneously, the maximum field was limited to 0.9 T due the mutual interaction.

In the vacuum environment of the dilution refrigerator, the maximum field is reduced by $\approx$20 % and the maximum sweep speed by a factor of 5. This is caused by the slightly higher temperature of 4.4 K instead of 4.2 K and less efficient thermalization

Fig. 4.10 Picture of the fully mounted vector magnet

Table 4.1 Benchmarks of the three different coils, which were immersed in liquid helium

	x-coil	y-coil	z-coil
I_{max}	15 A	14 A	18 A
B @ 10 A	0.9 T	0.8 T	1.1 T
$(dB/dt)_{\mathrm{max}}$	$\approx$1 T/s	$\approx$1 T/s	>10 T/s

of the coils. In liquid helium, any generated heat is directly mediated to the liquid helium bath, whereas in the vacuum environment of the DR, the heat has to diffuse to the copper thermalization of the coils, which creates a bottleneck in the thermal transport.

4.4 Current Leads

The current leads are the electrical link between the superconducting vector magnet and the room temperature connections outside the cryostat. To guarantee a stable operation, an equilibrium between the wire material, the diameter, and the length had to be found. Ideally, the material should be a very good electrical conductor but a very bad thermal conductor.

For the low temperature part of the cryostat, i.e. at temperatures below 77 K, high temperature superconductors (HTS) were used as current leads. Since superconductors are both perfect electrical conductors and very poor thermal conductors, they represent the material of choice. The HTS we used in the cryostat consisted of silver coated YBaCuO straps with a T_c of 90 K. The cross section of the straps was chosen to sustain 40 A at 77 K, thus, leaving a safety coefficient of two. The silver coating was needed to achieve a homogeneous temperature of the HTS along the strap. The drawback of these kind of superconductors is their fragility. Therefore, we terminated the HTS straps with low temperature superconductors of the NbTi type. This simplifies the soldering and unsoldering of the current leads from the vector magnet, which was necessary every time the sample is changed.
The high temperature part of the current leads was made of copper wires. Since the resistivity and thermal conductivity of copper varies with temperature, a design study was carried out to determine the optimal geometry. Yet, a too large diameter results in a thermal shortcut between stages of different temperatures, whereas a too small diameter could destroy the leads due to Joule heating. The same considerations can be made for the wire length L, since the heat conduction is proportional to $1/L$. Therefore, a very short wire will transmit a lot of heat into the dilution fridge, whereas a very long cable might not be able to remove the energy produced by Joule heating and the wire possibly melts.
To work out this optimization problem, the one dimensional heat equation was solved with the experimental boundary conditions. It is a inhomogeneous partial differential equation and given as [2]:

$$\frac{dT}{dt} - \frac{1}{c\rho}\frac{d}{dx}\left(\kappa(T)\frac{dT}{dx}\right) = \frac{\dot{q}}{c\rho} \tag{4.4.1}$$

$$\frac{dT}{dt} - \frac{1}{c\rho}\frac{d\kappa(T)}{dx}\frac{dT}{dx} - \frac{\kappa T}{c\rho}\frac{d^2T}{d^2x} = \frac{\dot{q}}{c\rho} \tag{4.4.2}$$

where κ is the thermal conductivity, c the specific heat capacity, ρ the density, T the temperature, and $\dot{q}$ the heating power per unit volume. The effect of the black body radiation was neglected since it is much smaller than the other parts of the equation within the temperature range of the experiment. The term on the right hand side is the source term and corresponds to the energy injected into the system. This energy is partly used to heat the wire (that is where the $\frac{dT}{dt}$ comes from) and is partly transported away, which gives rise to the $\frac{\kappa}{c\rho}\frac{d^2T}{dx^2}$ term. For a metal wire the power $\dot{q}$ due to Joule heating is given by:

$$\dot{q} = \frac{U \cdot I}{A \cdot L} = \frac{I^2 R}{A \cdot L} = \frac{I^2}{A^2}\frac{1}{\sigma(T)} \tag{4.4.3}$$

where I is the applied current, A is the cross section area, L the length of the wire, and σ the electrical conductivity of the metal.

Table 4.2 Results of the numerical optimization of the heat equation for the optimal wire length L and the optimal wire diameter D in the different temperature regions

	L (cm)	A (mm^2)
300–200 K	30	1.5
200–100 K	25	0.75
<100 K	25	0.5

The electrical conductivity $\sigma(T)$ of copper in the temperature range of 50 to 300 K can be modeled as [3]:

$$\frac{1}{\sigma(T)} = -3.204 \cdot 10^{-9} + 6.855 \cdot 10^{-11}\, T \quad [\Omega\mathrm{m}] \tag{4.4.4}$$

Another important parameter is the thermal conductivity $\kappa(T)$ of copper. In the range from 100 to 300 K it can be fitted by [4]:

$$\kappa(T) = 886 - 7.462 \cdot T + 0.045 \cdot T^2 - 1.2331 \cdot 10^{-4} \cdot T^3 + 1.267 \cdot 10^{-7} \cdot T^4 \tag{4.4.5}$$

and from 50 to 100 K by [4]:

$$\kappa(T) = 7051 - 277.4T + 4.69T^2 - 0.0368T^3 + 1.106 \cdot 10^{-4} T^4 \tag{4.4.6}$$

Since the electrical and thermal conductivity of copper increases for decreasing temperature, the diameter of the leads needs to be decreased to minimize heat leaks. For practical reasons the diameter reduction is done at two temperatures: 200 and 100 K. Using Eq. 4.4.2 in combination with Eqs. 4.4.3–4.4.6, the optimal parameters for the wire length L and the diameter D at the different temperature ranges were calculated. The optimization parameters were such that the created heat leak should be less than 1.7 W, which corresponds to ≈10% of the cooling power of the primary circuit at a ^{4}He flow rate of 3.6 l/min; and that the temperature increase during a steady state operation at 10 A remains smaller than 10 %. The results for the three different temperature regions are tabulated in Table 4.2.
The values given in Table 4.2 do not include the size of the thermalizations, which were chosen to be 20 cm at 200 K, 12 cm at 100K, and 16 cm at 50 K. They were realized by gluing copper litz wires onto the current leads with a mixture of araldite and silver powder.
In order to operate the vector magnet, six of these current leads (two for each coil) were fabricated. After having them installed together with the HTS, we tested the ensemble at 10 A per lead. During the test, the temperature of the outermost stage in the cryostat increased by about 50 K, whereas the temperature-increase of the 20 K stage was already below 1 K, so that the stable operation of the dilution fridge was guaranteed.

4.5 Sample Holder

The sample holder is the link between the sample and the cryostat. It consists of two parts, an exchangeable chip carrier and a fixed radiation shield, which is in direct contact with the mixing chamber of the cryostat. It is needed to block the 4 K radiation of the vector magnet and keeps the sample at mK temperatures. The sample holder was designed to have an independent vacuum, which protects the sample when heating up the cryostat to room temperature. A picture of the radiation shield is shown in Fig. 4.11.
The chip carrier was designed to have 24 DC strip lines and one 50 Ω matched broadband waveguide. It is connected via a 36 Pin PCI Express connector to the radiation shield, which, when it is closed, encapsulates the chip carrier. The chip carrier itself is made out of six copper/insulator layers, which are shown in Fig. 4.12.

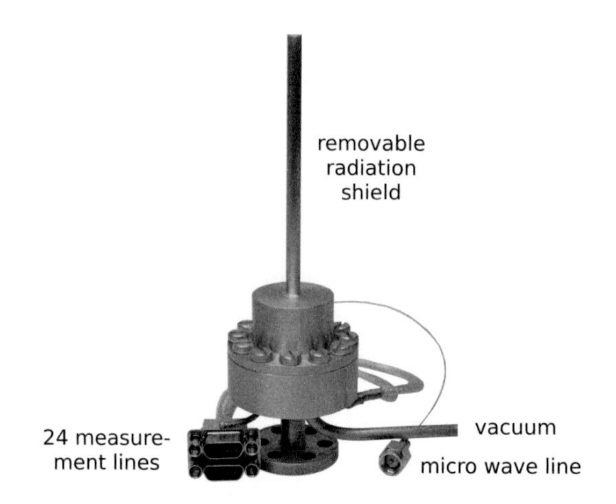

Fig. 4.11 Radiation shield with a independent vacuum, a feed through of 24 measurement lines and one mircowave line

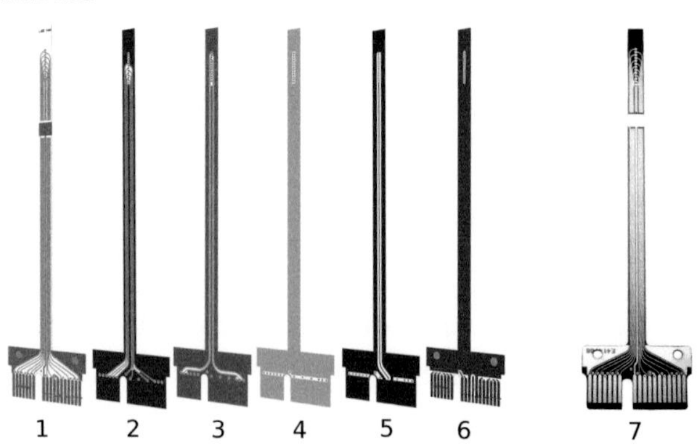

Fig. 4.12 (*1–6*) Layout of the chip carrier consisting of six independent layers. The *top three layers* contain the 24 DC strip lines and *three bottom layers* 50 Ω matched waveguide. (*7*) Picture of the sample holder

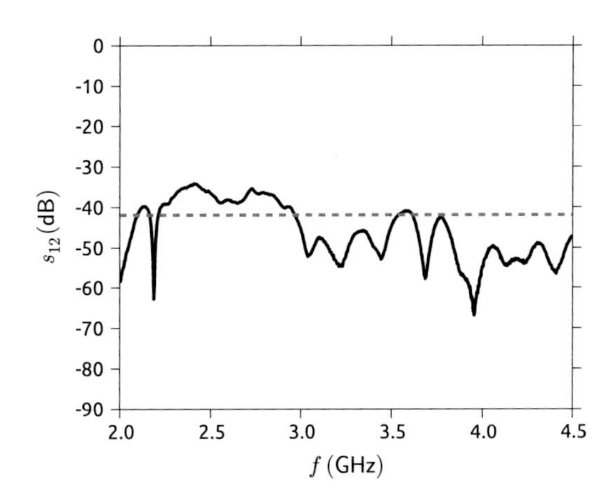

Fig. 4.13 Microwave transmission s_{12} measured from the SMA connector to the end of the waveguide using an Agilent E8362C vector network analyzer

The three top most layers contain the 24 DC strip lines, layer four, five, and six are used for the 50 Ω matched waveguide. The DC lines are soldered to two 12 pin Cannon connectors and the waveguide to a SMA terminated microcoax. Despite the 50 Ω matching the transmission s_{12} measured from the SMA connector to the end of the waveguide on the top layer is around -40 dBm (see Fig. 4.13). A large part of the attenuation is probably coming from reflections at the PCI Express interface. Insertion losses of the microcoax are about 10.5 dBm/meter at 1 GHz and have only a minor influence.

4.6 Filter

Most experiments exploring the quantum nature of matter are sensitive to external noise sources, which, if not properly attenuated, decrease the coherence time of a quantum state drastically. In general, there are three main noise source interfering with the experiment.

The first one is the noise generated by electro-magnetic radiation. It is produced by any wireless communication system and is in the order of a few Hz to a few GHz, e.g. Wifi, mobile phones, television, GPS, etc., or by improperly shielded power sources like any switching power supply or transformer.

The second noise source is the Johnson-Nyquist thermal noise, which is the electrical equivalent of Planck's blackbody radiation. The noise power in Watts is given by $P = k_B T \Delta f$, where k_B is the Boltzmann constant, T the temperature and Δf the frequency bandwidth. The magnitude of the noise is shown in Fig. 4.14.

The third noise source is vibrational noise, produced mainly by rotating parts, e.g. pumps. It is in the order of a few Hz to a few hundred Hz and can be minimized by vibrational low pass filters like a heavy stone ore metal plate and by reducing the amount of connectors from the sample to the amplifier.

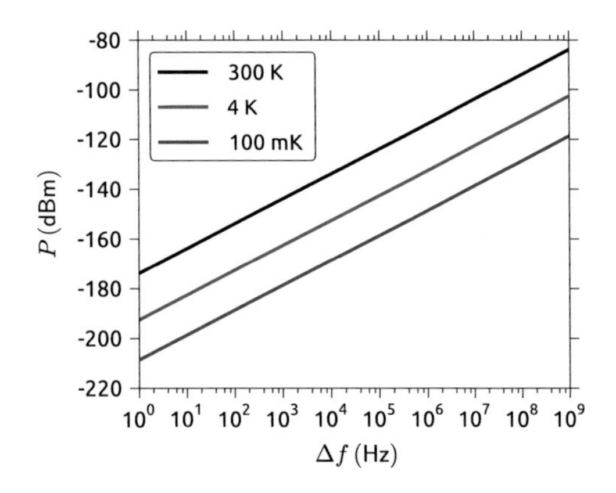

Fig. 4.14 Power of the Johnson-Nyquist noise as a function of the bandwidth at 300 K (*black*), 4 K (*red*) and 100 mK (*blue*). Notice that 0 dBm corresponds to 1 mW

4.6.1 Low Frequency Filters

To protect the experimental setup from electromagnetic radiation, every incoming and outgoing wire was shielded. Additionally, we tested low-pass filters, which can be mounted at the 4 K stage to further attenuate the remaining electromagnetic noise. They should have a negligible series resistance in order to be compatible with the electromigration (see Sect. 4.9.2). For this reason, we were looking for suitable pi-filters, consisting of two capacitors and one inductor. Their cut-off frequency f_0 should be around 1 MHz at cryogenic temperatures in order to have enough bandwidth for the electromigration technique. Since they will be mounted inside the cryostat, their size should of course be as small as possible. For testing purposes, we ordered several pi-filters with about equal size and cut-off frequency. The room temperature transfer function s_{12} is shown in Fig. 4.15. Their cut-off frequency is about 500 kHz with an initial attenuation of 20 dB/decade, which is increasing to 40 dB/decade at around 5 MHz. The kink in s_{12} is due to asymmetrical capacitors. In order to test their cryogenic compatibility, we performed ten temperature cycles from 300 to 77 K by repeatedly immersing them in liquid nitrogen. The final transfer function s_{12} after the tenth cycle at 77 K is shown in Fig. 4.16.
All devices show a shift of f_0 to higher frequencies at 77 K. This is due to a decreasing susceptibility ϵ_r of the dielectric with temperature. Only pi-filters from EMI Inc. with the X7R dielectric showed an acceptable temperature stability and were therefore selected for our setup.

4.6.2 High Frequency Filters

The attenuation of noise frequencies above 1 GHz requires a different type of filter since discrete filters like an LC-circuit become transparent due to parasitic effects [5] (see Figs. 4.15 and 4.16). Over the last decades, a diversity of solutions has been proposed. The most common high frequency filters are fine-grain metal powder

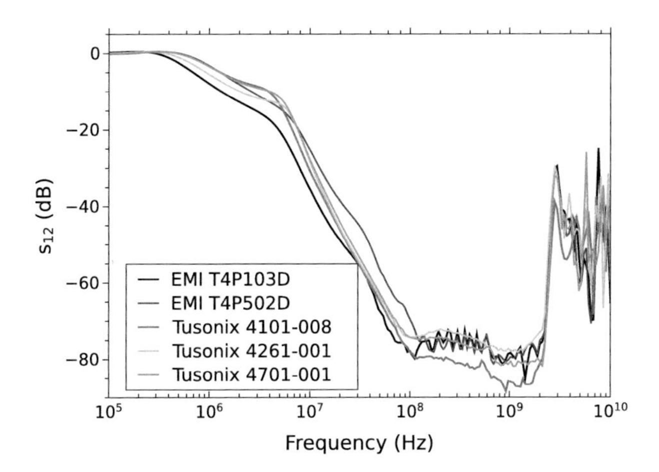

Fig. 4.15 Transfer function s_{12} at 300 K for several pi-filter measured with an Agilent E8362C vector network analyzer

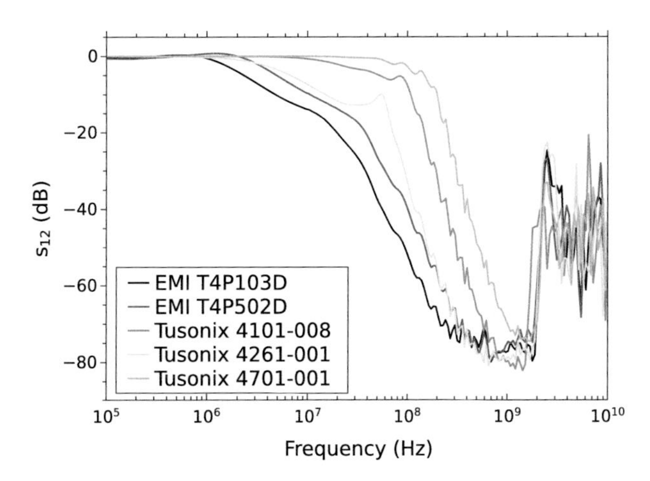

Fig. 4.16 Transfer function s_{12} at 77 K for several pi-filter measured with an Agilent E8362C vector network analyzer

filters [6–9], whose attenuation is based on skin-effect damping. Those filters are often bulky but have a very high performance. Moreover, thin coaxial cables, like mircocoax [10] or Thermocoax [11] have been tested. They are less space consuming but their attenuation is also smaller. A different approach involves lithographically fabricated meander lines, which work as distributed LRC filters [12–14]. Very recently, wires surrounded by Eccosorb, which is a microwave absorbing material, were testes under cryogenic conditions [6]. A nice summary of different filter types is given in [15].

In order to be space-efficient, we could use either the Thermocoax or Eccosorb filters. To compare the two filter techniques, we fabricated different measurement

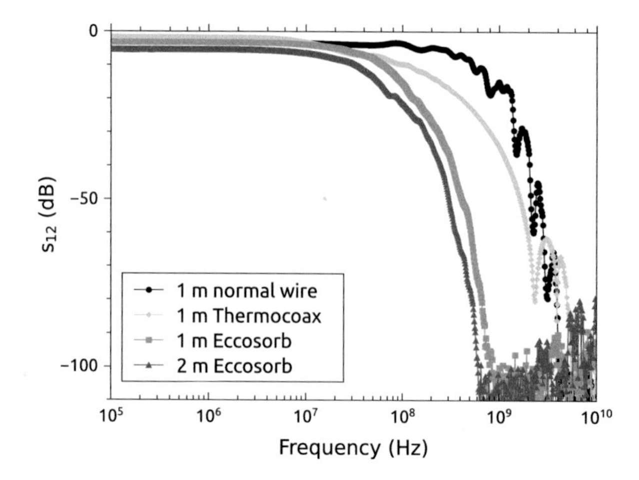

Fig. 4.17 Comparison of the attenuation of different filters. The Eccosorb filters were made out of Eccosorb-coated wires enclosed in a CuNi tube with 1.5 mm in diameter

lines, which were terminated by SMA connectors on both sides. Their attenuation was determined with the Agilent E8362C vector network analyzer. The results can be seen in Fig. 4.17. While at lower frequencies ($\approx$10 MHz) the attenuation is almost similar, the Eccosorb coated lines reach -70 dB attenuation already at around 600 MHz, the Thermocoax, however, only at around 2 GHz. As a comparison we measured also a 1 m long line without Eccosorb, which showed as expected the worst performance (see Fig. 4.17). Based on these results, we chose Eccosorb coated wires as high frequency attenuators. The final filter was made out of 24 superconducting wires made out of NbTi filaments embedded in a constantan matrix. They were coated with Eccosorb, and enclosed in a CuNi tube of 1.5 mm external diameter. The first meter of the tube is gradually thermalized from 300 K down to 40 mK, while the rest is thermalized to the 40 mK stage to attenuate all thermal noise sources. To be more space efficient, the very low temperature part of the filter was rolled up in a counterwind cylindrical coil. We chose the superconducting wires in order to keep the series resistance low, which is of paramount importance for the electromigration technique (see Sect. 4.9.2). The constantan matrix and the CuNi tube are needed to keep the heat leak from 300 K to 40 mK small. The attenuation of the final filter is shown in Fig. 4.17 (blue curve).

4.7 Signal Transducer

In Sect. 4.6, it was already pointed out that major noise sources at room temperature are electromagnetic radiation and vibrations. They couple to the experimental setup via ground loops, weak shielding or bad connectors. A way to curtail these problems is to use short cables and avoid connectors wherever it is possible. Therefore, we wanted to unify the commonly used switch box, amplifier, voltage divider, and low pass filters in one signal transducer. The development was done in close collaboration with Daniel Lepoittevin from the Néel Institute.

The signal transducer was designed to be compatible with the standard dilution fridge interface (12 pin Jaeger connector) and the batches of electromigration junctions, which have all a common source and gate, respectively (see Sect. 4.9.2). Due to the geometry of the 12 pin Jaeger connector, we ended up with 10 selectable signal injections lines (drains), one signal output line (source), and one gate. Every line can be grounded directly or via a 100 kΩ resistor. This prevents large discharge currents during the installation of the chip carrier in the cryostat, which are caused by a potential difference between the junctions and the dilution fridge. The drain and gate line have additional voltage dividers in order to increase the resolution of the data acquisition unit (see Sect. 4.8). In addition, an offset of ± 2.5 V or ± 5 V can be superimposed to the divided gate signal in order to shift the measuring range by keeping the resolution constant. To avoid sharp transitions between different offsets, a low pass filter with a time constant of 1 s is added to the summing amplifier. Furthermore, all inputs are equipped with a low pass filters to reject the incoming noise. Thereby the drain inputs have a cutoff frequency of 500 Hz and the gate input a cutoff frequency of 200 Hz. The higher value of the drain inputs was needed in order to transmit the lock-in signal, which can is modulated up to a few hundred Hz. In the following a more technical description to the signal transducer is given.

The signal transducer contains two built-in IV converters. The OPA129U (box 2 Fig. 4.18) is an ultra low input bias current amplifier. It has a current input bias of only 30 fA and is used for our actual measurements. It provides four selectable gains (R40-R43), which are 10^6–10^9. Parallel to R40-R43 are the capacitors C57-C60, which on the one hand prevent the amplifier from self-oscillation and on the other hand determine its bandwidth.

To adjust the offset of the OPA129U the circuit in box 3 Fig. 4.18 is added. It consists of a very stable current source (Ref200AU), which yields a current of $\pm 100\ \mu$A with a precision of 0.25 % for input voltages from 2.5–40 V. In the following, this current is transformed into a voltage via the resistors R48-R50 and amplified to give an offset compensation in the range of -30 to +30 mV. Moreover, it should be noticed that the input of the OPA129U is directly soldered to source line in order to minimize the electro-mechanical noise.

The fast feed back loop of the electromigration requires an amplifier with a large bandwidth. Therefore, a second IV converter (LT1028CS8) is mounted inside the signal transducer (box 1 Fig. 4.18). Its internal bandwidth is 75 MHz and its current input bias is 30 nA. Due to this large value, it must be disconnected during sensitive measurements since otherwise a huge part of the signal would be lost because of the input-leak current. Its gain is fixed to 10^3, which is the optimum range for the electromigration.

The switches to select the different drain terminals have 3 positions, ground, 100 kΩ via ground, and floating. The first two positions are used when connecting the sample to the cryostat, whereas the latter is used during the experiment. The polarizing resistor of the drain voltage divider was chosen to be only a fraction of the sample

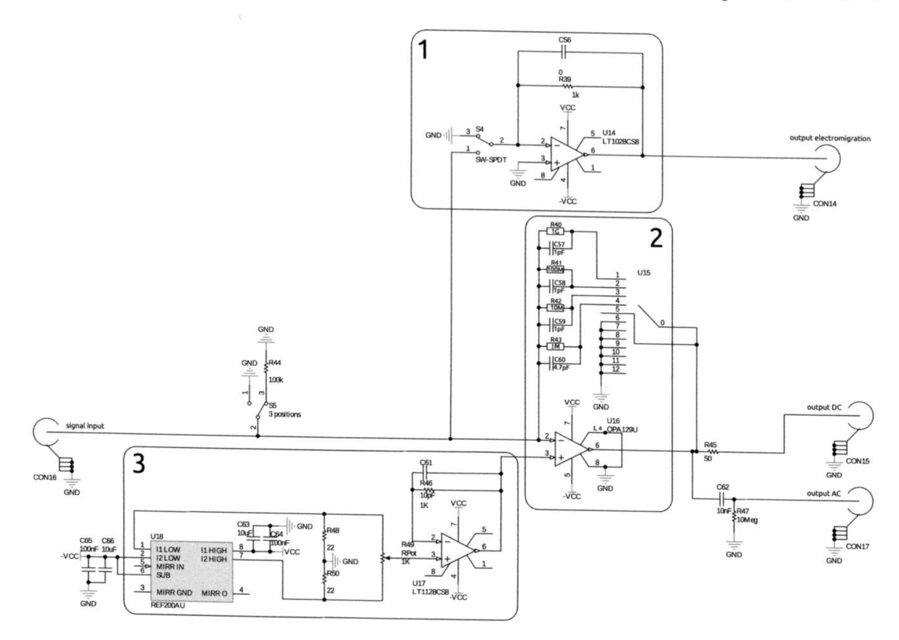

Fig. 4.18 Circuit diagram of the IV converter. Box (I) contains the fast current voltage converter with a bandwidth of 1 MHz and an amplification of 10^3. This IV converter is used for the electromigration only. Box (2) shows the ultra low noise IV converter with 4 selectable gains used for the electrical transport measurements. The offset of the IV converter in (2) is adjusted by the circuit embedded in (3)

resistance. To reject the input noise, an additional 500 Hz low pass filter was add to the drain input.

The gate-circuit divides the input voltage coming from the voltage source by up to 90 (box 3 Fig. 4.19). Afterward, an optional offset of ± 2.5 or ± 5 V is added to the divided signal using the summing amplifier (box 5 Fig. 4.19). The offset is created using the two voltage references in box 1 Fig. 4.19. The circuits in box 2 and 4 are unity gain buffer amplifiers, which transform the impedance of circuits 3 and 1 to almost zero Ω. Finally, the signal goes through a 30 Hz low pass filter (box 6 Fig. 4.19) in order to reject the low frequency noise. Figure 4.20 shows the final version of the signal transducer.

In order to analyze the performance of the high gain IV converter inside the signal transducer we were measuring its noise level with a Stanford SR760 FFT spectrum analyzer. To benchmark the results we did also measurements on the isolated IV converter and a commercial low noise IV converter Femto DLPCA-200. Therefore, we were able to verify the crosstalk to the surrounding electronics inside the signal transducer as well as the overall performance.

A scheme of the experimental setup for measuring the input and output noise is shown in Fig. 4.21a, b, respectively. Since the current noise of the inputs is very small, an additional amplifier had to be used.

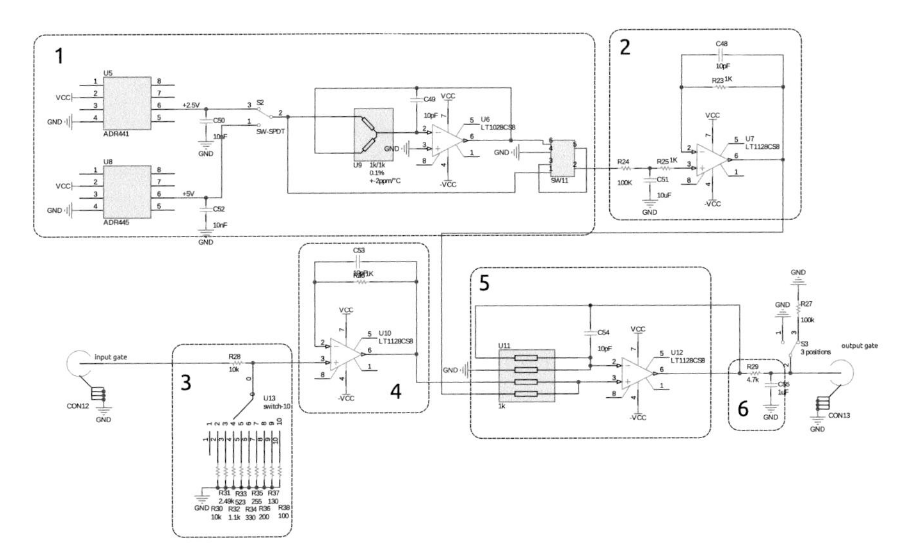

Fig. 4.19 Schematic of the gate circuit. Circuit diagram of the gate circuit. The input gate voltage is divided using the circuit in box 3. Subsequently an optional offset of 0, ±2.5 or ±5 V is generated using the circuit in box 1 and added to the divided signal using the summing amplifier in box 5. The circuit in 2 and 4 are used to match the output impedances of 3 and 1–5

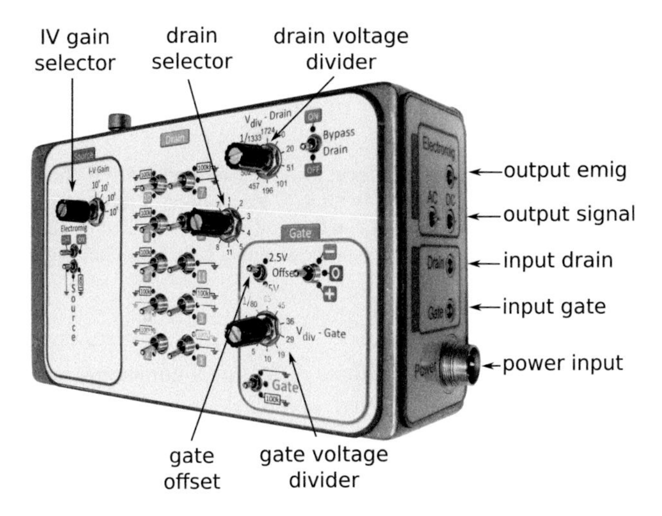

Fig. 4.20 Picture of the signal transducer

First, we disconnected the amplifier of Fig. 4.21a in order to acquire the background signal of the setup (see green curve in Fig. 4.21c). It was found to be 50 fA above the theoretical value of 128 fA ($\sqrt{4k_{\mathrm{B}}T/R}$), which is most likely due to additional noise

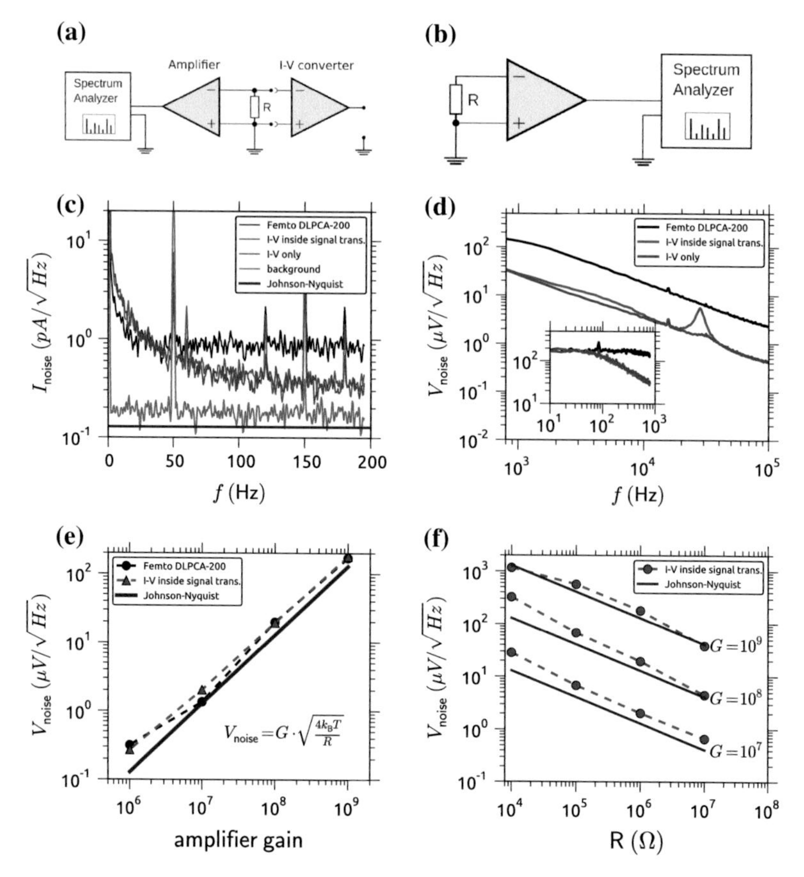

Fig. 4.21 **a** Setup for measuring the spectral input noise. The resistance R was chosen to be 1 MΩ. **b** Setup for measuring the spectral output noise with $R = 1$ MΩ. **c** Spectral input noise measurements for three different amplifiers versus frequency f. **d** Spectral output noise of three different amplifies versus frequency f. **e** Spectral output noise versus the amplifier gain. **f** Spectral output noise versus the input resistance for three different gains

of the second amplifier. Afterward, the three different IV converters were connected to analyze their input noise. The blue, red, and black curve in Fig. 4.21 correspond to the isolated IV converter, the IV converter of the signal transducer, and the Femto DLPCA-200, respectively. The noise level of the home-made IV converter shifted to 340 fA at 170 Hz, which is a 260 fA above the background. Since the blue and the red curve are almost identical, we can exclude any crosstalk between different parts of the signal transducer with the input IV converter. If, however, the Femto DLPCA-200 is connected to the spectrum analyzer we monitor noise level of around 900 fA at 170 Hz, which is 720 fA above the background and therefore more than twice the value of the OPA129U.

In Fig. 4.21d the noise levels measured at the outputs of the amplifiers are depicted. The frequency was varied from 0 to 100+kHz, the resistor is again 1 MΩ and the gain was fixed to 10^9 V/A. At frequencies below 100 Hz, the voltage noise is almost identical for all three IV converters. The different cutoff frequencies of the two I-V converters origin from a bandwidth of 100 Hz for the OPA129U and 1 kHz for the Femto at the same gain. Hence, about one order of magnitude less high frequency noise is collected but the lock-in frequency is limited values below 100 Hz at this gain. The peak in the red curve at around 26 kHz results from a parasitic LRC oscillator inside the signal transducer. Since this peak remains below the Femto noise level its influence is considered as negligible.
In Fig. 4.21e we compared the output noise of the Femto DLPCA-200 with the IV converter of the signal transducer for different gains at a frequency of 20 Hz and with a resistor of 1 MΩ. At this frequency, the two curves are almost identical and about 1.6 times higher than the theoretical value ($S_V = G \cdot \sqrt{4k_B T/R}$).
In Fig. 4.21f the output noise of the IV signal transducer was measured for different gains and resistances. The good agreement between the obtained data and the theoretical values shows the small value of extra noise added to the signal.

4.8 Real-Time Data Acquisition

The experimental setup required the control and read-out of multiple signals simultaneously. In a straight forward realization one could use several devices, linked one to another by a common ground. This, however, induces ground loops, which would be a major source of noise. Therefore, our motivation was to combine all tasks in one automation unit like a computer. However, in conventional computers the operating system is assigning priorities to different tasks. Thus, a task with low priority can be executed with a delay of several milliseconds. Additionally, the execution of a task with high priority is not guaranteed. Hence, the simultaneous control of different experimental parameters cannot happen in a synchronized way with precisions below several milliseconds.
For this reason, we were using an ADwin system instead of a standard PC (Fig. 4.22). It combines analog and digital inputs and outputs with a dedicated real-time processor and real-time operation system. It has a 16 bit output card with an integrated D/A converter. Its voltage range is ±10 V resulting in a step size of $20\text{V}/2^{16} = 305\ \mu V$. The input card, in contrary, has a resolution of 18 bit and an A/D converter with readout voltages ranging from −10 and +10 V at a resolution of $20\text{V}/2^{18} = 75\ \mu V$. An additional 14 bit input card with a clock frequency of 50 MHz was added to perform the electromigration using a fast feedback loop. All cards are controlled by a 300 MHz digital signal processor (DSP), which performs tasks with a precision of 3 ns. The response time in the feedback loop of the electromigration is 1.5 μs due to the execution of several lines of code.

Fig. 4.22 Picture of the ADwin automat showing the front panel with an 18 an 14 bit analog input card an a 16 bit analog output card

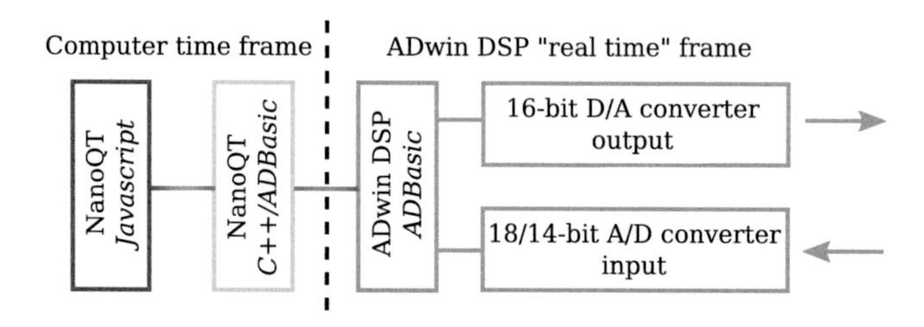

Fig. 4.23 Schematic representation of the different execution levels during the data acquisition. The user programmed Script is transcribed into different lines of C++ and ADBasic, the latter being the native language of the ADwin system. Those *lines* are send to the ADwin DSP who carries out the instructions at a frequency of 300 MHz

The ADwin is linked to a standard PC via an Ethernet connection and can be programmed using NanoQt (see Fig. 4.23). This is a home-made software, which was developed in our group by E. Bonet, C. Thirion and R. Picquerel. Its user interface is based on the JavaScript language and allows for the execution of user defined scripts.

4.9 Sample Fabrication

The device, which was studied in thesis, is a molecular spin-transistor. It consists of a single-molecule magnet, which is connected to source, drain, and gate terminals. The size of the molecule and therefore the characteristic dimension of the device was about 1 nm. Since the smallest dimensions, which can be created by electron beam lithography, are around 10 nm, other fabrication techniques were necessary.

Today, there are only a few techniques available to reliably connect a single molecule to metallic electrodes, such as a scanning tunneling microscopy [16], mechanical break junctions [17], and electromigrated break junctions [18]. Among those techniques, electromigration is the only one which can also implement an efficient gate to control the chemical potential of the molecule and therefore enables us to adjust the working point of the transistor.
The first step towards a molecular spin-transistor is the fabrication of a Nanowire with a well define weak point. Using electromigration in the next step enables us to craft a nanometer sized gap at the predefined breaking point of the nanowire. In the last step, we trap a molecule inside the nanogap to complete the transistor fabrication. In the following a more detailed explanation of the three fabrication steps is given.

4.9.1 Nanowire Fabrication

The nano fabrication of our devices was done using the clean room facilities of the Néel Institute. In order to reduce the number of external connections per transistor, a layout with 12 nanowires sharing a common source and gate was developed. An optical image of the layout is depicted in Fig. 4.24a. In Fig. 4.24b we can clearly see the 12 nanowires with their source in the middle of the image and the U-shaped gate underneath. It was already shown by [19] that back-gated single-molecule transistors show a very good gate response and are most compatible with the electromigration technique.

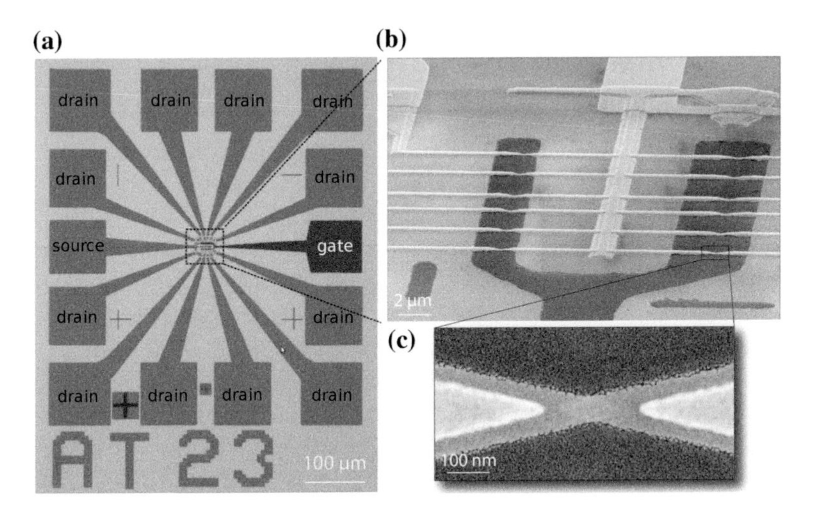

Fig. 4.24 **a** Layout of an array of 12 transistors sharing a common source and gate terminal. **b** Scanning electron microscope image showing the back gate (*grey*) as well as the common source and the different drain terminals. **c** Zoom showing the nanowire obtained by shadow mask evaporation

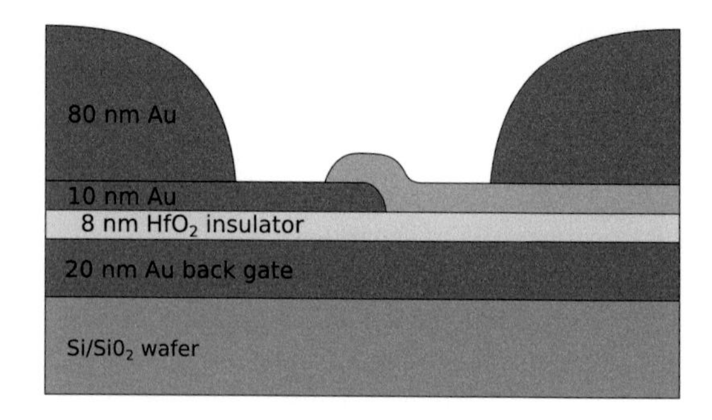

Fig. 4.25 Cross section of the nanowire and the predetermined breaking point. The titanium sticking layers under the gold electrodes are not shown

The first step in the device fabrication was the deposition of the back gate (grey electrode in Fig. 4.24b. In consists of a 20 nm thick gold layer, which was deposited onto a Si/SiO_2 wafer and a 3 nm Ti sticking layer using deep ultra violet optical lithography and metal evaporation. During this process, the contact pads as well as the U-shaped electrode were fabricated. To insulate the gate from the source and drain terminals, a 8 nm thick HfO_2 layer with a dielectric constant of $\approx$17 was deposited onto the gate by using atomic layer deposition. The thin oxide layer resulted also in the different color of the gate with respect to source and drain. Subsequently, we deposited the source and drain contact pads using ultra violet optical lithography and Ti/Au metal deposition. The most important part for the electromigration is the deposition of a nanowire with a predetermined breaking point. This step was done using electron beam lithography and shadow mask evaporation under different angles. A scanning electron microscope image of the constriction in the nanowire is shown in Fig. 4.24c. It has thickness of only 10 nm at the weakest point, whereas the nanowire itself is 80 nm thick. A schematic cross section of the nanowire and the predetermined breaking point is shown in Fig. 4.25.

4.9.2 *Electromigration*

In order to create a nanogap between the source and drain terminal, we made use of the electromigration technique at mK temperatures. The phenomenon of electromigration is known since a long time. Especially in the 1960s, it gained a lot of interest since it was found to be a reason for failure of micro-electronic devices [20, 21]. The phenomenon can be paraphrased as the diffusion of metal ions under the exposure of large electric fields. The force applied to the each metal ion can be written as [22]:

$$\boldsymbol{F} = Z * e\boldsymbol{E}, \tag{4.9.1}$$

where $Z*$ is the effective charge of the ion during the electromigration and can be decomposed into:

$$Z* = Z_{\mathsf{el}} + Z_{\mathsf{wind}}, \tag{4.9.2}$$

where Z_{el} can be seen as the nominal charge of the ion and Z_{wind} the momentum exchange effect between electrons and the ion, commonly also referred to as the electron wind [22]. In metals, only the latter contribution is responsible for the diffusion of the ion and not the electric field. Therefore, the diffusion happens in the direction of the electric current.

Our electromigration procedure is combination of the method of Park [18] and Strachan [23]. In order to limit the Joule heating during the electromigration, we polarize the break junction with a voltage instead of a current. The increasing resistance, which is expected during the migration of the metal, thus, leads to a power reduction (U^2/R) instead of a power increase (I^2R).

Furthermore, it was shown that a large series resistance leads to an increase of power dissipation during the electromigration [24–26], which results in larger gaps or even the complete destruction of the device. As already pointed out in Sect. 4.6.2, we were using superconducting wires inside the cryostat to reduce the total series resistance (120 Ω, measured from one connector outside the cryostat to another).

Moreover, we made us of the ADwin system to establish a fast feedback loop. It continuously reads-out the resistance of the wire and turns off the polarizing voltage within 10 μs. Since the typical time constant of the electromigration is in the order of 100 μs [27], we are able to control the size of the nanogap formation on the atomic level.

The conductance-voltage characteristic recorded during the electromigration typically looks like Fig. 4.26a. It shows a first decrease of the conductance due to Joule heating of the metal. The subsequent increase of the conductance is caused by a rearrangement of the metallic grain boundaries, which enlarge the average grain size and therefore reduced the scattering at the grain boundaries. The following sharp

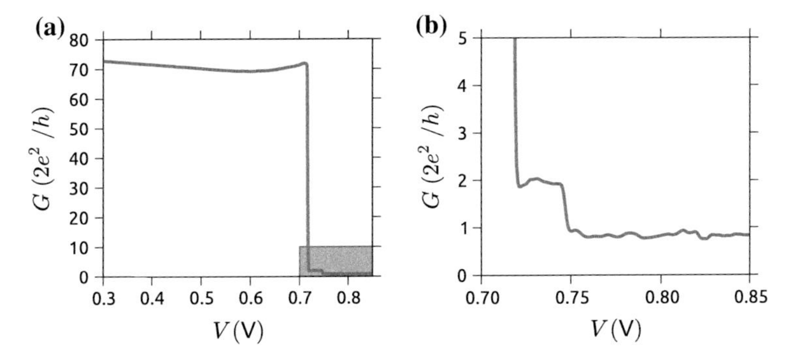

Fig. 4.26 **a** Conductance of the break junction during the electromigration. **b** Zoom into the *grey shaded region* of (**a**) showing quantized conductance steps

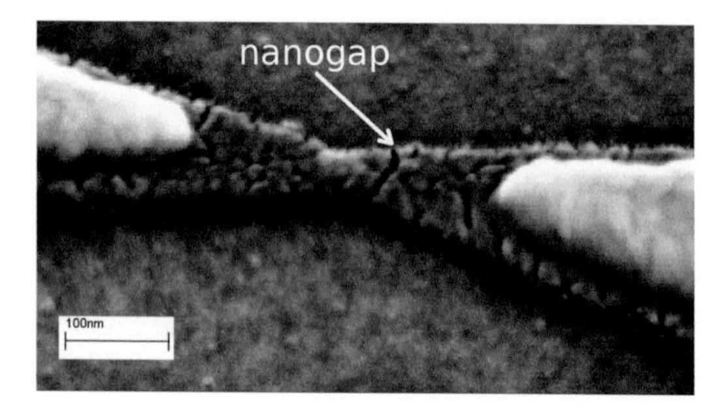

Fig. 4.27 Scanning electron microscope image of an electromigrated break junction

drop in the conductance curve is caused by the migration of the gold ions, leading to the formation of a nm sized gap. During the last seconds of the electromigration, we are often able to see quantized conductance steps, which arise from the current transport through the last remaining gold atoms.
A scanning electron microscope image of an electromigrated junction is presented in Fig. 4.27. It shows the predefined breaking point of the nanowire and a nanometer sized gap.

4.9.3 Fabrication of a Molecular Spin Transistor

Applying the procedures of Sects. 4.9.1 and 4.9.2 allows us to create a three terminal device with source, drain, and gate electrodes. In order to complete the fabrication of a molecular transistor, a single molecule needs to be trapped inside the nanogap, which was formerly created by electromigration.
In the first step, we cleaned the nanowires from organic residues using acetone and isopropanol, followed by an exposure to oxygen plasma for 2 min. Subsequently, we dissolved 3 mg of $TbPc_2$ crystals into 5 g dichlormethane and sonicated the solution at low power for 1 h. This ensures that the remaining $TbPc_2$ clusters are completely dissolved. Afterward, some droplets of the solution were deposited on the nanowire chip and blow dried with nitrogen.
In the next step, we glued the chip on the sample holder and established the electrical connections to the chip by microbonding aluminum wires. Subsequently, the sample was mounted inside a dilution refrigerator and cooled down to mK temperatures. Once the sample was cold, we started the electromigration to craft a nanometer gap into the nanowire. The heat created during this process enables the molecules to diffuse on the surface and therefore be trapped inside the gap. As a first indication if the fabrication procedure was successful, we measured the zero bias conductance through the device as a function of the gate voltage. If a nanometer size object was trapped inside the nanogap, it will create a quantum dot resulting in one or more Coulomb peaks (see Chap. 2). Yet, this is not a proof that we actually trapped a single $TbPc_2$ molecule. Especially when using electromigration, there are many

ways of creating a quantum dot. For example a gold nanoparticle or some organic residue, which was not completely removed during the cleaning procedure, would result in to the same transport signature when trapped inside the nanogap. In order to eliminate any doubt if the nanoparticle is a single $TbPc_2$ or not, we studied the magnetic properties of our device. As it will be shown Chap. 5, the $TbPc_2$ has a very unique magnetic signature, which can be used as a fingerprint of the molecule. In case we did not trap any or too many nanoparticles, we heated up the cryostat above 150 K and cooled it down again. This enables the surface diffusion of the molecules due to thermal activation and a subsequent retrapping at a different place. This procedure of warming up and cooling down was repeated up to ten times before changing the sample.

References

1. F. Pobell, J. Brooks, *Matter and Methods at Low Temperatures*, vol. 45, 2nd edn. (Springer, Heidelberg, 1992). ISBN 10 3-540-46356-9
2. W. Demtröder, *Experimentalphysik 1: Mechanik und Wärme (Springer-Lehrbuch) (German Edition)* (Springer, 2005). ISBN 354026034X
3. C. Boulder, V.J. Johnson, *A compendium of the properties of materials at low temperatures: (Phase 1-)*. Wright Air Development Division, Air Research and Development Command, U.S. Air Force (1961)
4. C.Y. Ho, R.W. Powell, P.E. Liley, *Thermal Conductivity of the Elements*, vol. 1. Amer Inst of Physics (1974)
5. M.M. Freund, T. Hirao, V. Hristov, S. Chegwidden, T. Matsumoto, A.E. Lange, Compact low-pass electrical filters for cryogenic detectors. Rev. Sci. Instrum. **66**, 2638 (1995)
6. S. Mandal, T. Bautze, R. Blinder, T. Meunier, L. Saminadayar, C. Bäuerle, Efficient radio frequency filters for space constrained cryogenic setups. Rev. Sci. Instrum. **82**, 024704 (2011)
7. A. Lukashenko, A.V. Ustinov, Improved powder filters for qubit measurements. Rev. Sci. Instrum. **79**, 014701 (2008)
8. F.P. Milliken, J.R. Rozen, G.A. Keefe, R.H. Koch, 50Ω characteristic impedance low-pass metal powder filters. Rev. Sci. Instrum. **78**, 024701 (2007)
9. J. Martinis, M. Devoret, J. Clarke, Experimental tests for the quantum behavior of a macroscopic degree of freedom: the phase difference across a Josephson junction. Phys. Rev. B **35**, 4682–4698 (1987)
10. D.C. Glattli, P. Jacques, A. Kumar, P. Pari, L. Saminadayar, A noise detection scheme with 10 mK noise temperature resolution for semiconductor single electron tunneling devices. J. Appl. Phys. **81**, 7350 (1997)
11. A.B. Zorin, The thermocoax cable as the microwave frequency filter for single electron circuits. Rev. Sci. Instrum. **66**, 4296 (1995)
12. D. Vion, P.F. Orfila, P. Joyez, D. Esteve, M.H. Devoret, Miniature electrical filters for single electron devices. J. Appl. Phys. **77**, 2519 (1995)
13. H. Courtois, O. Buisson, J. Chaussy, B. Pannetier, Miniature low-temperature high-frequency filters for single electronics. Rev. Sci. Instrum. **66**, 3465 (1995)
14. H. le Sueur, P. Joyez, Microfabricated electromagnetic filters for millikelvin experiments. Rev. Sci. Instrum. **77**, 115102 (2006)
15. K. Bladh, D. Gunnarsson, E. Hürfeld, S. Devi, C. Kristoffersson, B. Smalander, S. Pehrson, T. Claeson, P. Delsing, M. Taslakov, Comparison of cryogenic filters for use in single electronics experiments. Rev. Sci. Instrum. **74**, 1323 (2003)

16. J. Schwöbel, Y. Fu, J. Brede, A. Dilullo, G. Hoffmann, S. Klyatskaya, M. Ruben, R. Wiesendanger, Real-space observation of spin-split molecular orbitals of adsorbed single-molecule magnets. Nat. Commun. **3**, 953 (2012)
17. M.A. Reed, Conductance of a molecular junction. Science **278**, 252–254 (1997)
18. H. Park, A.K.L. Lim, A.P. Alivisatos, J. Park, P.L. McEuen, Fabrication of metallic electrodes with nanometer separation by electromigration. Appl. Phys. Lett. **75**, 301 (1999)
19. H. Park, J. Park, A.K. Lim, E.H. Anderson, A.P. Alivisatos, P.L. McEuen, Nanomechanical oscillations in a single-C60 transistor. Nature **407**, 57–60 (2000)
20. I.A. Blech, Direct transmission electron microscope observation of electrotransport in aluminum thin films. Appl. Phys. Lett. **11**, 263 (1967)
21. J.R. Black, Electromigration failure modes in aluminum metallization for semiconductor devices. Proc. IEEE **57**, 1587–1594 (1969)
22. K. Tu, Electromigration in stressed thin films. Phys. Rev. B **45**, 1409–1413 (1992)
23. D.R. Strachan, D.E. Smith, D.E. Johnston, T.H. Park, M.J. Therien, D.A. Bonnell, A.T. Johnson, Controlled fabrication of nanogaps in ambient environment for molecular electronics. Appl. Phys. Lett. **86**, 043109 (2005)
24. H.S.J. van der Zant, Y. Kervennic, M. Poot, K. O'Neill, Z. de Groot, J.M. Thijssen, H.B. Heersche, N. Stuhr-Hansen, T. Bjørnholm, D. Vanmaekelbergh, C.A. van Walree, L.W. Jenneskens, Molecular three-terminal devices: fabrication and measurements. Faraday Discuss. **131**, 347 (2006)
25. M.L. Trouwborst, S.J. van der Molen, B.J. van Wees, The role of Joule heating in the formation of nanogaps by electromigration. J. Appl. Phys. **99**, 114316 (2006)
26. T. Taychatanapat, K.I. Bolotin, F. Kuemmeth, D.C. Ralph, Imaging electromigration during the formation of break junctions. Nano Lett. **7**, 652–656 (2007)
27. K. O'Neill, E.A. Osorio, H.S.J. van der Zant, Self-breaking in planar few-atom Au constrictions for nanometer-spaced electrodes. Appl. Phys. Lett. **90**, 133109 (2007)

Chapter 5
Single-Molecule Magnet Spin-Transistor

One of the major motivations to study single molecule magnets (SMMs) is to design ultra dense data storage devices, where each bit of information is stored on the magnetization of a single molecule. However, due to the tiny magnetic moment of an SMM (few μ_B) and a size in the order of a nanometer, it is impossible to study isolated SMMs with standard magnetometers like a micro-squid.
Therefore, we were using a completely new type of detection device—a single-molecule magnet spin-transistor. It was fabricated using electromigration of a nanowire at mK temperatures (see Sect. 4.9). In this way, a nanometer sized gap was crafted between two very clean gold terminals, in which we trapped a single $TbPc_2$ molecule magnet. An artistic view of the device is shown in Fig. 5.1. By studying the electronic transport through the device as a function of the external magnetic field, we are able to read-out the electronic spin state of an isolated single-molecule magnet and the nuclear spin state of a single terbium ion. The latter will be briefly discussed at the end of this chapter and in more detail in Chaps. 6 and 7. This chapter will mainly focus on operation of the spin-transistor and how it can be exploited to study the electronic spin of a single-molecule magnet. The terminology "spin" when it is used alone will always refer to the electronic spin.

5.1 Mode of Operation

The first working molecular spin-transistor was fabricated 2012 in our group [1] and is referred to as sample A. Later on, I fabricated two other devices, which will be referred to as sample B and C. In order to explain the working principle of the molecular spin-transistor, we will schematically subdivide the device into three quantum systems, namely, a nuclear spin qubit, an electronic spin, and a read-out quantum dot (Fig. 5.1b).

(I) The **nuclear spin qubit** emerges from the atomic core of Tb^{3+} ion. It possesses a nuclear spin of $I = 3/2$ resulting in four different qubit states. Due the hyperfine

S. Thiele, *Read-Out and Coherent Manipulation of an Isolated Nuclear Spin*,
Springer Theses, DOI 10.1007/978-3-319-24058-9_5

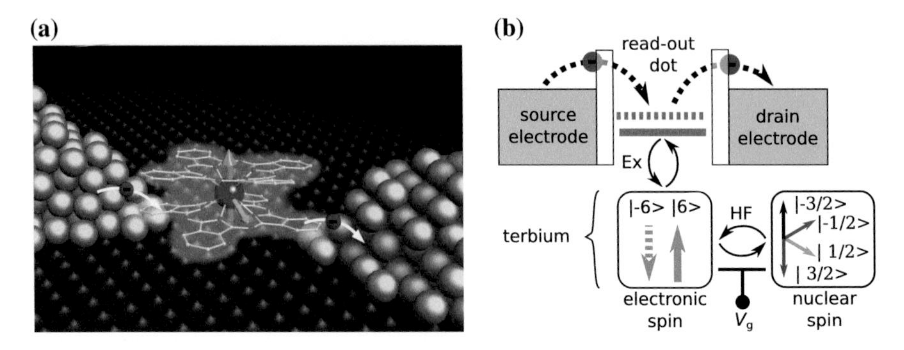

Fig. 5.1 **a** Artist view of the single-molecule magnet spin-transistor. The ligands of the $TbPc_2$ are tunnel coupled to the source and drain thus creating a quantum dot, which can be controlled by a back gate (not shown) underneath. In the center of the molecule is a Tb^{3+} ion possessing an electronic spin (*orange*) and a nuclear spin (*green*). **b** Simplified coupling scheme of the spin-transistor. It consists of three quantum systems: a nuclear spin qubit, an electronic spin and a read-out quantum dot. The nuclear spin is coupled with the electronic spin via the hyperfine interaction. This quantum mechanical link can be used to map the nuclear spin state onto the electronic spin, which amplifies the magnetic signal by $\approx 10^3$. Furthermore, the electronic spin is exchange coupled to the read-out quantum dot, which establishes the detection of the electronic spin and therefore nuclear spin qubit

interaction and the nuclear quadrupole moment, the degeneracy of the four levels is lifted, resulting into four unequally spaced levels (for further details see Sect. 3.7).

(II) The **electronic spin** arises from the terbium's 4f electrons. The intrinsic spin-orbit coupling and a strong ligand-field results in an electronic ground state doublet of $m_J = \pm 6$ and an easy axis of magnetization perpendicular to the ligand plane. This means that the electronic spin can be regarded as a two level system with its eigenstates $|\uparrow\rangle$ and $|\downarrow\rangle$. The degeneracy of the doublet is lifted by the hyperfine coupling to the nuclear qubit and splits each state into four levels, which are separated by approximately 2.5, 3.1 and 3.7 GHz [2]. For further details we refer to Sects. 3.5–3.7.

(III) The **read-out quantum dot** is created by the phthalocyanine (Pc) ligands. Their delocalized π-electron system is tunnel-coupled to the source and drain terminals, thus creating a conductive island. Furthermore, a finite overlap of the π-electron system with the terbium's 4f wave functions gives rise to an exchange coupling between the read-out dot and the electronic spin.
Using this device we were able to read-out the electronic spin state of the $TbPc_2$. Due to the exchange coupling between the read-out dot and the electronic spin, a slight modification in the read-out dot's chemical potential is created depending on whether the electronic spin points parallel or antiparallel to the external field. Since the position of the chemical potential with respect to the source and drain Fermi levels determines the conductance through the device, the two electronic spin states can be assigned to two different conductance values. Therefore, an electronic spin transition from $|\uparrow\rangle \rightarrow |\downarrow\rangle$ or $|\downarrow\rangle \rightarrow |\uparrow\rangle$ results in a conductance jump.

Furthermore, we can use the device to perform a single-shot read-out of the nuclear spin-qubit state. In contrary to the electronic spin detection, this is a two stage process, which takes advantage of the coupling between all three quantum systems. In the first stage, the nuclear qubit state is mapped onto the electronic spin using the hyperfine interaction. As already pointed in Chap. 3, the ligand field mixes the two electronic ground states, resulting in an anticrossing of $\Delta E \simeq 1\ \mu\mathrm{K}$ close to zero magnetic field. Sweeping the magnetic field slowly enough over such an anticrossing gives rise to the quantum tunneling of magnetization (QTM), which reverses the electronic spin according to the Landau-Zener probability. Due to the hyperfine interaction we get four instead of one anticrossing, which makes the magnetic field position of the QTM transition nuclear spin dependent (see Fig. 5.8a). In the second stage, we read-out the position of the QTM event through a jump in the read-out dot's conductance and establish in this way the detection of the nuclear spin-qubit state.
In the following sections, we will show step by step which experiments were conducted and what conclusions were drawn in order to derive to model explained above.

5.2 Read-Out Quantum Dot

The first experiments we performed after the electromigration of the nanowire were low-temperature electronic transport measurements. Those were used to check whether a nanometer object was trapped inside the nanogap. If so, we expected the object to behave as a quantum dot coupled to the source and drain terminals, which would result in the typical single-electron tunneling (SET) characteristics (see Chap. 2).
To check for the SET behavior we measured the conductance through the transistor as a function of the source-drain voltage V_{ds} and the gate voltage V_{g}. This way, a two dimensional map (stability diagram) like in Fig. 5.2a is obtained, where regions with high conductance are colored in red and the regions of low conductance are colored in blue. This stability diagram originated from device B. Figure 5.2a shows only one charge degeneracy point (CDP) within a wide gate voltage window. This is an indication of a relatively large charging energy, thus, the read-out dot must be very small. This is consistent with the claim that the quantum dot is created by the Pc ligands, but does not yet prove our model.
Furthermore, we observed a faint Kondo ridge to the left of the CDP, which indicated an odd number of electrons on the quantum dot and good coupling of the molecule to the source and drain terminals. The occurrence of a Kondo peak was observed in all three devices, indicating that a good coupling to the electrodes is probably a requirement for a functional molecular spin-transistor. The stability diagrams of the other two samples are shown in Fig. 5.2b and c for samples A and C respectively. They were measured for a smaller V_{ds} window in order to protect the devices from damage.

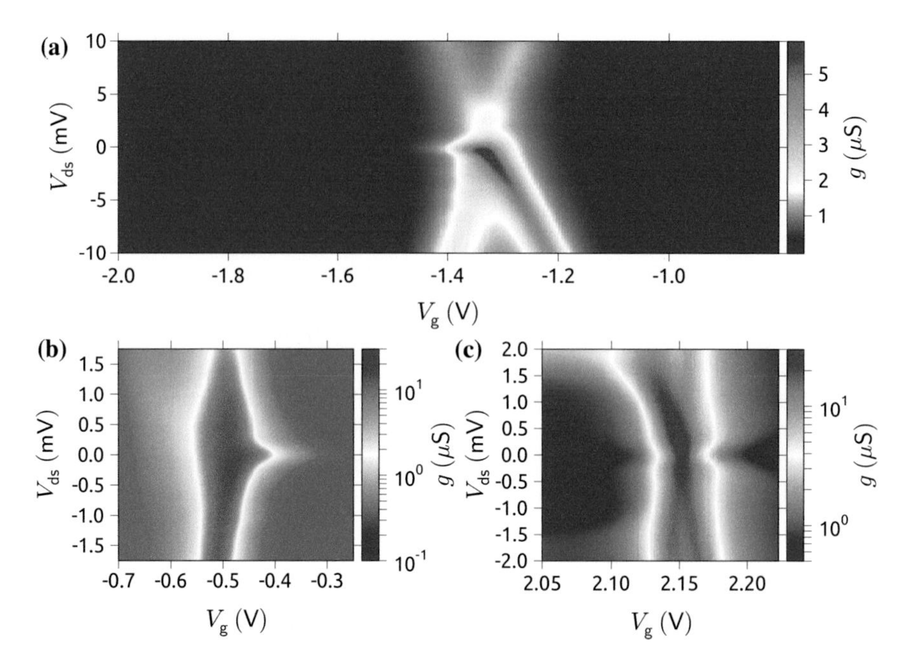

Fig. 5.2 Stability diagram of the read-out quantum dot for sample B (**a**), sample A (**b**) and sample C (**c**). They were measured by sweeping source-drain voltage V_{ds} at different gate voltages V_g while monitoring the conductance. They all show single-electron tunneling and a large Coulomb blockade effect, which was expected from electronic transport through a single molecule, tunnel-coupled to source and drain electrodes. Furthermore, a Kondo peak was observed for all devices, indicating a good coupling to the source and drain terminals. Note that the exotic appearance of the Kondo peak in (**c**) will be discussed in more detail in Sect. 5.4

5.3 Magneto-Conductance and Anisotropy

A first test to verify if the quantum dot, presented in the previous section, was coupled to the magnetic moment of the $TbPc_2$ molecule, is to study the conductance through the device as a function of the magnetic field. Since the magnetic moment of the terbium double-decker can be reversed with an external magnetic field, we expected to see a feature of this magnetization reversal in the electronic transport. To perform the magneto-conductance measurement, we fixed V_{ds} at zero Volt and V_g at a value close to the charge degeneracy point. That is where we expected the largest sensitivity to a magnetization reversal, as a slight variation of the quantum dot's chemical potential results in a strong modification of the conductance. Afterward, we swept the external magnetic field from negative to positive values (trace) and back again (retrace) while recording the conductance through the quantum dot.

As shown in Fig. 5.3a, by sweeping the magnetic field back and forth, we observed jumps in the read-out dot's conductance. Moreover, the magneto-conductance signal was hysteretic, which is the signature of an anisotropic magnetic object. Every

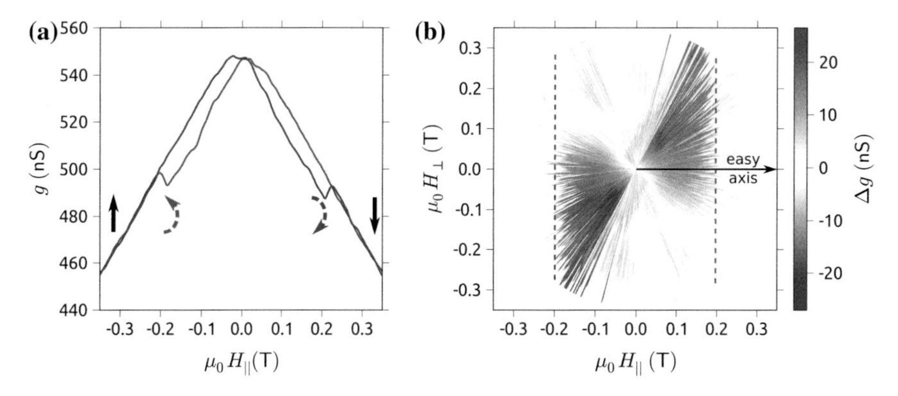

Fig. 5.3 **a** Trace (*blue*) and retrace (*red*) magneto-conductance signal of sample B as a function of $H_{||}$. The conductance jumps correspond to the reversal of the electronic spin carried by the $TbPc_2$ SMM. **b** Two dimensional magneto-conductance signal of sample B as a function of the external field recorded with an angular resolution of 0.5°. The magnetization reversal is shown as a sharp color change. The applied field to reverse the magnetization is smallest along $H_{||}$ and augments gradually with increasing angle. At an angle of 90° the magnetic field is applied in the hard plane and the magnetic moment cannot be reversed anymore

time this object reversed its magnetization the chemical potential of the read-out dot changed between two distinct values, giving rise to jumps in the conductance. The amplitude of the jump was about 3 % of the total conductance value and approximately the same for all three devices.

In order to find more proofs that those conductance jumps originated from the spin reversal of the $TbPc_2$ SMM, we investigated the angular dependence of those jumps. In Chap. 3 we pointed out that an isolated $TbPc_2$ molecule possesses a strong magnetic anisotropy, with an easy axis of magnetization perpendicular to the phthalocyanine plane. It was shown by spin resolved DFT calculations that this anisotropy is preserved even when the molecule is brought to contact with a metallic surface [3] and should therefore also be conserved in our spin-transistor configuration.

As it was shown in Fig. 5.3a the conductance through the read-out dot depends on the orientation of the electronic spin. While at negative $H_{||}$ the spin ground state is $|\uparrow\rangle$ and the excited state is $|\downarrow\rangle$, the Zeeman effect will inverse the energies of the two states at positive magnetic field. Therefore the observed conductance jump during the trace sweep at $B \approx 0.2$ T in Fig. 5.3a corresponds to the transition $|\uparrow\rangle \rightarrow |\downarrow\rangle$ and the jump during the retrace sweep at $B \approx -0.2$ T to the transition $|\downarrow\rangle \rightarrow |\uparrow\rangle$. We repeated the hysteresis measurement under different angles of the magnetic field, and thus scanned the magneto-conductance signal within a plane in the three dimensional vector space. Between two subsequent sweeps, the vector of the magnetic field was rotated by 0.5°. Notice that the specific orientation of the plane was chosen prior to the experiment in order to include the easy axis of magnetization. Subtracting the retrace signal from the trace signal at each angle resulted in Fig. 5.3b. The sharp color change from white to red/blue indicates the spin reversal. By looking at the angular dependence of the reversal it is evident that it becomes harder to flip the

spin, as we turn the magnetic field from $H_{||}$ towards $H_{\perp}$ since only the projection of the magnetic field onto $H_{||}$ is relevant. This behavior is a direct consequence of the molecule's magnetic anisotropy and therefore a strong evidence that the magnetic object is a $TbPc_2$ SMM.
The easy axis of magnetization is parallel to $H_{||}$, whereas the direction along $H_{\perp}$ is called hard axis. It lies within the hard plane, which is aligned parallel to the Pc ligands. Therefore, we can deduce the orientation of the molecule with respect to the experiment from Fig. 5.3b.

5.4 Exchange Coupling

In the previous section we stated that the origin of the magneto-conductance signal was due to a coupling between the read-out quantum dot and the molecule's electronic spin. In this section we are going to determine the strength of the coupling and discuss the possible origins.
To estimate the magnitude of the coupling we investigated the evolution of the Kondo peak of Fig. 5.2(a) (sample B) as a function of the bias voltage V_{ds} and the applied magnetic field B_z. Figure 5.4(a) shows that the Kondo peak is splitting linearly at a rate of 223 μV/T with augmenting B, which is expected for a spin 1/2 and a g-factor close to two.
By extrapolating the linear slopes at positive magnetic fields to negative fields, we found an intersection at approximately −210 mT, which is equal to a negative critical field B_c (see Sect. 2.4). This is in contrast to the classical spin 1/2 Kondo effect, where

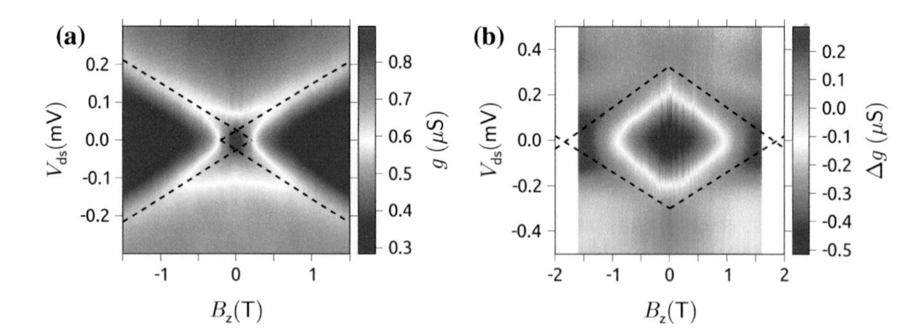

Fig. 5.4 **a** The conductance of sample B, at the *left side* of the charge degeneracy point, is measured as a function of the source-drain voltage V_{ds} and the external magnetic field B_z. It shows the linear evolution of the Kondo peak with respect to the magnetic field. The extrapolation of the linear slops shows an intersection at ±200 mT, which can be used to estimate the magnitude of the coupling between the read-out dot and the electronic spin. **b** Conductance measurement of sample C showing a linear decrease of the Kondo splitting with increasing magnetic field amplitude B_z. This feature is a signature of an antiferromagnetic coupling of the terbium's electronic spin to the read-out quantum dot

the B_C is always positive and linked to the Kondo temperature T_K via: $2g\mu_B B_C = k_B T_K$. In order to explain this finding, we used the analog to the underscreened spin 1 Kondo effect (see Sect. 2.4), where the antiferromagnetic coupling between the screened spin 1/2 and the electrons in the terminals is weakened by a ferromagnetic coupling to the unscreened spin 1/2, which decreases the critical field from finite values to almost zero Tesla [4]. In our device the negative B_C can be interpreted as a ferromagnetic coupling between the read-out dot and the terbium's electronic spin. Due to the larger magnetic moment of $9\mu_B$, the antiferromagnetic coupling to the leads is already destroyed at zero bias. To model the magnetic field behavior, we modified the above mentioned formula to [1]:

$$2g\mu_B B_C = k_B T_K + a\, g\mu_B J_z \tag{5.4.1}$$

where a is a negative for ferromagnetic coupling. From Fig. 5.4a we obtained the full width at half maximum of the Kondo peak at $B = 0$ T of 56 μV. Using the expression $eV = k_B T_K$, we get an estimated Kondo temperature to 650 mK. By inserting the Kondo temperature and $B_C = -210$ mT into Eq. 5.4.1 we extracted a coupling constant of $a = -200$ mT, indicating a strong ferromagnetic coupling.
The same experiments were performed on sample C (Fig. 5.4b) and sample A. Also in these two samples a splitting of the Kondo peak at zero magnetic field was observed. In contrary to sample A and B, sample C shows an antiferromagnetic coupling to the quantum dot since the splitting decreases for increasing B_z (see Fig. 5.4b). This behavior was modeled using a positive a in Eq. 5.4.1. Moreover, it demonstrated that the sign of the coupling constant a must be very sensitive to local deformations of the molecule, which are different from sample to sample. The coupling strengths of samples A, and C were extracted following the same procedure as explained above. Table 5.1 summarizes the three different values.
The modulus of the coupling is very large in all three samples, which makes the exchange interaction the most likely candidate. It has been demonstrated that it can attain values up to 7 T for a nitrogen atom inside a C_{60} [5] and it was presented that the $TbPc_2$ molecule shows an antiferromagnetic super-exchange coupling when deposited on a ferromagnetic surface [6].
Moreover, by considering the magnetic moment of the terbium ($9\mu_B$) and the average distance between the terbium ion and the electron on the phthalocyanine (0.7 nm), we can estimate the dipole-dipole interaction to be about 50 mT, which is smaller than the measured interaction and not sufficient to explain the coupling strength.

Table 5.1 Summary of the extracted coupling strengths of the electronic spin to the read-out quantum dot

Sample	A	B	C
a	−300 mT	−200 mT	+1.66 T

Yet, the exchange coupling is only possible if the read-out quantum dot and the terbium ion are geometrically very close to each other. This supports the assumption that the read-out quantum dot is created by the phthalocyanine ligands. Note that adding one electron to the read-out dot will not affect the charge state of the Tb ion, since this would require an oxidation or reduction of the terbium. It was shown by Zhu et al. [7] that up to the fifth reduction and second oxidation of the molecule, electrons are only added to the organic ligands of the double-decker, leaving the charge state and therefore the magnetic properties of the terbium ion untouched.

5.5 2D Magneto-Conductance of the Read-Out Dot

After having quantified the strength of the coupling between the read-out quantum dot and the electronic spin, we now investigate the magneto-conductance signal along two different directions. Using sample A, we measured the conductance through the read-out dot as a function of the magnetic field $H_\perp$ perpendicular to the easy axis of the $TbPc_2$ at four different parallel fields $H_{||}$ and vice versa (Fig. 5.5).
In order to assign the electronic spin state to a certain conductance value, we fitted the data to the empirical formula

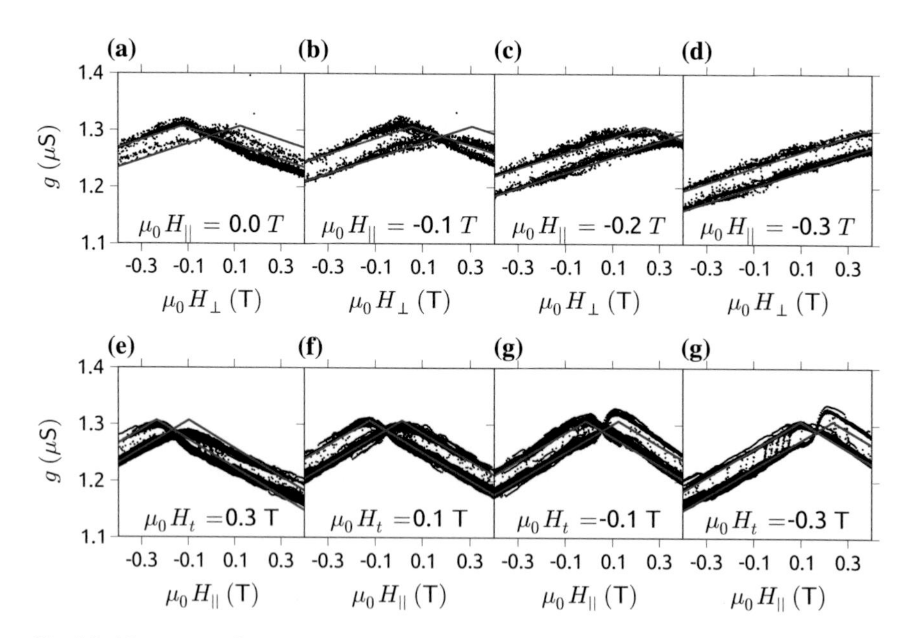

Fig. 5.5 Magneto-conductance signal of sample A as a function of the transverse field $B_\perp$ at four different parallel fields $B_{||}$ (**a–d**) and vice versa (**e–f**). The *red curve* is an empirical fit to the spin-down conductance and the blue curve to the spin-up conductance

$$g(B_{||}, B_t) = -\alpha|B_t - \beta B_{||} \pm \gamma/2| + g_0 \tag{5.5.1}$$

with $\alpha = 1.38 \times 10^{-7}$ S/T, $\beta = -1.8$, $\gamma = 0.25$ T and $g_0 = 1.307 \times 10^{-6}$ S. The fit to the spin-up conductance is depicted in blue, whereas the fit to the spin-down conductance was colored in red in Fig. 5.5. We observed that the difference between the two conductance values is constant over a large range in magnetic field but goes to zero at a particular combination of $H_{||}$ and $H_{\perp}$. To get a better visualization of this effect we simulated the two dimensional conductance map using Eq. 5.5.1 and the parameters extracted from the fits (see Fig. 5.6a). In the red area, the conductance was larger when the spin pointed up, whereas in the blue area, the conductance was larger when the spin pointed down. A remarkable feature is, however, the white stripe, indicating that the two conductance values were equal.
To get a deeper understanding of the origin of the magneto-conductance signal we used a semi-classical model to describe the read-out dot's chemical potential. We assumed that the read-out quantum dot possesses a spin $\boldsymbol{S}$, which is exchange coupled to the electronic spin $\boldsymbol{J}$ through $a\boldsymbol{SJ}$. The Hamiltonian of the read-out dot exposed to an external magnetic field $\boldsymbol{B}_{\text{ext}}$ is given by:

$$H = g\mu_B \boldsymbol{S}\boldsymbol{B}_{\text{ext}} + a\boldsymbol{SJ} = g\mu_B \boldsymbol{S}\left(\boldsymbol{B}_{\text{ext}} + \frac{a\boldsymbol{J}}{g\mu_B}\right) \tag{5.5.2}$$

with g the g-factor of the quantum dot and μ_B the Bohr magneton. In the semi-classical approach $\boldsymbol{J}$ is no longer an operator but a vector with $J_z = \pm 6\hbar$ and $max(J_x) = max(J_y) = \sqrt{7}\hbar$. A magnetization reversal of the electronic spin was modeled by changing $J_x \to -J_x$, $J_y \to -J_y$, $J_z \to -J_z$. Like in the experiment the external magnetic field was simulated to be in the y-z plane with $\boldsymbol{B} = (0, B_t, B_{||})$.

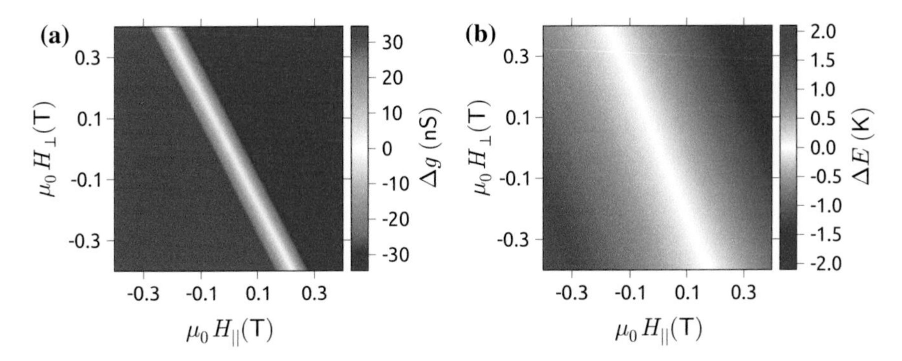

Fig. 5.6 **a** Fitted difference between the spin-down and spin-up conductance as function of the magnetic field parallel (||) and transverse ($\perp$) to the easy axis of the $TbPc_2$. **b** Zeeman energy difference of the read-out quantum dot with respect to the electronic spin up or down state, calculated using Eq. 5.5.3 and $a = 200$ mT. The qualitative agreement of the two plots shows that the magneto-conductance signal can be explained by a change of the read-out dot's Zeeman splitting ΔE_Z with respect to the electronic spin state of the terbium ion

The Hamiltonian of the quantum dot is now written as:

$$H = g\mu_B \left[S_y \left(B_\perp \pm \frac{aJ_y}{g\mu_B} \right) + S_z \left(B_{||} \pm \frac{aJ_z}{g\mu_B} \right) \right] \tag{5.5.3}$$

with S_y and S_z being the appropriate spin matrices for the spin $\boldsymbol{S}$ and the $\pm$ sign indicating the two different spin directions $|\uparrow\rangle$ and $|\downarrow\rangle$. Assuming $S = 1/2$ we can diagonalize the Hamiltonian for each electronic spin direction individually and calculate the difference of the Zeeman splittings $\Delta E_Z(|\uparrow\rangle) - \Delta E_Z(|\downarrow\rangle)$. Doing this at different $B_{||}$ and $B_\perp$ resulted in Fig. 5.6b. The two plots in Fig. 5.6 show a good qualitative agreement, especially the white region of zero sensitivity as well as the angle with respect to $B_{||}$ is very well reproduced. It shows that the origin of the magneto-conductance signal can be explained by a shift of the quantum dot's Zeeman splitting depending on whether the electronic spin points parallel or antiparallel to the external field.

5.6 Electronic Spin Relaxation

After having explained the coupling of the read-out quantum dot to the electronic spin, we now focus on the electronic spin only. In this section we investigate the relaxation behavior of the electronic spin at large magnetic fields.

From Fig. 5.3 we can already see that the spin relaxation at large magnetic fields is not exactly determined by the projection on $H_{||}$, which origins from the stochastic nature of the inelastic spin reversal. It requires an energy exchange with the thermal bath and the creation of a phonon.

In the case of an isolated terbium double-decker the energy exchange is mediated by the ligand field. In order to quantify this effect, we will use a model taken from Abragam and Bleaney [8].

Therein, we assume a two-level spin-system whose energies are separated by $\hbar\omega$ and which is in contact with a phonon bath of temperature T. Then, the transition rates between state $|1\rangle$ and $|2\rangle$ are given by the Einstein coefficients of absorption and emission:

$$w_{1\to 2} = B\rho_{ph}, \tag{5.6.1}$$

$$w_{2\to 1} = A + B\rho_{ph} = B\rho_{ph} exp\left(\frac{\hbar\omega}{k_B T}\right) \tag{5.6.2}$$

where ρ_{ph} is the phonon density, B the coefficient of stimulated emission or absorption and A the coefficient of spontaneous emission. If the spin-system is out of thermal equilibrium, it will return to it in a characteristic time τ:

$$\frac{1}{\tau} = w_{1\to 2} + w_{2\to 1} \tag{5.6.3}$$

which under substitution of Eqs. 5.6.1 and 5.6.2 results in:

$$\frac{1}{\tau} = B\rho_{\mathrm{ph}}\left[exp\left(\frac{\hbar\omega}{k_B T}\right) + 1\right] \tag{5.6.4}$$

The phonon density of a three dimensional crystal is given as:

$$\rho_{\mathrm{ph}} = \underbrace{\frac{3}{2\pi^2}\frac{\omega^2}{v^2}}_{\text{density of states}} \underbrace{\frac{\hbar\omega}{exp\left(\frac{\hbar\omega}{k_B T}\right) - 1}}_{\text{average phonon energy}} \tag{5.6.5}$$

Hence, inserting this expression into Eq. 5.6.4 gives:

$$\frac{1}{\tau} = \frac{3\hbar\omega^3}{2\pi^2 v^3} B\, coth\left(\frac{\hbar\omega}{2k_B T}\right) \tag{5.6.6}$$

The lattice vibrations couple not directly to the terbium ion, instead they modulate the ligand field. To take this indirect interaction into account we develop the ligand field in powers of strain [9]:

$$V = V^{(0)} + \epsilon V^{(1)} + \epsilon^2 V^{(2)} + \ldots \tag{5.6.7}$$

where the first term on the right is just a static term and the second and third term correspond to first and second order corrections respectively. Applying Fermi's golden rule:

$$w_{i\to j} = \frac{2\pi}{\hbar^2}\left|\langle i|\, H^{(1)}\, |j\rangle\right|^2 f(\omega) \tag{5.6.8}$$

where $H^{(1)}$ is the first order perturbation Hamiltonian and $f(\omega)$ the normalized line-shape function. Inserting $2\rho v^2\epsilon^2 = \rho_{\mathrm{ph}} d\omega$ and integrating over all frequencies results in $w_{i\to j} = \frac{2\pi}{\hbar^2}\frac{\rho_{\mathrm{ph}}}{2\rho v^2}$. When we compare this expression with Eq. 5.6.1, we get $B = \frac{\pi}{2\pi\hbar\rho v^2}\left|V^{(1)}\right|^2$, with ρ being the density of the material. Hence, using Eq. 5.6.6 we get:

$$\frac{1}{\tau} = \frac{3}{2\pi\hbar\rho v^5}\left|V^{(1)}\right|^2 \omega^3\, coth\left(\frac{\hbar\omega}{2k_B T}\right) \tag{5.6.9}$$

In the case of $TbPc_2$, the energy difference between the two spin states is $\hbar\omega = g\mu_B \Delta m_j \mu_0 H_{||}$. Furthermore, if $\hbar\omega \gg 2k_{\mathrm{B}}T$ the hyperbolic cotangent is close to unity and the characteristic relaxation times is proportional to $(\mu_0 H_{||})^3$:

$$\frac{1}{\tau} \propto \alpha(\mu_0 H_{||})^3 \tag{5.6.10}$$

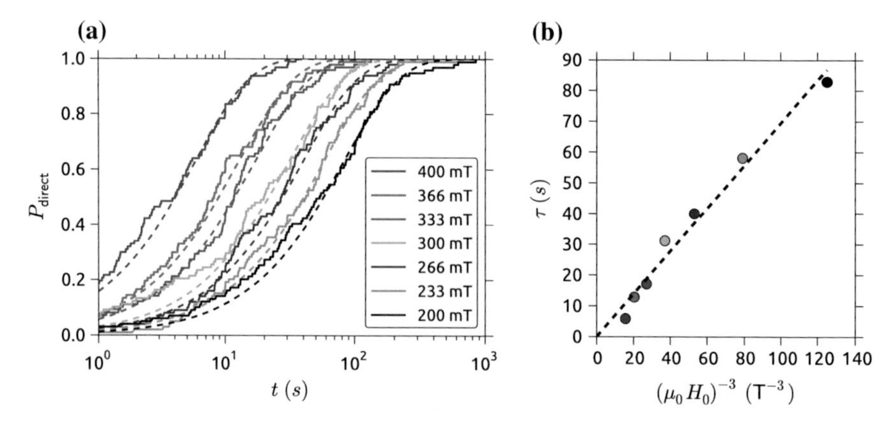

Fig. 5.7 **a** The solid lines represent the relaxation probability P_{direct} as a function of the waiting time t at different magnetic fields H_0, which are indicated in the legend. The dashed lines are a fit to $P_{\text{direct}} = 1 - exp(t/\tau)$. **b** Characteristic relaxation time τ extracted from the fits of (a) as a function of H_0

In order to verify if this model is correct within the limit of an isolated molecule, we performed the following experiment. We prepared the spin in its ground state by applying a large negative magnetic field of $\mu_0 H_{||} = -600$ mT. Afterward, we initialized the spin in its excited state by sweeping the magnetic field at 50 mT/s to $+\mu_0 H_0$, which was ranging from 200 to 400 mT. If a magnetization reversal occurred before reaching $+\mu_0 H_0$, the initialization was repeated. If, however, the spin was properly initialized in its excited stated, we recorded the time necessary to relax back into its ground state. We repeated this procedure 100 times at each H_0 and plotted the waiting times in a normalized histogram. Integrating the latter led to the extraction of the relaxation probability P_{direct} as a function of the waiting time t (see Fig. 5.7a). Subsequently each curve was fitted to the function $P_{\text{direct}} = 1 - exp(t/\tau)$ in order to obtain the characteristic relaxation time τ at each H_0. By plotting every τ as a function of $(\mu_0 H_0)^{-3}$, a straight line can be fit to the data.
This experiment is another evidence that the observed conductance jumps were indeed due to the relaxation of the electronic spin. Furthermore, the single electronic-spin quantum-system is coupled to the ligand field, which makes it behave as a classical two level system.

5.7 Quantum Tunneling of Magnetization

In the previous section, the relaxation of the magnetic moment of a single $TbPc_2$ due to a direct transition was discussed. However, the quantum nature of single molecule magnets allows for a second type of spin reversal, which is called quantum tunneling of magnetization (QTM). It was first discovered by Friedman and Thomas in 1996

[10, 11] as they measured the hysteresis loop of a Mn_{12} SMM. Henceforward, it has been extensively studied by different groups on clusters or arrays of single molecule magnets [12, 13]. Nevertheless, measuring the phenomenon on an single molecule level is quite exclusive and was first presented in 2013 using a $TbPc_2$ spin valve coupled to a carbon nanotube [14].
Before explaining the experiment, we want to recall the Zeeman diagram of the $TbPc_2$ electronic ground state doublet (see Fig. 5.8a). It shows that each electronic state was split into four levels due to the hyperfine coupling. All lines with the same slope correspond to the same electronic spin state and all lines with the same color correspond to the same nuclear qubit state. Our main focus is directed on the avoided level crossings, highlighted by colored rectangles. They were induced due to off-diagonal terms in the ligand field Hamiltonian and mix the electronic spin-up $|\uparrow\rangle$ and spin-down $|\downarrow\rangle$ state (see Sect. 3.6). However, the horizontal separation of the anticrossings is determined by the hyperfine coupling between the terbium's electronic and nuclear spin.
By applying an external magnetic field parallel to the easy axis of the molecule, we move along the lines of the Zeeman diagram. Every time we pass by one of those anticrossings, the molecule's electronic spin is able to reverse due to a process which is referred to as the quantum tunneling of magnetization (QTM). The probability of the reversal P_{LZ} is given by the Landau-Zener (LZ) formula [15, 16]:

$$P_{\mathrm{LZ}} = 1 - exp\left[-\frac{\pi\Delta^2}{2\hbar g_{\mathrm{J}}\Delta m_{\mathrm{J}}\mu_0 dH_{||}/dt}\right]. \tag{5.7.1}$$

Since this process is only allowed in the close vicinity of the anticrossing, the electronic spin can tunnel only at four distinct magnetic fields (see Fig. 5.8). Therefore, the detection of the four QTM transitions would be the final evidence that the magnetic object, which is coupled to the quantum dot, is without a doubt a single $TbPc_2$ SMM.
In order to find experimental evidence of this process, we biased the spin-transistor at $V_{\mathrm{ds}} = 0$ and set V_{g} to a value in the vicinity of the charge degeneracy point. Afterward, the external magnetic field was swept from -60 to 60 mT and back while measuring the conductance through the quantum dot. The recorded magneto-conductance signal of four selected sweeps is depicted in Fig. 5.8b. It shows conductance jumps at four different magnetic fields with an amplitude of 3 % of the total conductance. The change from one conductance value to another originated from the electronic spin reversal. To demonstrate that these reversals were caused by a QTM transition, we recorded the magneto-conductance signal for several thousand sweeps. For each electronic spin reversal, we determined the magnetic field of the resulting conductance jump. By plotting the positions of all detected jumps in a histogram we obtained Fig. 5.8c and d for samples A and C respectively. More details on the data analysis are given in Sect. 6.1. We observed four nonoverlapping peaks, whose maxima coincide with the magnetic field of the four anticrossings, which is a direct evidence that the magnetic object coupled to the read-out quantum dot is a single terbium double-decker SMM. Moreover, the experiment establishes the electronic detection of the

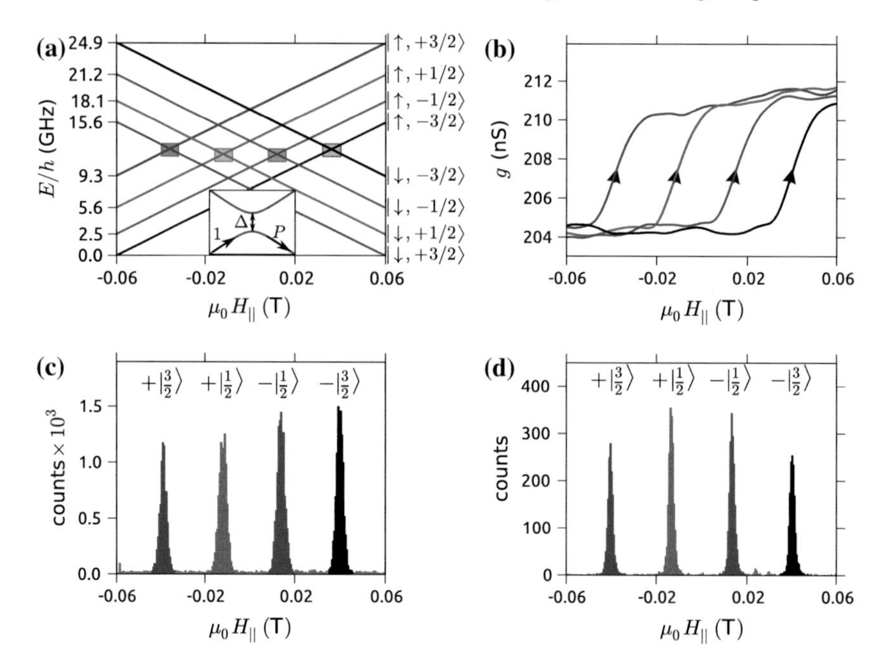

Fig. 5.8 **a** Zeeman diagram of the $TbPc_2$ molecular magnet, focusing on the isolated electronic spin ground state doublet $m_J = \pm 6$, as a function of the external magnetic field $H_{||}$ parallel to the easy-axis of magnetization. Both electronic spin states $|\uparrow\rangle$ and $|\downarrow\rangle$ are split into four energy levels due to a strong hyperfine interaction with the Tb nuclear spin. The ligand field induces off-diagonal terms in the spin Hamiltonian leading to avoided level crossings (colored rectangles and inset), where quantum tunneling of magnetization (QTM) is allowed. Note that for each QTM event the nuclear spin is preserved. Therefore, the positions in magnetic field $H_{||}$, where the electronic spin reversal happens, yields the nuclear spin states $|-3/2\rangle$, $|-1/2\rangle$, $|+1/2\rangle$ or $|+3/2\rangle$. **b** Magneto-conductance measurement of the read-out quantum dot. The electronic spin reversal results in a conductance jump of about 3 % of the signal. **c** Histogram of all recorded conductance jumps measures on sample C. Is shows four nonoverlapping peaks originating from the QTM transitions at the avoided level crossings. They are used as a fingerprint to identify the $TbPc_2$ single molecule magnet and establish the read-out of a single nuclear spin since they link the magnetic field of the conductance jump to each nuclear qubit state. **d** Histogram similar to (**c**) measured on sample A

nuclear spin qubit since the position of each conductance jump becomes nuclear spin dependent.

In the following we present the tunnel probability P_{QTM} as a function of the sweep rate $dH_{||}/dt$ using sample A and C. Focusing on the QTM probability averaged by the four anticrossings, we swept the magnetic field back and forth from −60 to +60 mT. Moreover, by limiting the magnetic field amplitude to 60 mT we could also suppress direct transitions whose characteristic time was extrapolated to 53 min at this field using Fig. 5.7b. For each measurement, we checked for a conductance jump indicating the QTM of the spin. By repeating this protocol 100 and 1000 times for each sweep rate and counting the amount of the detected QTM transitions, we were

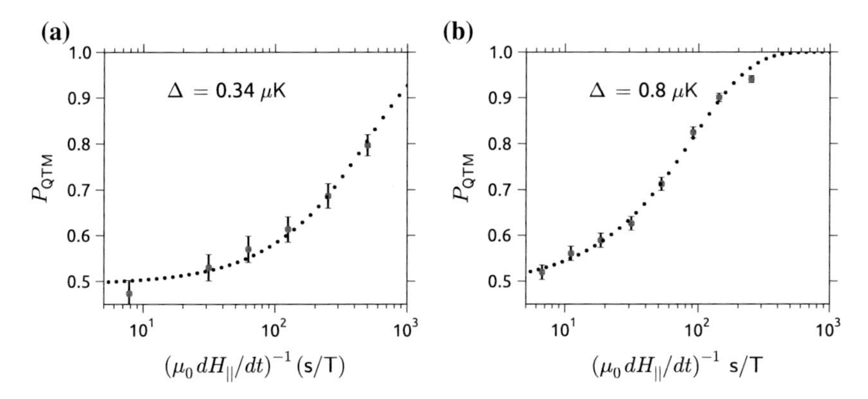

Fig. 5.9 Probability of observing a quantum tunneling of magnetization P_{QTM} of a single spin as a function of the magnetic field sweep rate $\mu_0 dH/dt$ for sample A (**a**) and C (**b**). The experimental results (*red dots*) were fitted to the function $P_{\mathrm{QTM}} = 1 - A\,exp(B/dH_{||}/dt)$

able to extract the tunnel probability P_{QTM} as function of $dH_{||}/dt$ for samples A and C, respectively (see Fig. 5.9).

The results show an exponential increase of the tunnel probability with decreasing sweep rate. Fitting the data to the function $P = 1 - A\,exp(B/dH_{||}/dt)$ enabled us to extract a tunnel splitting of $\Delta = 0.34\ \mu\mathrm{K}$ for sample A and $\Delta = 0.8\ \mu\mathrm{K}$ for sample C. Both values are close to the value of 1 μK determined by Ishikawa et al. [2]. However, there is a striking deviation from Eq. 5.7.1, the tunnel probability P_{QTM} appears to converge to 50 % at high sweep rates for both samples. This implies that there must be a second process, different from the QTM, causing a reversal.

In order to learn more about the additional transition, we determined the correlation between subsequent measurements. Since the tunnel process is a random event, its correlation will vanish leaving only the additional transition for the analysis. To calculate the autocorrelation function C_n we applied the following algorithm. A spin reversal in measurement i was saved as $x_i = 1$ and no spin reversal resulted in $x_i = -1$. Subsequently the autocorrelation function was determined as

$$C_n = \frac{\sum_{i=0}^{N-n}(x_i - \bar{x})(x_{i+n} - \overline{x})}{\sqrt{\sum_{i=0}^{N-n}(x_i - \bar{x})^2}\sqrt{\sum_{i=n}^{N}(x_i - \bar{x})^2}} \tag{5.7.2}$$

with N being the total number of measurements and $\overline{x}$ mean value. Figure 5.10 shows the result of this calculation up to $n = 1000$. In order to be truly random, the correlation function must be below the $2/\sqrt{N}$ limit (red-dotted line), which corresponds to the 95 % fidelity of a random event. Since this is true, apart from some exceptions, we concluded that the additional reversal process is a random event as well and has no magnetic origin.

In the following we studied the number of transitions as a function of the source-drain offset voltage in order to analyze if additional spin reversals might be activated by the tunnel current. Therefore, we swept the magnetic field back and forth between −60

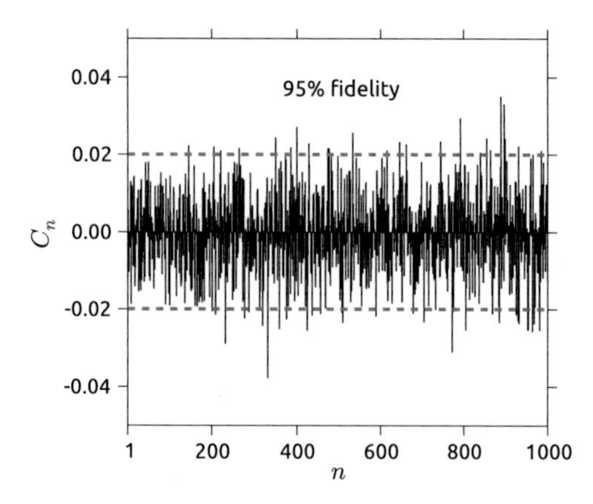

Fig. 5.10 The autocorrelation function C_n between QTM events, which were separated by n measurement (*black line*) and the 95 % fidelity threshold indicating the randomness of the autocorrelation function (*red dotted line*). Since C_n is most of the time below the threshold we could reason that QTM transitions happened at random, as expected from a quantum tunneling process

to 60 mT at 50 mT/s while gradually increasing the source-drain voltage. Every spin reversal recorded during this measurement is marked as a black point in Fig. 5.11a. It illustrates that the four peaks, corresponding to the four different nuclear spin states, were broadened with increasing bias voltage, and the appearance of additional noise in between the peaks was observed. Increasing the offset above 350 μV led to total loss of the signal. Dividing Fig. 5.11a into ten intervals of 40 μV (corresponding to 800 measurements), and integrating the number of reversals in each interval, yielded Fig. 5.11b. It shows a continuous increase of spin reversals with augmenting bias, which demonstrates the activation of the spin reversal due to the tunnel current. We suppose that the mechanism is similar to the one presented by Heinrich et. al [17], where tunnel electrons having an energy larger than the Zeeman splitting of a manganese atom are able to flip its spin. The difference in our case, is that the 4f electrons of the Tb are not directly exposed to the tunnel current as it is the case of the 3d electrons in the manganese. Therefore, we belief that the effect is less pronounced and thus less efficient.

To find out more about the activation probability, we subtracted a histogram of 1000 measurements at 300 μV offset from a histogram acquired at zero offset (see Fig. 5.11c). The result still exhibits four peaks, albeit broadened, which suggests that the activated spin reversal had a higher probability in the vicinity of the four avoided level crossings and therefore at smaller energy gaps.

Using this results, we can explain the convergence of the QTM probability to 50 %. In order to obtain a decent signal to noise ratio, lock-in amplitudes of 250 μV were necessary. From Fig. 5.11a we see that those amplitudes are already sufficient to activate the spin reversal around the avoided level crossing. This, in turn, would lead to additional transitions at each avoided level crossing, which resemble QTM events. A definite answer, however, requires theoretical modeling and is left as a future project.

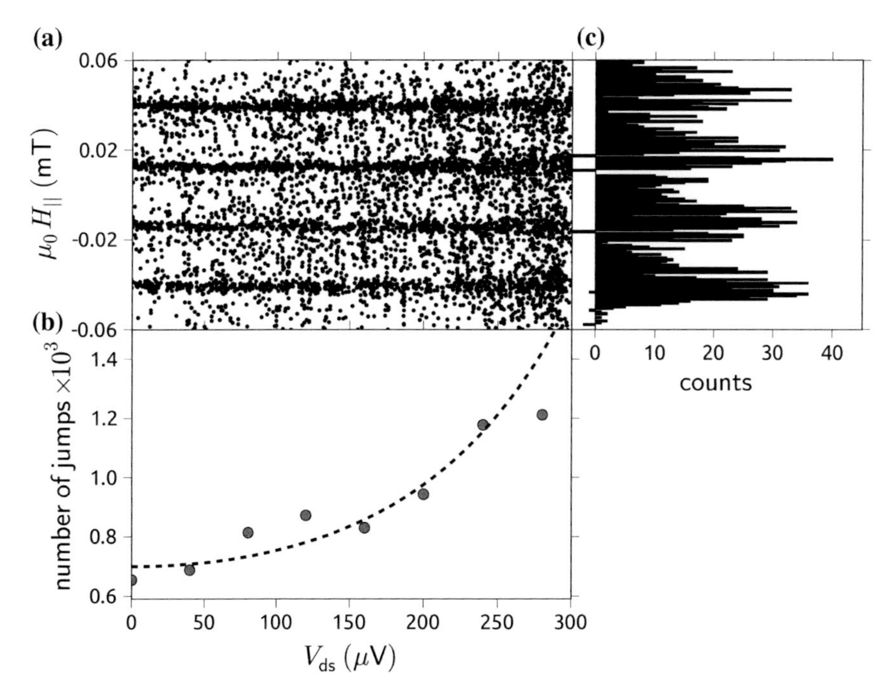

Fig. 5.11 **a** Electronic spin reversal as a function of the magnetic field $H_{||}$ and the offset bias voltage V_{ds}. **b** Integrated number of spin reversals of (**a**) for intervals of 800 sweeps. The steady increase of reversals with augmenting V_{ds} demonstrated the an activation of the reversal due to the tunnel current and is assumed to be the reason of the offset of the QTM probability. **c** Difference between a histogram of 1000 measurements taken at 300 μV offset and zero offset. The perceptibility of the four peaks shows that the activation probability is inverse proportional to the level splitting between up and down and therefore larges in the vicinity of the anticrossing

5.8 Summary

In this chapter we were able to show that an single $TbPc_2$ molecular magnet was trapped in between two gold contacts, allowing for the electronic read-out of the molecule's spin via a quantum dot. A close investigation of the coupling between the spin and the read-out quantum dot suggested that the latter was created by the organic ligands of the molecule. We presented a schematic model, which was able to describe the mode of operation of a single-molecule magnet spin-transistor, i.e., the read-out of the electronic spin. Furthermore, we presented a study of the electronic spin relaxation at large magnetic fields ($B > 200$ mT) and could extract a field dependent relaxation time $\tau(B_{||}) = 0.7(\mu_0 H_{||})^{-3}$ T^3s. In the end, we investigated the quantum tunneling of magnetization of the electronic spin. From the experiments we extracted a tunnel splitting of 0.34 and 0.8 μK for samples A and C, which was in the same order of magnitude as the theoretical values given by Ishikawa et al. [2]. Moreover, we were able to identify the four individual QTM transitions, which is the strongest evidence that nano-object under investigation was as single $TbPc_2$ SMM. In the next chapter we will use those transitions to perform a time-resolved read-out of the nuclear qubit state.

References

1. R. Vincent, S. Klyatskaya, M. Ruben, W. Wernsdorfer, F. Balestro, Electronic read-out of a single nuclear spin using a molecular spin transistor. Nature **488**, 357–360 (2012)
2. N. Ishikawa, M. Sugita, W. Wernsdorfer, Quantum tunneling of magnetization in lanthanide single-molecule magnets: bis (phthalocyaninato) terbium and bis (phthalocyaninato) dysprosium anions. Angew. Chem. (International ed. in English) **44**, 2931–2935 (2005)
3. L. Vitali, S. Fabris, A.M. Conte, S. Brink, M. Ruben, S. Baroni, K. Kern, Electronic structure of surface-supported bis (phthalocyaninato) terbium (III) single molecular magnets. Nano Lett. **8**, 3364–3368 (2008)
4. N. Roch, S. Florens, T.A. Costi, W. Wernsdorfer, F. Balestro, Observation of the underscreened kondo effect in a molecular transistor. Phys. Rev. Lett. **103**, 197202 (2009)
5. N. Roch, R. Vincent, F. Elste, W. Harneit, W. Wernsdorfer, C. Timm, F. Balestro, Cotunneling through a magnetic single-molecule transistor based on N@C_{60}. Phys. Rev. B **83**, 081407 (2011)
6. A. Lodi Rizzini, C. Krull, T. Balashov, J.J. Kavich, A. Mugarza, P.S. Miedema, P.K. Thakur, V. Sessi, S. Klyatskaya, M. Ruben, S. Stepanow, P. Gambardella, Coupling single molecule magnets to ferromagnetic substrates. Phys. Rev. Lett. **107**, 177205 (2011)
7. P. Zhu, F. Lu, N. Pan, D.P. Arnold, S. Zhang, J. Jiang, Comparative electrochemical study of unsubstituted and substituted bis (phthalocyaninato) rare earth (iii) complexes. Eur. J. Inorg. Chem. **2004**, 510–517 (2004)
8. A. Abragam, B. Bleaney, Electron Paramagnetic Resonance of Transition Ions (Oxford Classic Texts in the Physical Sciences) (Oxford University Press, USA, 2012). ISBN:0199651523
9. R. Orbach, Spin-lattice relaxation in rare-earth salts. Proc. R. Soc. A: Math. Phys. Eng. Sci. **264**, 458–484 (1961)
10. J.R. Friedman, M.P. Sarachik, R. Ziolo, Macroscopic measurement of resonant magnetization tunneling in high-spin molecules. Phys. Rev. Lett. **76**, 3830–3833 (1996)
11. L. Thomas, F. Lionti, R. Ballou, D. Gatteschi, R. Sessoli, B. Barbara, Macroscopic quantum tunnelling of magnetization in a single crystal of nanomagnets. Nature **383**, 145–147 (1996)
12. W. Wernsdorfer, N. Chakov, G. Christou, Determination of the magnetic anisotropy axes of single-molecule magnets. Phys. Rev. B **70**, 1–4 (2004)
13. M. Mannini, F. Pineider, P. Sainctavit, C. Danieli, E. Otero, C. Sciancalepore, A.M. Talarico, M.-A. Arrio, A. Cornia, D. Gatteschi, R. Sessoli, Magnetic memory of a single-molecule quantum magnet wired to a gold surface. Nat. Mater. **8**, 194–197 (2009)
14. M. Urdampilleta, S. Klyatskaya, M. Ruben, W. Wernsdorfer, Landau-Zener tunneling of a single Tb^{3+} magnetic moment allowing the electronic read-out of a nuclear spin. Phys. Rev. B **87**, 195412 (2013)
15. L. Landau, Zur Theorie der Energieubertragung II. Phys. Sov. Union **2**, 46–51 (1932)
16. C. Zener, Non-adiabatic crossing of energy levels. Proc. R. Soc. A: Math. Phys. Eng. Sci. **137**, 696–702 (1932)
17. A.J. Heinrich, J.A. Gupta, C.P. Lutz, D.M. Eigler, Single-atom spin-flip spectroscopy. Science (New York, N.Y.) **306**, 466–469 (2004)

Chapter 6
Nuclear Spin Dynamics—T_1

The detection and manipulation of nuclear spins has become an important multidisciplinary tool in science, reaching from analytic chemistry, molecular biology, to medical imaging and are some of the reasons for a steady drive towards new nuclear spin based technologies. In this context, recent breakthroughs in addressing isolated nuclear spins opened up a new path towards nuclear spin based quantum information processing [1–4]. Indeed, the tiny magnetic moment of a nuclear spin is well protected from the environment, which makes it an interesting candidate for storage of quantum information [5, 6]. On this account, we are going to investigate an isolated nuclear spin using a single-molecule magnet spin-transistor in regard to its read-out fidelity and lifetime, which are important figures of merits for quantum information storage and retrieval.

6.1 Signal Analysis

The experimental results in this chapter were obtained via electrical transport measurements through a three terminal single-molecule magnet spin-transistor and by using two different samples to demonstrate the reproducibility of the data. The spin-transistors were placed into a dilution refrigerator with a base temperature of 150 mK for sample A and 40 mK for sample C. Each device was surrounded by a home-made three-dimensional vector magnet and biased at $V_{\mathrm{ds}} = 0$ V. The gate voltage V_{g} was adjusted in order to shift the chemical potential of the read-out dot slightly above or below the source-drain Fermi level, resulting in the highest sensitivity of the device. From Sect. 5.1, we know that the conductance of the read-out dot at given V_{ds} and V_{g} depends on the direction of the electronic spin. By sweeping the external magnetic field parallel to the easy axis of the $TbPc_2$, we induced reversals of the terbium's electronic spin at the four avoided level crossings due to a quantum tunneling of magnetization (QTM). These reversal result in jumps of the read-out dot's conduc-

S. Thiele, *Read-Out and Coherent Manipulation of an Isolated Nuclear Spin*,
Springer Theses, DOI 10.1007/978-3-319-24058-9_6

tance. Since the magnetic field of the QTM transition is nuclear spin dependent, each conductance jump can be assigned to the nuclear spin qubit state. In the following, we will describe the data treatment in order to automatize the read-out process and explain in detail how we measured the lifetime T_1 of a single nuclear spin.

Figure 6.1a displays the raw data of five different measurements, including four sweeps where the electronic spin reversed due to a QTM transition and one sweep without a reversal. The conductance jump was evoked by a shift of the read-out dot's chemical potential due to the exchange coupling to the terbium's electronic spin (see Chap. 5).

In order to read-out the nuclear spin state, we had to analyze if, and where a conductance jump occurred during the magnetic field sweep. Therefore, the raw data were passed through a filter, which computed the first derivative with an adjustable smoothing over N data points. The output of the filtered signals from Fig. 6.1a are displayed

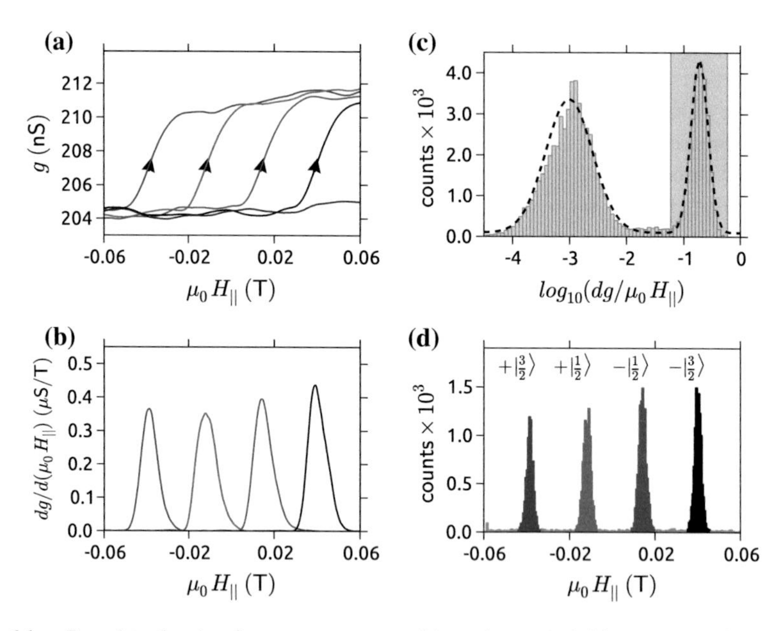

Fig. 6.1 **a** Raw data showing four measurements with a spin reversal (*blue*, *green*, *red*, and *black curve*) and one measurement without a spin reversal (*purple curve*). The conductance jump was induced by a shift of the read-out dot's chemical potential due to the exchange coupling to the electronic spin. **b** Filtered signal of (**a**), which is similar to a smoothed first derivative. Data including a reversal are transformed into peaks whose maxima indicate the respective jump position, whereas sweeps without a reversal are strongly suppressed. **c** Histogram of the maximum amplitudes of all filtered sweeps. Measurements without a spin reversal (*left peak*) can be separated from measurements containing a reversal (*right peak*) by a threshold (*yellow rectangle*). **d** Histogram of the jump positions of 75,000 measurement whose filtered maxima were within the *yellow rectangle* of (**c**). The four peaks originate from conductance jumps in the vicinity of the four anticrossings and allow for the unambiguous attribution of each detected conductance jump to a nuclear spin qubit state. The plot was generated using sample C, notice that sample A shows identical characteristics (compare Fig. 6.9a)

in Fig. 6.1b. The signal, which did not show a jump, is strongly suppressed by the filter. However, the sweeps, which contained a conductance jump, are transformed into peaks, whose maxima indicated at which magnetic field the jumps occurred.
To obtain a good statistical average, we measured the conductance signal of 75,000 magnetic field sweeps. Plotting the maximum amplitudes of all filtered data in a histogram gave rise to Fig. 6.1c. It shows that the jump amplitudes are divided into two distinct peaks, separated by more than two orders of magnitude. The left peak, corresponding to small amplitudes, originates from all measurements without a reversal; whereas the right peak, corresponding to large amplitudes, finds its origin in sweeps including a spin reversal. To sort out the measurements with spin reversals from the rest of the data we defined a threshold indicated by the yellow rectangle. If the maximum amplitude of the filtered signal lied within this rectangle, the sweep was considered to contain a QTM transition and the position of the jump was stored in an array. Subtracting the inductive field delay of the coils from the jump positions and plotting them into a histogram results in Fig. 6.1d. It shows that the conductance jumps happened almost exclusively in the vicinity of the four avoided level crossings, corresponding to the four nuclear spin states. Hence, we can unambiguously assign a nuclear qubit state to each detected jump. The width of the four peaks is determined by the lock-in time constant and the electronic noise of the setup, which leads to a broadening much larger than the intrinsic linewidth. The error induced by our nuclear spin read-out procedure is mainly due to inelastic electronic spin reversals (grey data point in Fig. 6.1), which were misinterpreted as QTM events and is estimated to be less than 5 % for sample A and less than 4 % for sample C.

6.2 Relaxation Time T_1 and Read-Out Fidelity F

After being able to read out the state of an isolated nuclear spin qubit, we are going now one step further by recording the real-time trajectory of an isolated nuclear spin. Using sample A, we present measurements, obtained by sweeping the magnetic field up and down between ±60 mT at 48 mT/s (2.5 s per sweep), while recording the conductance through the read-out quantum dot (see Fig. 6.2a). As explained in Sect. 6.1, we can assign each conductance jump to a certain nuclear spin qubit state and, due to the fixed frequency of the magnetic field ramp, to a certain time (see Fig. 6.2b).
By sweeping the magnetic field faster than the relaxation time, we obtained a real-time image of the nuclear spin trajectory. The first 2000s of this trajectory are shown in Fig. 6.3. The grey dots illustrate the position of the recorded conductance jumps. If the jump occurred within a window of ±7 mT around the avoided level crossing (indicated by colored bars), it was assigned to the corresponding nuclear spin state. If, however, a jump was recorded outside this window, the measurement was rejected. The black line shows the assigned time evolution of the nuclear spin state.
Figure 6.4a shows a magnified region of the nuclear spin trajectory including 170s of data. In order to access the nuclear spin relaxation time T_1, we performed a bit by

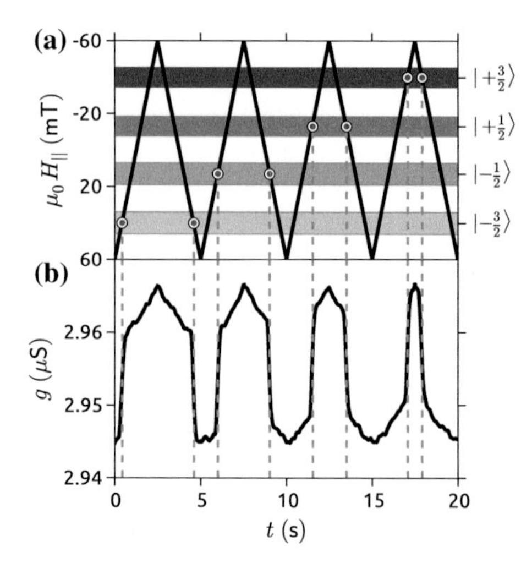

Fig. 6.2 Protocol to measure the nuclear spin trajectory. **a** The magnetic field $B_{||}$ is swept *up* and *down* between ±60 mT at a constant rate of 48 mT/s, corresponding to 2.5 s per sweep. **b** Each detected conductance jump can therefore be assigned to a certain nuclear spin state at a certain time *t*

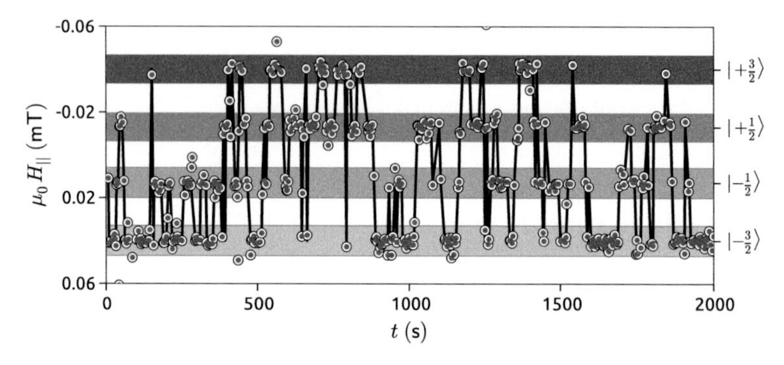

Fig. 6.3 First 2000 s of the nuclear spin trajectory measured using sample A. The *grey dots* illustrate the recorded conductance jump. If the jump was found inside a window of ± 7 mT (*colored stripes*) around one of four peaks of Fig. 6.1d it was assigned to the corresponding nuclear spin state, otherwise the measurement was rejected. Using this data analysis results in the single nuclear spin trajectory, shown as a *black line*

bit post-processing of this data. Therefore, we extracted the different dwell times, i.e. the time the nuclear spin remained in a certain state before going into another state. Plotting these dwell times for each nuclear spin state in separate renormalized histograms yielded the black data points of Fig. 6.4b–e.

A further fitting to an exponential function $y = exp(-t/T_1)$ gave the nuclear spin dependent relaxation times $T_1 \simeq 13$ s for $m_I = \pm 1/2$ and $T_1 \simeq 25$ s for $m_I = \pm 3/2$ for sample A. The perfect exponential decay indicated that no memory effect is present in the system. Furthermore, the obtained lifetimes were an order of magnitude

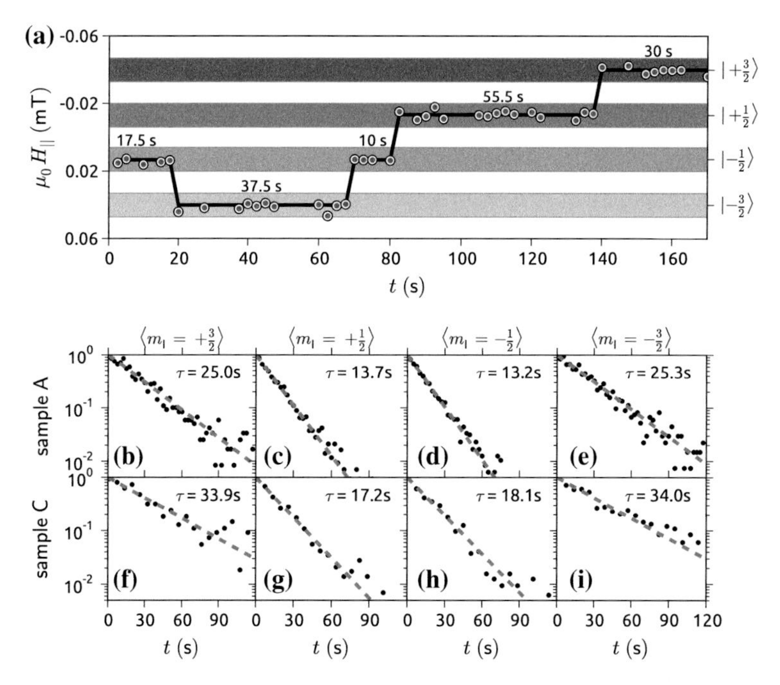

Fig. 6.4 **a** Zoom of the nuclear spin trajectory (*black curve*), which was obtained from the detected conductance jumps (*grey dots*). Every time the nuclear spin qubit changes over to a new state, we determined the dwell time in this state (*black numbers*). **b–e** Plotting the dwell times for each nuclear spin state in separate histograms led to the *black data points*. A further fitting to the exponential function $y = exp(-t/T_1)$ (*red dotted line*) yielded the relaxations times T_1 for each nuclear spin qubit state of sample A. **f–i** The relaxation times T_1 for sample C are obtained analog to sample A

larger than the measurement interval, which denotes that the same quantum state could be measured multiple times without being destroyed by the measurement process. Such a detection scheme is referred to as a quantum nondemolition (QND) read-out. Instead of demolishing the quantum system, it will only project the system onto one of its eigenstates [7]. Notice that superposition states will be destroyed by this projection.

Usually the Hamiltonian of the entire system can be written as $H = H_0 + H_\mathrm{M} + H_\mathrm{I}$, with H_0 being the Hamiltonian of the quantum system under study, H_M the Hamiltonian of measurement system and H_I the interaction Hamiltonian between the two systems. In order to perform a real QND measurement, it has been shown that the commutator between the measured variable q and the interaction Hamiltonian must be zero: $[q, H_\mathrm{I}] = 0$ [7, 8]. In our experiment the measurement variable is I_z and the interaction is described by the hyperfine Hamiltonian $H_\mathrm{hf} = A\boldsymbol{I}\boldsymbol{J}$. The latter possesses terms of $A/2(I_+J_- + I_-J_+)$, which do not commute with I_z. This can be seen as a deviation of the ideal QND measurement. However, the Hamiltonian $A/2(I_+J_- + I_-J_+)$, accounting for flip-flop processes of the nuclear and the electronic spin represents only a weak perturbation, as it would cause additional tunnel events at all crossings in Fig. 5.8a, not marked by colored rectangles. Since

Fig. 5.8c shows only four peaks, it demonstrates that the perturbation is negligible and the deviation from an ideal QND measurement must be small. An important point to notice is that performing a QND measurement is equivalent to initialize the nuclear spin in the measured state.
The read-out fidelities F are obtained by calculating the probability to stay in a certain nuclear spin qubit state during the time necessary to measure it. Due to the QTM probability of 51.5 %, two subsequent measurements were separated by ≈5 s in average resulting in fidelities of $F(m_I = \pm 3/2) \approx exp(-5/25.2\text{ s}) \approx 82\,\%$ and $F(m_I = \pm 1/2) \approx exp(-5/13.2\text{ s}) \approx 69\,\%$ for sample A. By repeating this measurement on sample C (see Fig. 6.4f–i), we obtained values of $T_1 \approx 17$ s for $m_I = \pm 1/2$ and $T_1 \approx 34$ s for $m_I = \pm 3/2$, which are comparable to sample A and shows the high reproducibility of the experiment and the excellent isolation of the nuclear spin in molecular spin-transistor devices, which is promising for future device architectures.
Due to a new vector magnet (see Sect. 4.3), which was designed for larger sweep rates (>200 mT/s), the measurement interval for the experiments with sample C could be reduced to 1.2 s. Given the rather identical QTM probability of 52 % for sample C, two subsequent measurements are separated by 2.31 s in average, leading to fidelities of $F(m_I = \pm 3/2) \approx exp(-2.31/34\text{ s}) \approx 93\,\%$ and $F(m_I = \pm 1/2) \approx exp(-2.3/17\text{ s}) \approx 87\,\%$. These values are comparable to fidelities given by Robledo et al. [9] who measured a single nuclear spin of a nitrogen vacancy center.
The limitation of our read-out fidelity comes from the currently rather slow detection rate of 0.5 measurements per second with respect to experiments on other nuclear spin qubits, which make use of the much faster electron spin resonance (ESR). Since the magnetic field cannot be stabilized at one of the anticrossing in the Zeeman diagram, $|+6\rangle$ and $|-6\rangle$ remains the only available basis and therefore flipping the electronic spin of the $TbPc_2$ involves a $\Delta m_J = 12$. This makes the ESR process highly improbable for this system. Nevertheless, we tried to flip the electronic spin sending microwaves with the transition frequency $m_I = -1/2 \longleftrightarrow m_I = 1/2$, while sweeping the magnetic field around 0 mT. However, the expected additional transition at $B = 0$ T was not observed so far. Another possibly to perform the ESR measurements is to use the transition $m_J = 6 \rightarrow m_J = 5$. Unfortunately, the transition frequency of around 12 THz is hard to access, and our coaxial cables are not suited to guide such high frequencies.
A mid term solution would be to design special vector magnets consisting of two types of coils: larger vector magnets similar to the ones presented in Chap. 4, to generate the static magnetic field and very small coils, used to generate high frequency magnetic field ramps. In this way the detection rate could be speed up by a factor of 10–100. The long term approach, however, is to find SMMs having a strong hyperfine coupling and allowing for electronic spin transitions $\Delta m_J = \pm 1$. In an easily accessible frequency range [2 GHz → 10 GHz], these SMMs would be perfect candidates for ESR detection implementation. The resulting speed up in measurement time by 2 orders of magnitude could lead to fidelities close to 1.

6.3 Quantum Monte Carlo Simulations

In order to perform a more quantitative analysis of the nuclear spin lifetime and the involved relaxation process we wanted to make use of computational techniques. However, to do a proper quantum mechanical simulation we needed to include the coupling of the nuclear spin to a thermal bath, which requires methods that go beyond the usual solution of the Schrödinger equation. There are currently two widely used approaches to simulate such quantum trajectories. In the usual approach the master equation is written for a reduced density matrix ρ_A [10]. It computes the ensemble average of the time evolution of ρ_A. An equivalent approach is the so-called Monte Carlo wavefunction method [11–13], which calculates the stochastic evolution of the atomic wavefunction using a quantum Monte Carlo (QMC) algorithm. It can be shown that the ensemble average of the master equation is analogue to the time average of the QMC technique. However, the latter could be adapted more easily to our experimental conditions and was therefore our method of choice. The following algorithm was developed in cooperation with Markus Holzmann from the LPMCC in Grenoble.

6.3.1 *Algorithm*

In the following we are briefly discussing the Monte Carlo wavefunction algorithm. Notice that the complete QMC code is shown in appendix C.
Suppose the wave function of the isolated system $|\Psi\rangle$ is entirely described by the Hamiltonian H_0, and all the influence of the environment on the time evolution of the system can be described in terms of a non-Hermitian operator H_1:

$$H_1 = -\frac{i\hbar}{2}\sum_m C_m^\dagger C_m \tag{6.3.1}$$

where $C_m (C_m^\dagger)$ is an arbitrary relaxation (excitation) operator. In the following, we assume that the environment can be modeled as a bosonic bath. Furthermore, we allow only transitions of the nuclear spin, which obey $|\Delta m| = 1$, as expected from the nuclear spin transition. Thus, we get only two contributions in the Hamiltonian H_1, namely:

$$C_1^{i,j} = \sqrt{\Gamma_{i,j}(1 + n(\omega_{i,j}, T))}\, \delta_{i,j+1} \tag{6.3.2}$$

which accounts for relaxations between the state i and j, and

$$C_2^{i,j} = \sqrt{\Gamma_{i,j}(n(\omega_{i,j}, T))}\, \delta_{i+1,j} \tag{6.3.3}$$

which accounts for excitations between the state i and j in terms of their energy differences $\omega_{i,j}$ and relaxation rates $\Gamma_{i,j}$. Notice, both are symmetric in i, j and

$\omega_{i,j} = |\omega_i - \omega_j|$. Both, C_1 and C_2 have the dimension $1/\sqrt{\text{time}}$. The function $n(\Delta\omega, T) = \left(1 + exp(\frac{\hbar\Delta\omega}{k_B T})\right)^{-1}$ is the Bose-Einstein distribution, which takes the density of the bosonic bath into account; and ($\Gamma_{0,1}$, $\Gamma_{1,2}$, and $\Gamma_{2,3}$) are the state dependent transition rates, with 0, 1, 2 and 3 being the ground state, first, second, and third excited state. The effective Hamiltonian is the sum of H_0 and H_1

$$H = H_0 - \frac{i\hbar}{2}\sum_{m=1}^{2} C_m^\dagger C_m \tag{6.3.4}$$

Notice that H_1 is non-Hermitian, since its eigenvalues are imaginary. To obtain the nuclear spin trajectory, we have to calculate the time evolution of the wavefunction, which is done in the following three steps.

Step I

In the first step we calculate the wavefunction after a small time step δt. Therefore, we make use of the classical Schrödinger equation.

$$\frac{d\tilde{\Psi}}{\delta t} = -\frac{i}{\hbar}(H_0 + H_1)\Psi$$
$$\tilde{\Psi}(t + \delta t) = exp\left(-\frac{i}{\hbar}H_1\delta t\right) exp\left(-\frac{i}{\hbar}H_0\delta t\right)\Psi(t)$$

Here, we have neglected an error of δt^2, in which case H_0 and H_1 are not commuting. Furthermore, we chose δt in a way that $\left|\frac{i}{\hbar}H_1\delta t\right| \ll 1$. Thus, the term $exp(-\frac{i}{\hbar}H_1\delta t)$ can be written in a first order Taylor series expansion $exp(-\frac{i}{\hbar}H_1\delta t) \approx 1 - \frac{i}{\hbar}H_1\delta t$. Since we are only interested in the amplitude of the wavefunction, the term $exp\left(-\frac{i}{\hbar}H_0\delta t\right)$ will be neglected in the following. It adds only a phase term to the wavefunction and can be reintroduced at any point in the calculation if necessary. Hence, the amplitude of the wavefunction after a time step δt is:

$$\tilde{\Psi}(t + \delta t) = \left(1 - \frac{i}{\hbar}H_1\delta t\right)\Psi(t) \tag{6.3.5}$$

Step II

In the second step we calculate the transition probability from one state to another. As mentioned before the Hamiltonian H_1 is non-Hermetian and therefore the wavefunction is not normalized. Up to an error of δt^2 we can write:

$$\langle\tilde{\Psi}(t+\delta t)|\tilde{\Psi}(t+\delta t)\rangle = \langle\Psi(t)|1 - \frac{i}{\hbar}\delta t(H_1 + H^{\dagger}) + O(\delta t^2)|\Psi(t)\rangle$$
$$= 1 - \delta p \tag{6.3.6}$$

with

$$\delta p = \frac{i}{\hbar}\delta t\ \langle\Psi(t)|\left(H_1 - H_1^{\dagger}\right)|\Psi(t)\rangle \tag{6.3.7}$$

Since Eq. 6.3.6 is only a first order approximation, we have to adjust δt to assure that $\delta p \ll 1$. Moreover the term δp can be written as the sum of the relaxation and excitation probability: $\delta p = \delta p_{\mathrm{rel}} + \delta p_{\mathrm{exc}}$, because we have only allowed those two transitions in our model, where

$$\delta p_{\mathrm{rel}} = \delta t\ \langle\Psi(t)|\left(C_1^{\dagger}C_1\right)|\Psi(t)\rangle$$
$$\delta p_{\mathrm{exc}} = \delta t\ \langle\Psi(t)|\left(C_2^{\dagger}C_2\right)|\Psi(t)\rangle$$

Step III

In the third step we will account for the random evolution of the wavefunction, which will introduce the nonreversibility of a transition. At this point the wavefunction is at a bifurcation point and could evolve in three different directions:

1 the systems stays in the same state and nothing happens,
2 a relaxation in an energetically lower state occurs,
3 the systems is excited in an energetically higher state.

In order to decide which of three events is happening, we draw and uniformly distributed pseudo-random number $\epsilon = [0, 1]$. If $\epsilon > \delta p$, no quantum jump occurs and we will renormalize the wavefunction:

$$\Psi(t+\delta t) = \frac{\tilde{\Psi}(\delta t)}{\sqrt{1-\delta p}} \tag{6.3.8}$$

If however $\epsilon < \delta p$, the system undergoes a quantum jump. If furthermore $\epsilon < \delta p_{\mathrm{rel}}$, we are relaxing the system according to:

$$\Psi(t+\delta t) = \frac{C_1\tilde{\Psi}(t)}{\sqrt{\delta p_{\mathrm{rel}}/\delta t}} \tag{6.3.9}$$

On the other hand if $\epsilon > p_{\mathrm{rel}}$, we excite the system using the following expression:

$$\Psi(t+\delta t) = \frac{C_2\tilde{\Psi}(t)}{\sqrt{\delta p_{\mathrm{exc}}/\delta t}} \tag{6.3.10}$$

The denominator in Eqs. 6.3.9 and 6.3.10 accounts for the normalization of the wave-function.

6.3.2 Including the Experimental Boundaries

In order to simulate the experimentally obtained nuclear spin trajectory of Fig. 6.3 using the algorithm of Sect. 6.3, we had to introduce some slight modifications.
The read-out of the nuclear spin happens due to a QTM event only once per measurement cycle and therefore at finite time steps t_{measure}. Furthermore, sweeping the magnetic field back and forth to measure these QTM events implicates that each nuclear spin qubit state is probed at a different time during the sweep. Moreover, the QTM transition of the electronic spin occurred with a probability of 51.5 %, and, as a consequence, reversed the order of the ground and excited states of the qubit.
To simulate this experimental conditions appropriately, the computation cycles of duration δt were grouped into five time intervals Δt_i as shown in Fig. 6.5, with $\sum_i \Delta t_i = t_{\mathrm{measure}}$ and t_{measure} being the time needed for one magnetic field sweep. The individual Δt_i were chosen in a way, that at the end of each interval, the magnetic field would have been at one of the four anticrossings corresponding to $m_\mathrm{I} = -3/2$, $-1/2$, $1/2$, or $3/2$ respectively. Hence, we checked every Δt_i if the nuclear spin qubit was in the appropriate state to allow for a QTM transition (question marks in Fig. 6.5). If so, we drew a second random number $\epsilon_2 = [0, 1]$ to simulate the probabilistic nature of the transition. An ϵ_2 which was smaller than the QTM probability P_{QTM}, was interpreted as a QTM event. However, an ϵ_2 that was larger than P_{QTM}, resulted in no QTM transition. Moreover, every time the QTM happened, we saved the nuclear spin state and reversed the nuclear qubit ground stated and its excited states, just like in the experiment. Once we finished the simulation of interval 5, corresponding to the end of a field sweep, we computed the time intervals in reversed order (5, 4, 3, 2, 1), which is equivalent to sweeping back the magnetic field to its initial value.

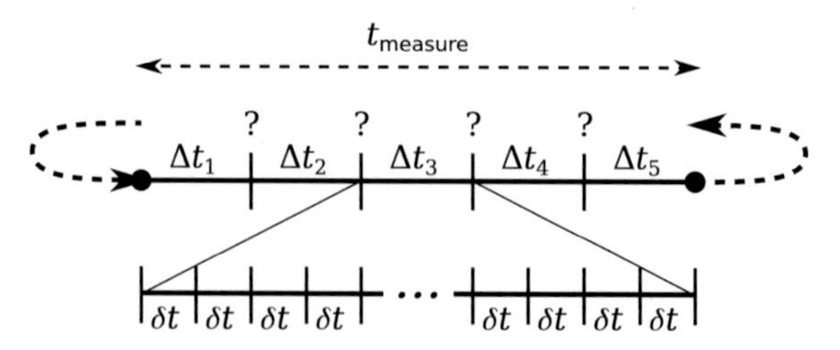

Fig. 6.5 In order to include the experimental boundaries into our simulations, the computation cycles of duration δt were grouped into intervals of Δt_i, where the sum of all Δt_i corresponds to the time needed to sweep the magnetic field during the trace or retrace measurement. At the end of each interval Δt_i, corresponding to a certain magnetic field, we checked if the nuclear spin was in the appropriate state to allow for a QTM transition. If so, the QTM event was accepted with the probability P_{QTM}, leading to the storage of the nuclear spin state and an inversion of the ground state and the excited states

6.4 Comparison Experiment—Simulation

6.4.1 Relaxation Mechanism

In the following, the computational results obtained with the algorithm of Sect. 6.3 are compared with experimental data from sample A, in order to extract further information about the underlying physics of the relaxation process. The parameters used to perform the simulation are listed in Table 6.1. The temperature T, which corresponds to the electron temperature of sample A, the measurement period of 2.5 s, and the QTM probability of 51.5 % were taken as fixed parameters. Only the transition rates Γ_{01}, Γ_{12} and Γ_{23} were varied in order to obtain the best fit to the experimental data shown in Fig. 6.6a–d.
Computing the trajectory for 2^{24} Monte Carlo time steps and following the procedure of Sect. 6.2 to extract the lifetime T_1 gave rise to the data displayed in Fig. 6.6e–f. The first striking feature, which can be extracted from this comparison, is that the difference in T_1 between the $\pm 3/2$ states and the $\pm 1/2$ states is nicely reproduced by simulation. An explanation for this observation can be given by looking at the Hamiltonian H_1 (Eq. 6.3.1), containing the relaxation and excitation operators C_1 and C_2. If the nuclear spin is in the $|\pm 1/2\rangle$ state, the two operators C_1 and C_2 contribute to the relaxation and excitation process. If, however, the nuclear spin is in the $|\pm 3/2\rangle$ state, one of the operators becomes zero, no matter what the electronic

Table 6.1 Input parameters for the quantum Monte Carlo algorithm introduced in Sect. 6.3.

t_{measure}	δt	P_{QTM}	T	Γ_{01}	Γ_{12}	Γ_{23}
2.5 s	$\frac{2.5}{60}$ s	51.5 %	150 mK	1/41 s^{-1}	1/82 s^{-1}	1/90.2 s^{-1}

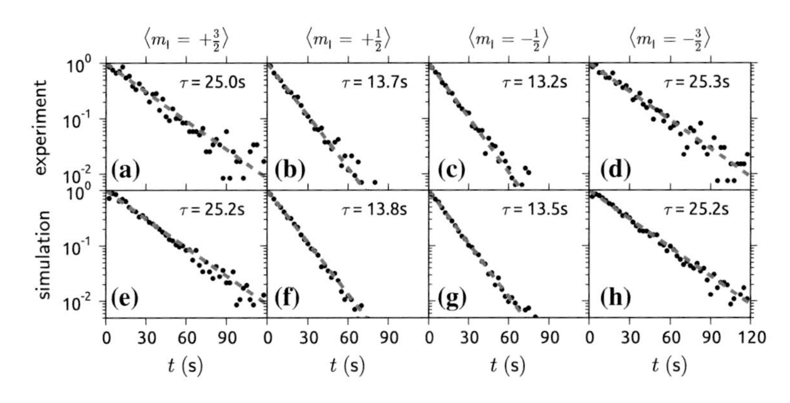

Fig. 6.6 **a–d** Experimental data for sample A taken from Fig. 6.4. **e–f** Computed data points using the parameters of Table 6.1 and the algorithm of Sect. 6.3. The red dotted line in each subplot is a fit to an exponential function $y = exp(-t/T_1)$, yielding the relaxation time T_1 for each nuclear spin state

spin state is, resulting in a smaller transition rate and therefore a larger T_1. A more descriptive explanation can be given by considering the number of transition paths. If the nuclear spin is in the ground or most excited state ($m_I = \pm 3/2$), there is only one way to change its state—excitation **or** relaxation, whereas if the nuclear spin is in an intermediate state ($m_I = \pm 1/2$) it has two escape paths—excitation **and** relaxation. Since the lifetime is roughly inversely proportional to the number of transition paths, if the rates for each part were equal, the T_1's show a difference of approximately two. The exact ratio depends of course on the temperature and the individual transition rates.

In the next step we wanted to reveal the dominant relaxation mechanism, which could be caused by spin-lattice interactions and nuclear spin diffusion. The latter mechanism was found to be very weak in bulk terbium [14] and can, hence, be neglected for rather isolated and nonaligned SMMs. Concerning the spin-lattice relaxation mechanism, we examined closer the $\Gamma_{i,j}$'s derived by fitting the results of QMC simulations to experimental data. Depending on its proportionality to the nuclear level spacing $\omega_{i,j}$ we can distinguish between three types of mechanisms.

1 The Korringa process, in which conduction electrons polarize the inner lying s-electrons. Since these couple with the nuclear spins via contact interaction, an energy exchange over this interaction chain is established, leading to $\Gamma_{i,j} \propto |\langle i|I_x|j\rangle|^2$ [15].
2 The Weger process, which suggests that the spin-lattice relaxation is dominated by the intra-ionic hyperfine interaction and the conduction electron exchange interaction [16]. It is a two-stage process, where the energy of the nucleus is transmitted to the conduction electrons via the creation and annihilation of a Stoner excitation. This process is similar to the Korringa process but results in $\Gamma_{i,j} \propto |\langle i|I_x|j\rangle|^2 \omega_{i,j}^2$.
3 The magneto-elastic process, which leads to a deformation of the molecule due to a nuclear spin relaxation, yields $\rightarrow \Gamma_{i,j} \propto |\langle i|I_x|j\rangle|^2 \omega_{i,j}^4$ [17].

The term $|\langle i|I_x|j\rangle|^2$ arises from the fact, that only rotations of the spin perpendicular to the z-directions are responsible for longitudinal transitions [18]. A comparison between the $\Gamma_{i,j}$'s and the different mechanisms is shown in (Fig. 6.7a). The almost perfect agreement with the Weger process suggests that the dominant relaxation process is caused by the conduction electrons. Since they are exchange coupled to the Tb electronic spin which in turn is hyperfine coupled to the nuclear spin, an energy and momentum exchange via Stoner excitations could be possible.

This implies that by controlling the amount of available conduction electrons per unit time the relaxation rates $\Gamma_{i,j}$ can be changed. Hence, an electrically control of T_1 by means of the bias and gate voltages is possible. To verify this conclusion, we measured the relaxation time T_1 of $m_I = \pm 3/2$ as a function of the tunnel current through the quantum dot. The result in Fig. 6.7b shows a decrease of the lifetime by a factor of three while increasing the current by 100 %. This finding could be interesting to speed up the initialization of the nuclear spin in its ground state prior to a quantum operation by inducing a fast relaxation to the ground state through a series of current pulses.

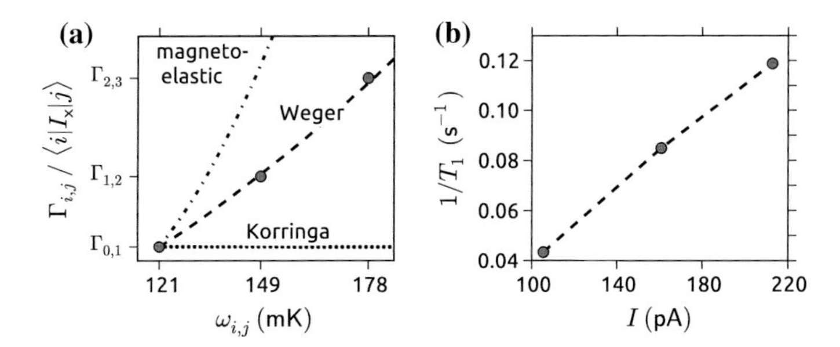

Fig. 6.7 **a** The transition rates $\Gamma_{i,j}$, derived by fitting the results of QMC simulations to experimental data, exhibit a quadratic dependence on the nuclear spin level spacing $\omega_{i,j}$. This behavior is expected from a Weger relaxation process, in which the nuclear spin is coupled via virtual spin waves to conduction electrons. **b** The decrease of lifetime with increasing current is probably due to an increase of electrons tunneling through the read-out dot in addition to an increase of temperature

Another experiment which shows the coupling of the nuclear spin to the electrons was carried out by measuring the nuclear spin temperature as a function of the applied bias voltage. Since we are dealing with a single nuclear spin, the physical quantity temperature has only a meaning if we are speaking of time averages.

The read-out fidelity of the nuclear spin state is rapidly decreasing for increasing bias voltages as shown in Fig. 5.11a. Therefore we developed the protocol shown in Fig. 6.8a in order to measure the nuclear spin at bias voltages beyond 300 μV.

In a first step, the source-drain voltage V_{ds} was rapidly increased from 0 V to values of 1 and 2 mV, at which we waited for 6 s. The tunnel current through the quantum dot at 1 mV was about 1 nA. Extrapolating Fig. 6.7 to 1 nA results in a T_1 of 1.49 s, which was four times smaller than the waiting time and therefore long enough to thermalize the nuclear spin. During this time, the magnetic field was at -60 mT, leading to ground state of $m_I = -3/2$ (see Fig. 6.8b).

Afterward, V_{ds} was decreased to 0 V in order to probe the nuclear spin state by sweeping the magnetic field to $+60$ mT and back while checking for a QTM transition. Repeating this procedure 6000 times for 0, 1, and 2 mV led to the black histograms of Fig. 6.8c–e, showing the four peaks corresponding to the four nuclear spin states. Integrating each peak over a window of ± 7 mT around the maximum and normalizing the outcome led to the nuclear spin population, which was subsequently fitted the to Boltzmann distribution (red dotted line in Fig. 6.8c–e). From the fitting parameters, we obtained the time average nuclear spin temperature.

As shown in Fig. 6.8f, the temperature is increasing monotonically with augmenting V_{ds}, which demonstrates the coupling of the nuclear spin to the electronic bath. During the experiment the temperature of the cryostat was stable at 150 mK, suggesting that the increase of the time average nuclear spin temperature is caused by an energy exchange with the electrons tunneling through the read-out quantum dot. A deeper analysis, however, is quite difficult since the local Joule heating of the device is unknown. More insights to this topic might be provided by Clemens Winkelmann et al., working at the Néel institute. They started a three year project, dedicated to investigate the heat conduction through a single molecule inside a breakjunction using local thermometers.

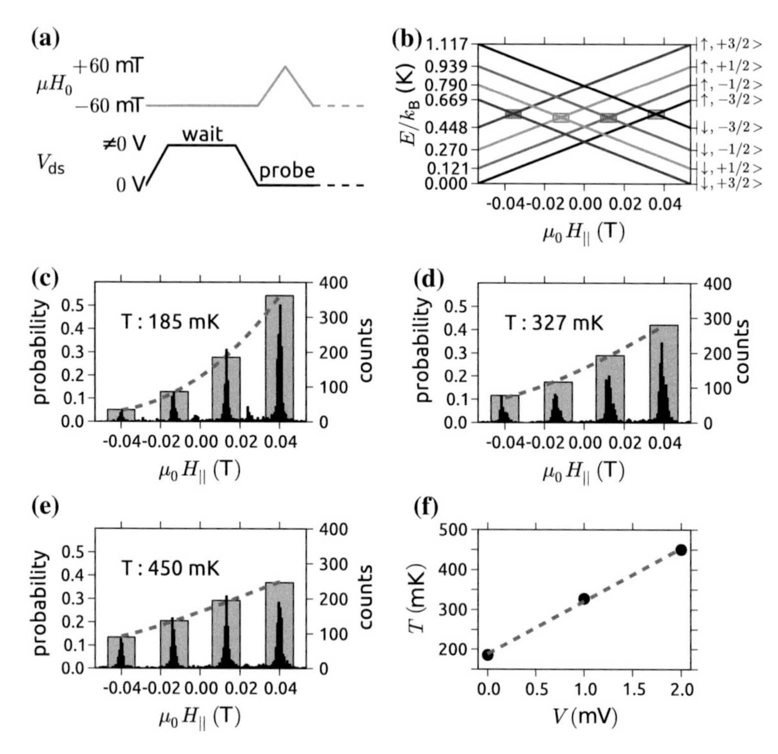

Fig. 6.8 **a** Protocol to measure the time average nuclear spin temperature. The source-drain voltage V_{ds} is rapidly increased to a finite value, at which we thermalized the nuclear spin. Afterward, V_{ds} is brought by to zero in order to probe the nuclear spin state by sweeping the magnetic field from −60 to 60 mT and back. **b** Zeeman diagram of the nuclear spin. During the waiting period in (**a**) the external magnetic field is at −60 mT, making $m_I = -3/2$ the ground state of the nuclear qubit. **c–e** Histogram of 6000 sweeps at 0 mV (**c**), 1 mV (**d**) and 2 mV (**e**) V_{ds} offset. The *grey bars* show the time average population of the nuclear spin and were obtained by integrating each peak over a window of ±7 mT around its maximum. Fitting the population to a Boltzmann distribution (*red dotted curve*) allowed for the extraction of the time average nuclear spin temperature T. **f** Fitted temperatures of (**c–e**) versus the applied source-drain voltage V_{ds}

6.4.2 Dynamical Equilibrium

Measuring the nuclear spin trajectory by sweeping the magnetic field up and down leads to an inversion of the nuclear qubit ground state and the excited states at every QTM transition. Since this inversion period is smaller than T_1, the time-average population of the nuclear spin converges to a dynamical equilibrium, which is far from the thermal Boltzmann distribution. Plotting the data obtained from nuclear spin trajectory in a histogram, and integrating over each of the four peaks, reveals the average population within this dynamical equilibrium (see Fig. 6.9a). It shows that the probability for being in each state is not 25 %, but slightly larger for $m_I = \pm 1/2$ compared to $m_I = \pm 3/2$. The time-average population obtained by the QMC

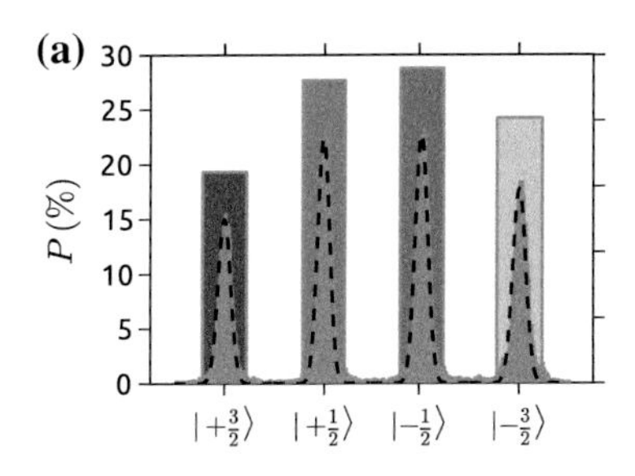

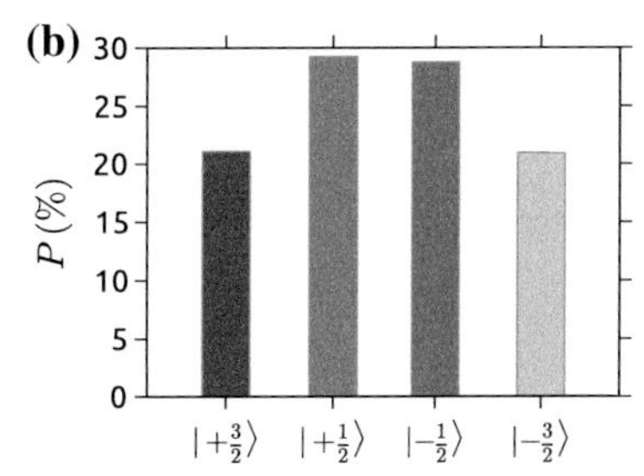

Fig. 6.9 **a** Histogram of the data obtain during the measurement of the nuclear spin trajectory of sample A. The *grey bars* correspond to the integral over each peak, revealing the time-average population of each nuclear spin. **b** Time-average population simulated using the parameters of Table 6.1 and the algorithm of Sect. 6.3. The higher probability of $m_\text{I} = \pm 1/2$ with respect to $m_\text{I} = \pm 3/2$ comes from the difference in the transition rates $\Gamma_{0,1}$ and $\Gamma_{2,3}$, and the periodical inversion of the ground state and the most excited states due to a QTM transition. For more details see text

simulations shows the same feature (see Fig. 6.9b), which allows for an explanation within the framework of the QMC model.
We found that the shape of the time-average population, in the case where the measurement time t_measure is smaller than T_1, is mainly governed by the individual transition rates $\Gamma_{i,j}$. As shown in Table 6.1, $\Gamma_{0,1}$ is much smaller than $\Gamma_{2,3}$, which causes a faster transition from the most excited state into the second excited state than from the first excited state into the ground state. Due to this asymmetry, and the periodic inversion of the ground state and the excited states, we are actively pumping the population into $m_\text{I} = \pm 1/2$ states. Notice that for equal $\Gamma_{i,j}$'s the time-average population would be 25 % for each state.

6.4.3 Selection Rules

During the analysis of the nuclear spin trajectory, we observed transitions with $\Delta m_\text{I} \neq \pm 1$. In order to clarify if this effect arose from a finite time resolution, i.e. multiple $\Delta m_\text{I} = \pm 1$ transitions between two subsequent measurements or additional transition paths, allowing for $\Delta m_\text{I} \neq \pm 1$, we compared experimental and simulated data. By counting the number of transitions corresponding to $\Delta m = 0$, ± 1, ± 2 and ± 3 and normalizing them with respect to the total amount of transitions, we obtained the red histogram in Fig. 6.10. Repeating this protocol for the simulated nuclear spin trajectory gave rise to the grey histogram.
The good agreement with the experimental data supports our assumption that the nuclear spin can only perform quantum jumps, which change its quantum number by one since the computational model allowed only for such transitions. All higher orders of Δm_I are therefore multiple transitions of $\Delta m_\text{I} = \pm 1$, which were not resolved due to the finite time resolution.

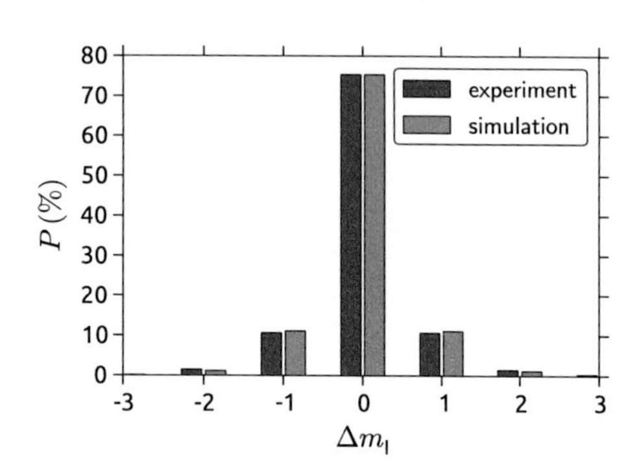

Fig. 6.10 Histogram of all transitions observed in the experiment (*red*) and simulation (*grey*). Transitions with $\Delta m_I \neq \pm 1$ correspond to multiple transitions of $\Delta m_I = \pm 1$, which were not resolved due to the finite time resolution

6.5 Summary

In this chapter we presented the dynamical evolution of the nuclear spin. Making use of the single-molecule magnet spin-transistor as a detection device, we recorded the real-time nuclear spin qubit trajectory over many days. Using a post treatment of the experimental data, we could extract the relaxation time T_1 for each nuclear spin state individually. Repeating this measurement on a second sample confirmed that the lifetime T_1 was in the order of a few tens of seconds, showing that the nuclear spin is well protected in the our devices. In order to perform a more sophisticated analysis of the experimental data, we developed a quantum Monte-Carlo code to numerically retrace the nuclear spin evolution. Fitting the simulation to the experimental data led to the extraction of the otherwise hardly accessible state dependent relaxation rates of the nuclear spin. These were found to depend strongly on type of relaxation, which enabled us to identify that the nuclear spin relaxation is dominated by an energy exchange with the electrons tunneling through the read-out quantum dot. An experimental confirmation of this conclusion was found in the tunabilty of the nuclear spin lifetime T_1 with respect to the tunnel-current. Additional evidence of the coupling between the nuclear spin and the tunnel electrons could be found through an increase of the nuclear spin temperature with augmenting tunnel current. Moreover, the experiments shed light on the read-out fidelities of the nuclear qubit, which were better than 69 and 87 % for sample A and C respectively, and are important figures of merit toward single-molecule magnet based quantum bits.

References

1. M.V.G. Dutt, L. Childress, L. Jiang, E. Togan, J. Maze, F. Jelezko, A.S. Zibrov, P.R. Hemmer, M.D. Lukin, Quantum register based on individual electronic and nuclear spin qubits in diamond. Science (New York, N.Y.) **316**, 1312–1316 (2007)
2. P. Neumann, J. Beck, M. Steiner, F. Rempp, H. Fedder, P.R. Hemmer, J. Wrachtrup, F. Jelezko, Single-shot readout of a single nuclear spin. Science (New York, N.Y.) **329**, 542–544 (2010)

3. R. Vincent, S. Klyatskaya, M. Ruben, W. Wernsdorfer, F. Balestro, Electronic read-out of a single nuclear spin using a molecular spin transistor. Nature **488**, 357–360 (2012)
4. J.J. Pla, K.Y. Tan, J.P. Dehollain, W.H. Lim, J.J.L. Morton, F.A. Zwanenburg, D.N. Jamieson, A.S. Dzurak, A. Morello, High-fidelity readout and control of a nuclear spin qubit in silicon. Nature **496**, 334–338 (2013)
5. B.E. Kane, A silicon-based nuclear spin quantum computer. Nature **393**, 133–137 (1998)
6. P.C. Maurer, G. Kucsko, C. Latta, L. Jiang, N.Y. Yao, S.D. Bennett, F. Pastawski, D. Hunger, N. Chisholm, M. Markham, D.J. Twitchen, J.I. Cirac, M.D. Lukin, Room-temperature quantum bit memory exceeding one second. Science (New York, N.Y.) **336**, 1283–1286 (2012)
7. V. Braginsky, F. Khalili, Quantum nondemolition measurements: the route from toys to tools. Rev. Mod. Phys. **68**, 1–11 (1996)
8. V.B. Braginsky, Y.I. Vorontsov, K.S. Thorne, Quantum nondemolition measurements. Science (New York, N.Y.) **209**, 547–557 (1980)
9. L. Robledo, L. Childress, H. Bernien, B. Hensen, P.F.A. Alkemade, R. Hanson, High-fidelity projective read-out of a solid-state spin quantum register. Nature **477**, 574–578 (2011)
10. C. Cohen-Tannoudji, Frontiers in laser spectroscopy. in *Les Houches Summer School Proceedings* (1975)
11. J. Dalibard, Y. Castin, K. Mølmer, Wave-function approach to dissipative processes in quantum optics. Phys. Rev. Lett. **68**, 580–583 (1992)
12. K. Mølmer, Y. Castin, J. Dalibard, Monte Carlo wave-function method in quantum optics. J. Opt. Soc. Am. B **10**, 524 (1993)
13. K. Mølmer, Y. Castin, Monte Carlo wavefunctions in quantum optics. Quantum Semiclassical Opt. J. Eur. Opt. Soc. Part B **8**, 49–72 (1996)
14. N. Sano, J. Itoh, Nuclear magnetic resonance and relaxation of 159 Tb in ferromagnetic terbium metal. J. Phys. Soc. Jpn. **32**, 95–103 (1972)
15. J. Korringa, Nuclear magnetic relaxation and resonnance line shift in metals. Physica **16**, 601–610 (1950)
16. M. Weger, Longitudinal nuclear magnetic relaxation in ferromagnetic iron, cobalt, and nickel. Phys. Rev. **128**, 1505–1511 (1962)
17. N. Sano, S.-I. Kobayashi, J. Itoh, Nuclear magnetic resonance and relaxation of Dy 163 in ferromagnetic dysprosium metal at low temperature. Prog. Theoret. Phys. Suppl. **46**, 84–112 (1970)
18. M. McCausland, I. Mackenzie, Nuclear magnetic resonance in rare earth metals. Adv. Phys. **28**, 305–456 (1979)

Chapter 7
Nuclear Spin Dynamics—T_2^*

Nuclear spin qubits are interesting candidates for quantum information storage due to their intrinsically long coherence times. In the last chapter, we investigated the relaxation time T_1 of a single nuclear spin and the read-out fidelity. Thus, having demonstrated four out of five DiVincenzo criteria, we turn now to the coherent manipulation of the nuclear qubit, which will complete the list.
To perform such a manipulation on a nuclear spin, large resonant AC magnetic fields are necessary. To be able to address spins individually, those AC fields are usually generated by driving large currents through nearby microcoils [1]. Yet, in order to reduce the parasitic cross talk to the read-out quantum dot and the Joule heating of the device, the maximum amplitude of the magnetic field is limited and rarely exceeds a few mT [2].
To avoid those problems, especially the Joule heating, a manipulation by means of an electric field is advantageous, in particular for scalable device architectures. Since the electric field is unable to rotate the nuclear spin directly, an intermediate quantum mechanical interaction is necessary, which transforms the electric field into an effective magnetic field. Such interactions are for example the spin-orbit coupling [3, 4], the g-factor modulation [5], or the hyperfine interaction [6].
In this chapter we will show how the latter can be used to perform coherent rotations of the nuclear spin, which are up to two orders of magnitude faster than state of the art micro-coil approaches.

7.1 Introduction

7.1.1 Rabi Oscillations

Any two level spin qubit system is characterized by its two spin orientations $|\uparrow\rangle$ and $|\downarrow\rangle$. To visualize such a system, people make use of the Bloch sphere representation.

S. Thiele, *Read-Out and Coherent Manipulation of an Isolated Nuclear Spin*,
Springer Theses, DOI 10.1007/978-3-319-24058-9_7

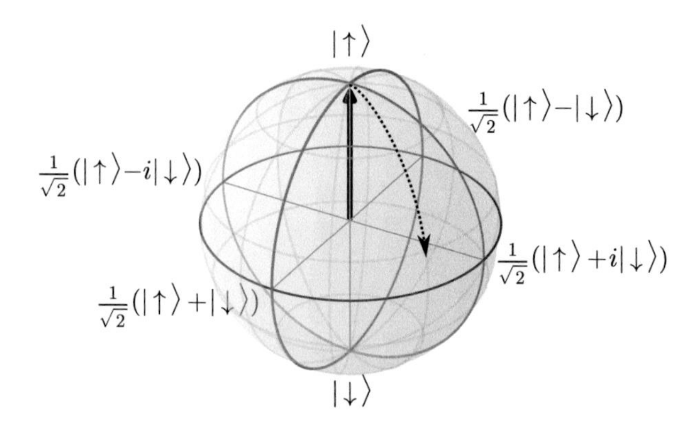

Fig. 7.1 Bloch *sphere* representation of a two level spin qubit system. The north and south pole of the sphere correspond to the two eigenstates $|\uparrow\rangle$ and $|\downarrow\rangle$, whereas the qubit state is indicated as a vector, which can be at any point on the surface. A coherent manipulation of the qubit is shown as a rotation of the vector on the sphere. Note that the trajectory of the rotation was chosen arbitrary and has no further meaning

Therein, the qubit state is symbolized as a Bloch vector, pointing from the origin of the sphere towards its surface. Moreover, the two eigenstates $|\uparrow\rangle$ and $|\downarrow\rangle$ correspond to the north and south pole of the sphere and any linear superposition $a|\uparrow\rangle + b|\downarrow\rangle$ is depicted as a point on the sphere's surface. To complete the picture, any coherent manipulation of the qubit can be illustrated as a rotation of the Bloch vector around the sphere (Fig. 7.1).
In order to manipulate the two level spin qubit, we first have to lift its degeneracy. This can be done by applying a static magnetic field B_Z along the z-axis. The Hamiltonian accounting for this effect is the Zeeman Hamiltonian (see Sect. 3.3):

$$H_Z = \hbar\omega_Z\sigma_Z \tag{7.1.1}$$

with $\hbar\omega_Z = g\mu B_Z$ being the separation between the ground state and the first excited state and σ_Z the Pauli spin operator, which performs a quantum mechanical operation that can be thought of a precession of the spin around the z-axis.
Now, the actual manipulation of the spin qubit requires an AC magnetic field in x- or y-direction. Without loss of generality, we assume that the magnetic field of magnitude $2B_1$ is applied along the x-axis. Decomposing the term into two counter-rotating parts as shown in Fig. 7.2 will simplify the calculation.

$$\boldsymbol{B}_R = B_1\left(cos(\omega t)\boldsymbol{e}_x + sin(\omega t)\boldsymbol{e}_y\right) \tag{7.1.2}$$
$$\boldsymbol{B}_L = B_1\left(cos(\omega t)\boldsymbol{e}_x - sin(\omega t)\boldsymbol{e}_y\right) \tag{7.1.3}$$

Furthermore, we assume $\boldsymbol{B}_R$ will rotate in sense with the nuclear spin precession and $\boldsymbol{B}_L$ in the opposite sense. In the frame work of the rotating wave approximation,

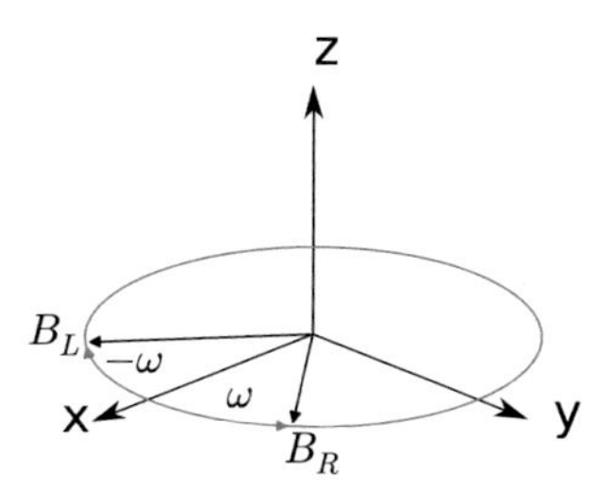

Fig. 7.2 Decomposition of the AC magnetic field $\boldsymbol{B} = 2B_1 cos(\omega t)\boldsymbol{e}_x$ into two counter-rotating parts

one can show that near the resonance ($\omega \simeq \omega_z$), the counter-rotating part can be neglected [7] and the time dependent part becomes:

$$H_{AC} = \hbar\Omega \left(cos(\omega t)\sigma_x + sin(\omega t)\sigma_y\right) \tag{7.1.4}$$

with $\hbar\Omega = g\mu B_1$ and σ_x and σ_y are the Pauli spin matrices, accounting for rotations around x and y. The qubit Hamiltonian H, including both contributions $H_Z + H_{AC}$, is given as:

$$H = \hbar\omega_z\sigma_z + \hbar\Omega \left(\sigma_x cos(\omega t) + \sigma_y sin(\omega t)\right) \tag{7.1.5}$$

To simplify the equation the following equality is applied [7]:

$$\sigma_x cos(\omega t) + \sigma_y sin(\omega t) = e^{-i\omega t\sigma_z}\sigma_x e^{i\omega t\sigma_z} \tag{7.1.6}$$

resulting in:

$$H = \hbar\omega_z\sigma_z + \hbar\Omega e^{-i\omega t\sigma_z}\sigma_x e^{i\omega t\sigma_z} \tag{7.1.7}$$

In order to eliminate the phase factors $e^{\pm i\omega t\sigma_z}$, we perform a unitary transformation of $U = exp(i\omega t\sigma_z)$. Physically, this can be understood as switching from the laboratory frame to the frame rotating around the z-axis with the frequency ω. To write the Hamiltonian in its usual way, we introduce $\Delta = \omega_z - \omega$, being the detuning between the MW frequency and the qubit level spacing.

$$H = \frac{\hbar\Delta}{2}\sigma_z + \frac{\hbar\Omega}{2}\sigma_x \tag{7.1.8}$$

To visualize the enormous advantage of the rotating frame approximation, we calculated the evolution of the qubit wavefunction exposed to an AC magnetic field in x-direction in the laboratory frame (see Fig. 7.3a) and in the rotating frame (see Fig. 7.3b). We assumed that the qubit was at $t = 0$ in the $|\uparrow\rangle$ state (grey vector). To compute the trajectory on the Bloch sphere, we used the Qutip [8, 9] master equation solver. Therein, the wavefunction $|\Psi\rangle = a|\uparrow\rangle + +b|\downarrow\rangle$ is calculated at different times steps, in which the expectation values σ_x, σ_y, and σ_z were evaluated. The Python code using the Qutip library to generate Fig. 7.3 is presented in appendix D.

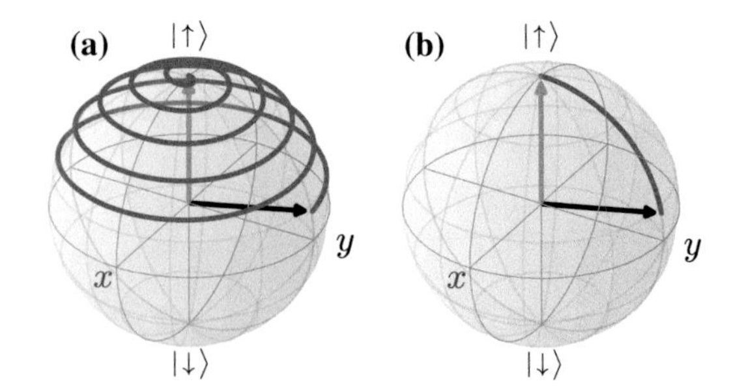

Fig. 7.3 The trajectory of a spin qubit, initialized in the $|\uparrow\rangle$ state (*grey arrow*) at $t = 0$, was computed in the laboratory frame (**a**) and the rotating frame (**b**), while being exposed to an AC magnetic field in x-direction at resonance frequency ($\Delta = 0$)

Note that in the rotating frame, the magnetic field in z-direction is proportional to Δ and therefore zero at the resonance frequency, whereas it is B_z in the laboratory frame. The big advantage of the rotating frame is that all fields are static, which allows for an easy superposition of the different components. Hence, at $\Delta \neq 0$, the Bloch vector rotates around a vector of angle $\theta = arctan(\Omega/\Delta)$ with respect to the z-axis (see Fig. 7.4), and the frequency of the precession is simply given by:

$$\Omega_R = \sqrt{\Delta^2 + \Omega^2} \tag{7.1.9}$$

with Ω_R being the Rabi frequency.
To actually measure the precession trajectory on the Bloch sphere, as presented in Fig. 7.5a, MW pulses with different duration τ are applied (see Fig. 7.5b). Before each pulse, the qubit is initialized in the $|\uparrow\rangle$ state. The following pulse is rotating the spin with the frequency Ω_R around an axis given by Ω and Δ. After the duration τ, the expectation value of σ_z of the qubit is measured. By plotting the expectation value versus the pulse duration τ, we obtain Rabi oscillations as shown in Fig. 7.5c. The amplitude and the frequency of the oscillations strongly depends on the detuning and power of the MW. Note that the largest amplitude is found at the resonance.

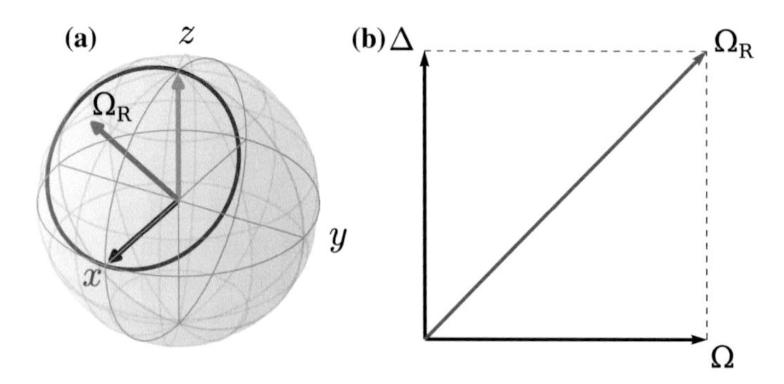

Fig. 7.4 **a** Spin precession around an effective magnetic field in the rotating frame. **b** The precession frequency Ω_R is given by $\Omega_R = \sqrt{\Delta^2 + \Omega^2}$

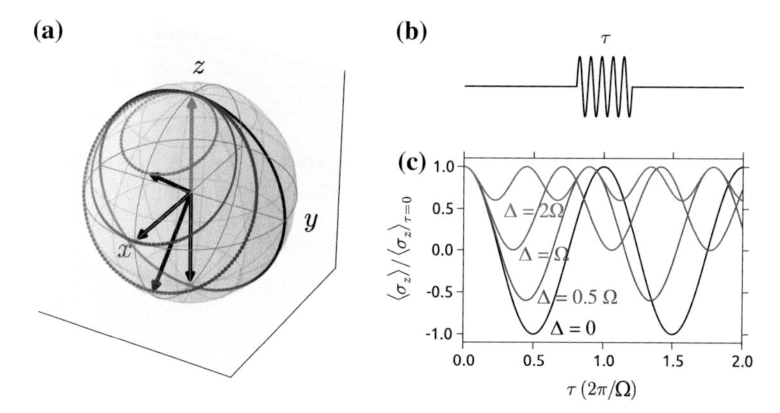

Fig. 7.5 **a** Trajectory of the Bloch vector in the rotating frame at different detunings $\Delta = 0$ (*black curve*), $\Delta = 0.5\Omega$ (*blue curve*), $\Delta = \Omega$ (*red curve*), and $\Delta = 2\Omega$ (*blue curve*). The spin was initialized in the ground state (*grey vector*) and exposed to pulses with different duration τ (**b**). The expectation value $\langle\sigma\rangle_Z$ was evaluated at the different pulse durations and different detunings resulting in the Rabi oscillations (**c**) whose amplitude is largest at the resonance frequency ($\Delta = 0$)

Moreover, the Rabi frequency at $\Delta = 0$ is $\Omega_R = g\mu B_1/\hbar$. For an electronic spin μ is the Bohr magneton, however, for the nuclear spin $\mu = \mu_N$, the nuclear magneton, which is 2000 times smaller than μ_B. Hence, to manipulate a nuclear spin with the same speed as an electron spin, three orders of magnitude larger magnetic fields are necessary. The usual approach to generate AC magnetic makes use of on-chip microcoils, which are in the vicinity of the qubit. Yet, the parasitic cross talk to the quantum dot and the Joule heating of the entire sample limit the magnetic fields to a few mT [2]. To circumvent these problems, a manipulation could be performed by means of an electric field. Especially the Joule heating is tremendously reduced, which is of major importance for scalable device architectures. Since the electric field is not able to rotate the spin directly, an intermediate quantum mechanical interaction is necessary to transform the electric field into an effective magnetic field. Such interactions are for example the spin-orbit coupling [3, 4], the g-factor modulation [5], or the hyperfine interaction [6]. In order to manipulate a nuclear spin, the latter seems the most suited and will therefore be in the focus of the next chapter.

7.1.2 Hyperfine Stark Effect

The origin of the hyperfine coupling is exlained as a dipolar coupling between the nuclear magnetic moment μ_I, and the orbital magnetic moment μ_L and spin magnetic moment μ_S of the electron respectively. Notice, there exsists a second, although smaller contribution, which origins from the nonzero probability density of s-electrons at the core. It is referred to as the Fermi contact interaction and only relevant for s-shell electrons.

The Hamiltonian describing the hyperfine interaction is formulated as:

$$H_{\mathrm{hf}} = A\boldsymbol{I}\boldsymbol{J} \tag{7.1.10}$$

with A being the hyperfine constant, I nuclear spin, and J the electronic angular momentum. From the nuclear spin's point of view, Eq. 7.1.10 can be rewritten as an effective Zeeman Hamiltonian:

$$H_{\mathrm{hf}} = g_{\mathrm{N}}\mu_{\mathrm{N}}\boldsymbol{I}\boldsymbol{B}_{\mathrm{eff}}(A, J) \tag{7.1.11}$$

with $\boldsymbol{B}_{\mathrm{eff}}(A, J)$ being the effective magnetic field operator. The terms $I_+J_- + I_-J_+$, which account for electron-nuclear spin-flip transitions, can be neglected since $m_{\mathrm{J}} = \pm 5$ levels are separated by 600 K. Therefore $\boldsymbol{B}_{\mathrm{eff}}(A, J)$ can be associated with an ordinary magnetic field at the center of the nucleus.
In order to create an effective AC magnetic field, $\boldsymbol{B}_{\mathrm{eff}}(A, J)$ needs to be modulated periodically. If this modulation is done by means of an electric field, we referred to it as the hyperfine Stark effect. In analogy to the ordinary Stark effect, which describes the modification of the electronic levels under an external electric field, the hyperfine Stark effect deals with the shift of the nuclear energy levels.
One of the first experimental evidence of this effect was given by Haun et al. [10]. They investigated the shift of the hyperfine transition $|F = 4, m_{\mathrm{F}} = 0\rangle \longleftrightarrow |F = 3, m_{\mathrm{F}} = 0\rangle$ for the ^{133}Cs ground state (see Fig. 7.6). In their measurements they observed a quadratic dependence of the level splitting on the electric field. Since the level shift remains small compared to the hyperfine splitting, an explanation of this behavior can be given by first order perturbation theory. If e is the electric charge of the electron, $\mathcal{E}$ the electric field, and r component of the vector connecting the nucleus and the electron along $\mathcal{E}$, the perturbation is given by $er\mathcal{E}$. Since this is a odd-parity term and the atomic ^{133}Cs ground states are of well defined parity, all first order perturbation terms are zero. The first nonzero elements occur in second order of perturbation and contain $(er\mathcal{E})^2$, which gives rise to the quadratic Stark shift.

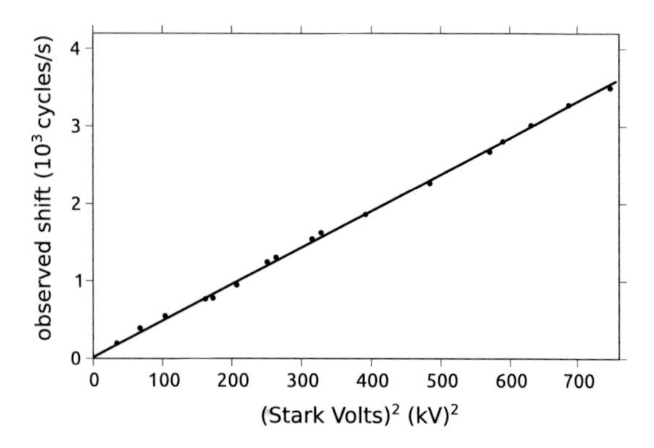

Fig. 7.6 Shift of the $F = 4, m_{\mathrm{F}} = 0 \longleftrightarrow F = 3, m_{\mathrm{F}} = 0$ transition frequency of the ^{133}Cs ground state as a function of the square of the applied voltage

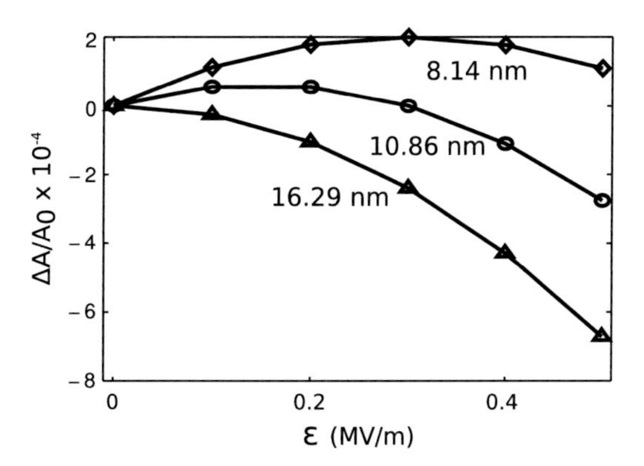

Fig. 7.7 Electric field response of the hyperfine constant at different distances between the impurity and the silicon interface. Adapted from [12]

In 1998, Kane applied the idea of the hyperfine Stark effect on ^{31}P nuclear spin qubits in silicon. He suggested that by using local gates at each qubit, the different nuclear spins can be tuned in and out of resonance independently [11]. This way, he established the individual addressability of nuclear spin qubits unsing only a global microwave field.

To show the feasibility of Kane's idea, Rahman et al. evaluated the hyperfine Stark shift of a ^{31}P impurity near the silicon interface withing the framework of the tight binding theory [12]. Since the interface breaks the symmetry around impurity, the wavefunctions of the ^{31}P are modified, resulting in states with mixed parity. Therefore, the first order perturbation terms are nonzero, giving rise to a change of the hyperfine splitting which is linear in $\mathcal{E}$. In their model, this modification is expressed as a change of the hyperfine constant $\Delta A/A_0$, which was found to be up to $\approx 10^{-3}$ at electric fields of 1 MV/m (see Fig. 7.7).

Now we turn to the $TbPc_2$ SMM. From Sect. 3.7 we know that the hyperfine constant of the Tb^{3+} inside the molecule is $A = 24.9$ mK [13]. Using Eqs. 7.1.10 and 7.1.12 we obtain an effective magnetic field at the nucleus of:

$$\boldsymbol{B}_{\text{eff}}(A, J) = \frac{AJ}{g_{\text{N}}\mu_{\text{N}}} = 313\ T \tag{7.1.12}$$

which is two orders of magnitude larger than the usual laboratory fields. Assuming we could periodically modify the hyperfine constant A by 1/1000, we would be able to generate AC magnetic field of ± 313 mT. Since the orientation of the quantization axis of the molecule with respect to the electric field is not well determined, the effective magnetic field will have components in the x- and z-direction. However, in terms of oscillating fields only the component in x-direction is able to rotate the nuclear spin, whereas the z-component induces additional decoherence. Moreover, we can predict a linear response to an external magnetic field, since the phthalocyanine ligands break the inversion symmetry of the Tb^{3+}, analog to the ^{31}P impurities at the interface. Therefore the first, instead of the second harmonic, of the oscillating electric field must be matched to the nuclear transition frequency.

In the following sections we will demonstrate how we used the hyperfine Stark effect to perform a coherent manipulation of a nuclear spin. Additionally, we will compare the experimental results to a more profound theoretical model.

7.2 Coherent Nuclear Spin Rotations

In this section we are presenting the first experimental evidence of a coherent single nuclear spin manipulation by means of an electric field. As pointed out in the previous sections, the hyperfine Stark effect is used as a mediating quantum mechanical process to transform an oscillating electric field into an AC magnetic field. This procedure can be viewed as the AC extension of Kane's proposal form 1998, and will allow for the generation of large amplitude local magnetic fields without the inconvenience of using large AC currents through close by microcoils.
In order to simplify the problem, we will focus on the nuclear spin subspace containing only the $|+3/2\rangle$ and $|+1/2\rangle$ qubit states. By assigning the $|+3/2\rangle$ and $|+1/2\rangle$ states the Bloch vectors pointing to the north and south pole of the Bloch sphere respectively, we can use the theory that was presented in Sect. 7.1.1 to explain the quantum manipulation. However, in this subspace the operator I_x becomes $\sqrt{3}\sigma_x$, I_y becomes $\sqrt{3}\sigma_y$, and I_z becomes σ_z, with $\sigma_{x,y,z}$ being the corresponding Pauli spin 1/2 matrices. Note that the other two nuclear spin subspaces would have worked as well.

7.2.1 *Frequency Calibration*

The coherent manipulation of the nuclear spin qubit requires the knowledge of the exact level spacing between the $|+3/2\rangle$ and $|+1/2\rangle$ states. This frequency depends of course on the electrostatic environment due to the hyperfine stark effect. A first indication of the approximate position of the resonance frequency could be found in the work of Hutchison and Ishikawa [13, 14] who gave values of 2.3 and 2.5 GHz. In conventional NMR experiments, the nuclear spins start to absorb a notable amount of microwave power at the resonance frequency, which can be detected by a change of the reflection or transmission of the microwave signal. In case of a single nuclear spin, this signal is much to small to be detected. Therefore, we developed our own protocol, which is sensitive to an increase of the relaxation rate if the two nuclear spin transition is in resonance to the frequency of the applied AC electric field.
The schematic of the protocol is shown in Fig. 7.8a. First, the nuclear spin was initialized by sweeping the magnetic field $\mu_0 H_{||}$ from negative to positive values (purple curve) while checking for a QTM transition at one of the 4 avoided level crossings (see colored rectangles (b)). Subsequently, we applied a MW pulse of duration $\tau = 1$ ms. The final state is then detected by sweeping back the external

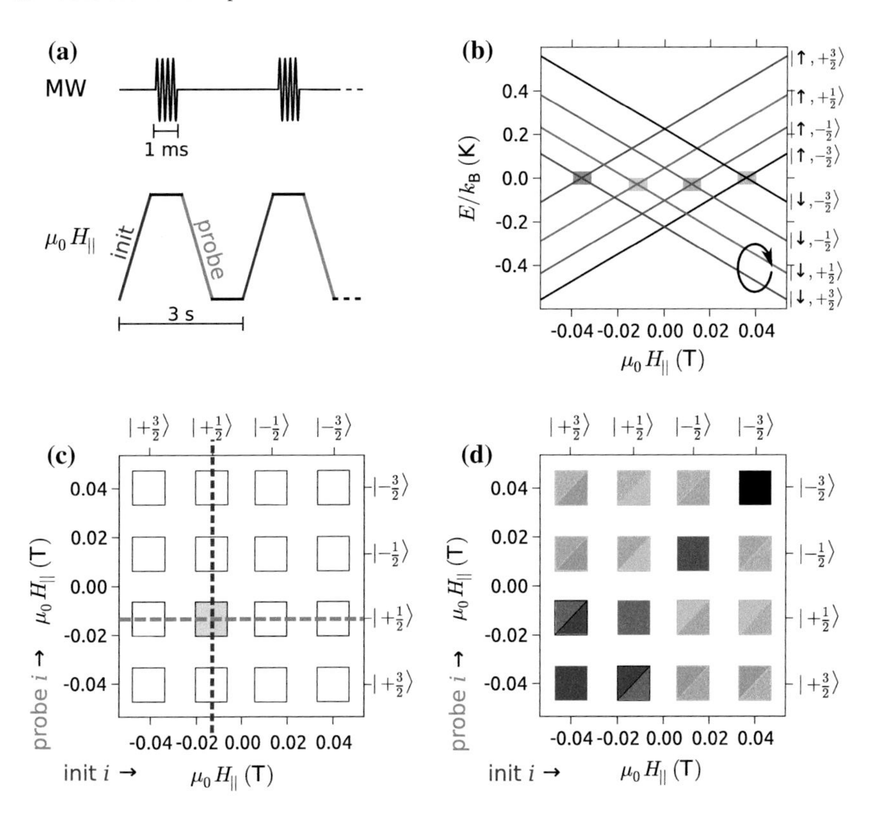

Fig. 7.8 **a** Measurement protocol to find the resonance frequency. To initialize the nuclear spin, the magnetic field $\mu_0 H_{||}$ is swept from negative to positive values (*purple curve*), while checking for QTM transition. Subsequently, we kept $H_{||}$ constant (*black curve*) and applied a microwave (MW) pulse of 1 ms. In the end, the final state is probed by sweeping back $H_{||}$ to negative values (*orange curve*). One measurement cycle has a duration of 3 s and is therefore much faster than T_1. **b** If the microwave was in resonance with the two lowest nuclear qubit levels, a transitions between $m_I = +3/2 \longleftrightarrow m_I = +1/2$ could be induced at positive $H_{||}$. **c** Schematic showing the construction of a 2D matrix to visualize the transitions. In the shown example the nuclear spin was initialized in the $|+1/2\rangle$ state (*vertical line*) and probed in the same state (*horizontal line*) giving to an element on the diagonal line. **d** Full 2D matrix. The elements on the diagonal, having only one color, correspond to measurements, where the nuclear spin state was not change between the initialization an the probe sweep. If the microwave was in resonance with the $m_I = +3/2 \longleftrightarrow m_I = +1/2$ transition, increased offdiagonal will appear, as indicated by blue-green rectangles. The other offdiagonals will also appear due to relaxation processes but with much less intensity

magnetic field in a time scale faster than the measured relaxation times of both nuclear spin states. The entire sequence is rejected when the initial or final state was not detected due to a missing QTM transition. A full cycle had a duration of 3 s and was therefore much faster than T_1 so that thermal relaxation processes were observed only every 6–11 measurements, depending on the nuclear spin state.

The MW was pulsed because of less heating of the device with respect to a continuous irradiation. However, the pulse width should be larger than the dephasing time T_2^*, which was expected to be smaller than 1 ms, in order to avoid accidental full coherent rotations in the Bloch sphere, which preserve the nuclear spin state. If the MW frequency was in resonance with the two lowest nuclear qubit levels at positive $H_{||}$, a transition between $m_I = +3/2 \longleftrightarrow m_I = +1/2$ could be induced (see Fig. 7.8b) resulting in an increased relaxation rate of the two states.

To visualize the relaxation rate, we constructed a two-dimensional matrix as follows. The detected nuclear spin state during the initialization determined the column of the matrix, whereas the probed nuclear spin stated determined the row. An example is given in Fig. 7.8c, where the nuclear spin was initialized (vertical line) and probed (horizontal line) in the $|1/2\rangle$ state, giving rise to an element on the diagonal of the matrix. Notice that the diagonal is going from the lower left to the upper right corner. By repeating this procedure several hundred times, we gathered enough data points to plot the 2D matrix (see Fig. 7.8d). Since the relaxation time is much longer than the measurement cycle, most elements are on the diagonal of the matrix. If, however, the MW was in resonance with the $m_I = 3/2 \longleftrightarrow m_I = 1/2$ transition, increased offdiagonal elements will appear, as indicated by blue-green rectangles. Other off-diagonal elements were also observed due to thermal relaxation processes but with much less intensity. Scanning the frequency, in steps of 2 MHz, from 2.3 to 2.5 GHz led to the results presented in Fig. 7.9a. When the microwave frequency hit the resonance of the nuclear qubit transition at 2.45 GHz, we obtained a matrix as shown in Fig. 7.9b, in which off-diagonal elements for the expected transition are clearly observed.

If, however, the microwave power was chosen too large or the pulse width was set too long, the device suffered from heating of the nuclear spin states, resulting in additional off-diagonal elements. In contrary to the resonant condition, the thermal heating affected all four nuclear spin states and can easily be distinguished (see Fig. 7.10).

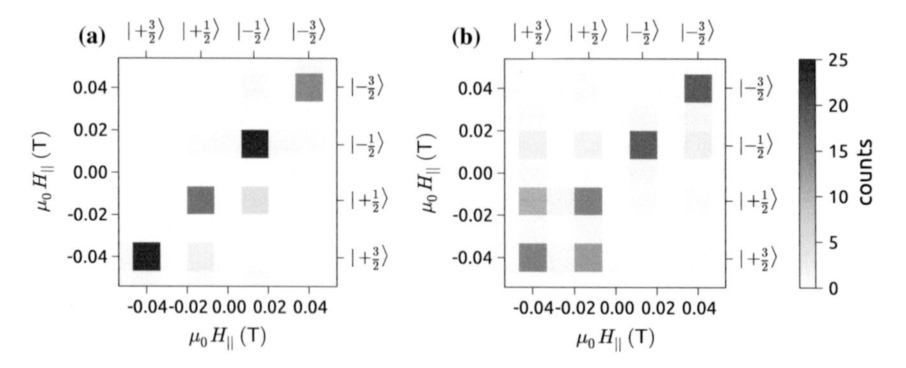

Fig. 7.9 Matrix similar to Fig. 7.8 for 400 sweeps when the microwave frequency was off resonance (**a**) and on resonance (**b**) with the $m_I = 3/2 \longleftrightarrow m_I = 1/2$ transition

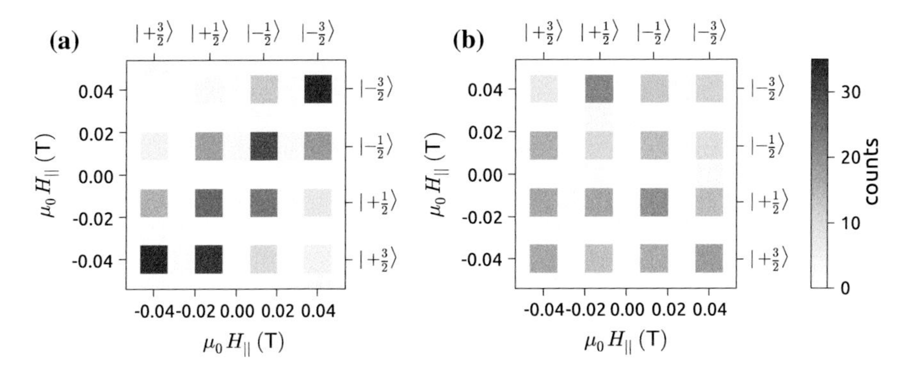

Fig. 7.10 Matrix similar to Fig. 7.8 for 1000 measurement at two different powers, off-resonant to the nuclear qubit transition

7.2.2 *Rabi Oscillations*

After having found the resonance frequency between the two lowest lying nuclear qubits, we started performing measurements for a fixed frequency as a function of the MW pulse duration τ. To initialize the nuclear spin qubit in its $|+3/2\rangle$ ground state, the external magnetic field is swept back and forth between -75 and 75 mT/s at 100 mT/s (see Fig. 7.11a) until a QTM transition is measured at -38 mT, which is the signature of the $|+3/2\rangle$ qubit state (Fig. 5.7.1a). Using a Rhode & Schwarz SMA100A signal generator, a MW pulse of duration τ is then applied while keeping the external field constant (Fig. 7.11a). The resulting state is detected by sweeping back the external magnetic field in a time scale faster than the measured relaxation times of both nuclear spin states. The sequence was rejected when the final state was not detected due to a missing QTM transition. In order to get a sufficient approximation of the nuclear spin qubit expectation value, the procedure was repeated 100 times for each pulse duration, resulting in coherent Rabi oscillations, as presented in Fig. 7.11b, c for two different microwave powers. The visibility of the measurements presented in Fig. 7.11b, c is $\sim$50 %.
From Eq. 7.1.9 we see that the Rabi frequency is proportional to the amplitude of the effective magnetic field for zero detuning Δ. Assuming that increasing the microwave power will increase the effective magnetic field, we should observe a monotonic increase of the Rabi frequency with the microwave power. To investigate this behavior we measured the frequency of the Rabi oscillation Ω_R at different injection powers P (see Fig. 7.12). The result shows a linear dependence of Ω_R with $\sqrt{P}$ above 2 mW of injection power. For smaller powers, however, we found a deviation from the this linear curve. One reason could be a nonlinearity in the hyperfine Stark effect or a slight gate voltage drift during the 5 days needed to perform this experiment. Indeed, we will see in the following that the Rabi frequency is extremely sensitive to modifications of the gate voltage because of the Stark effect.

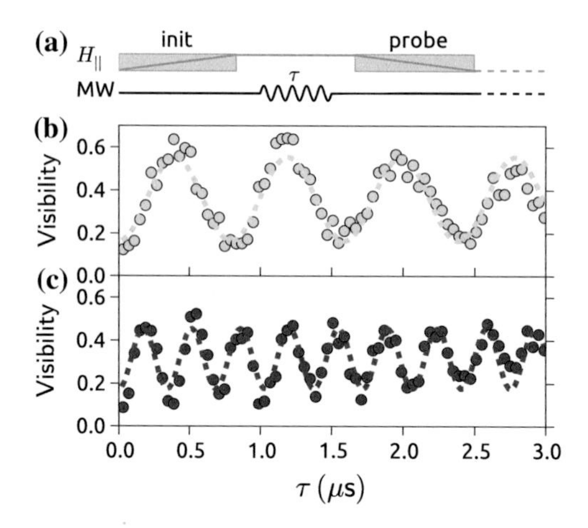

Fig. 7.11 Rabi oscillations of a single nuclear spin qubit. **a** Time dependent external magnetic field $H_{||}$ and pulse sequence generated to observe Rabi oscillations between the two lower states of the nuclear spin qubit having a resonant frequency ν_0. The nuclear spin is first initialized by detecting a conductance jump while sweeping up $H_{||}$ (init sequence). A subsequent MW pulse of frequency ν_0 and duration τ is applied, modifying periodically the hyperfine constant A. It induces an effective oscillating magnetic field resulting in coherent manipulation of the two lower states of the nuclear spin qubit. Finally, $H_{||}$ is swept down to probe the final state of the nuclear spin qubit. **b** Rabi oscillations obtained by repeating the above sequence 800 times for each τ, for two different MW powers, $P_{\rm MW} = 1$ mW and $P_{\rm MW} = 1.58$ mW for the red and violet measurements

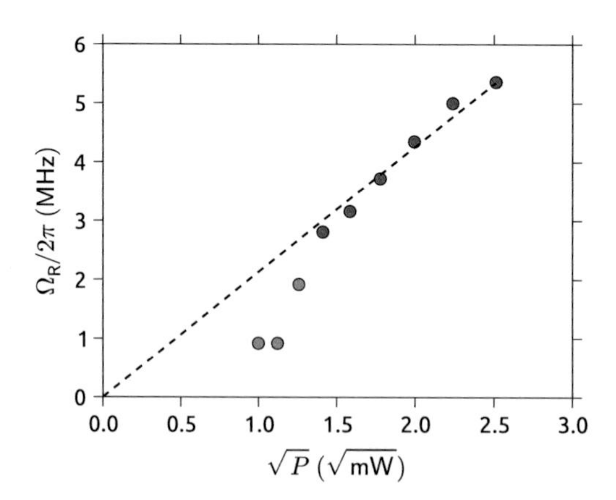

Fig. 7.12 Rabi frequency $\Omega_R/2\pi$ as a function of the microwave power P

7.3 Experimental Discussion of the Hyperfine Stark Effect

7.3.1 DC Gate Voltage Induced Hyperfine Stark Effect

We now present and discuss the study of the visibility of the Rabi oscillations as a function of the applied MW frequency at three different gate voltage values (Fig. 7.13a). As expected from theory (compare Fig. 7.5), the visibility of the Rabi oscillations was largest at the resonant frequency ν_0 and decreases for increasing detuning $\Delta = |\nu - \nu_0|$. However, a clear dependence of the nuclear qubit resonance frequency on the gate voltage is also observed in Fig. 7.13a. This effect can be attributed to the static HF Stark shift, due to the additional electric field induced by the gate voltage, which shows our ability to tune the HF constant A between the electronic spin and the nuclear spin qubit. Notice that only the z-component of the effective magnetic field will modify the level splitting. Applying a gate voltage offset of 10 mV and 16 mV resulted in a shift of $\Delta\nu_0$ =1.72 and 7.03 MHz respec-

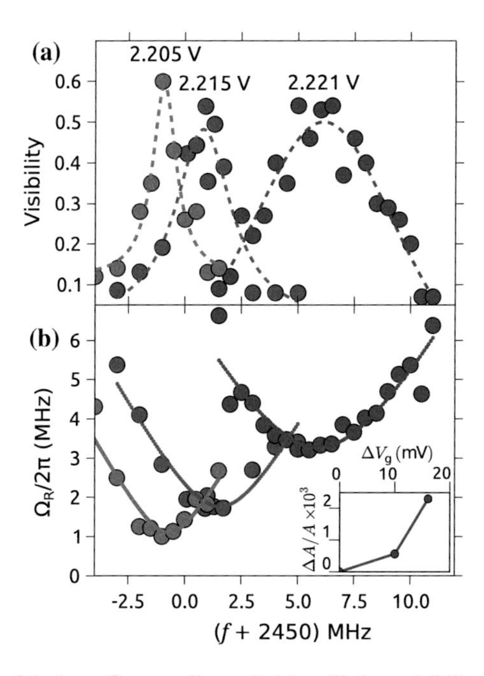

Fig. 7.13 Stark shift of the hyperfine coupling. **a** Rabi oscillations visibility measured at different MW frequencies for three different gate voltages V_g. The resonance shift of the nuclear spin qubit frequency ν_0 is caused by a modification of the hyperfine coupling A due to V_g induced Stark shift. **b** Rabi frequencies Ω_R corresponding to the visibility of (**a**). The continuous lines are fit to the experimental points following the theoretical expression of the Rabi frequency dependence (see main text). The magnitude of the effective magnetic field induced by the oscillating hyperfine constant A due to Stark shift reaches a few hundreds of mT, resulting in Rabi frequencies up to several MHz

tively. Converting this frequency shift into a change of the hyperfine constante gives $\Delta A/A = 5.6 \times 10^{-4}$ for $\Delta V_g = 10$ mV and $\Delta A/A = 2.3 \times 10^{-3}$ for $\Delta V_g = 16$ mV. (see inset Fig. 7.13). Those values can be compared with calculations presented in Sect. 7.4. There we estimate an order of magnitude of $\Delta A/A = 10^{-3}$ for an electric field of 1 mV/nm. The conversion of the back gate voltage into an electric field can be done using the simple formula E = V/d, a gate oxide thickness of 7 nm, and the screening factor of 0.2 (which is a typical value for devices created by electromigration). Doing so, we obtain $\Delta A/A = 2.9 \times 10^{-4}$ for $\Delta V_g = 10$ mV and 4.6×10^{-4} for $\Delta V_g = 16$ mV. Those values are smaller but in the same order of magnitude as the experimental values and therefore within the error bar of the theoretical model. But most importantly, these results show our ability to control the resonance frequency of a single nuclear spin qubit by means of an electric field only.

7.3.2 *AC Induced Hyperfine Stark Effect*

We turn now to the estimation of the effective AC magnetic field. To do so, the Rabi frequency Ω_R was measured for the three different gate voltage as a function of the detuning Δ (Fig. 7.13b). The horizontal evolution of the minimum of the Rabi oscillations as a function of the MW frequency is induced by the DC Stark shift as explain in Sect. 7.3.1. By further fitting the measurements to the function $\Omega_R/2\pi = \sqrt{(\Delta/2\pi)^2 + (\sqrt{3} g_N \mu_N B_x/h)^2}$, with g_N being the nuclear g-factor ($\approx$1.354 for Tb [15]), μ_N the nuclear magneton, we can extract the effective magnetic field in the x-direction B_x. Astonishingly, the data of Fig. 7.13 gives values of $B_x = 62, 98$ and 183 mT for $V_g = 2.205, 2.215$ and 2.221 V, which are up two orders of magnitude higher than magnetic fields created by on-chip micro-coils. In order exclude that those magnetic fields where produced by currents in the vicinity of the spin we were considering the following cases.

(I) The magnetic field could have been generated by the magnetic field component emitted by the microwave antenna itself. Assuming a minimal distance of 10 μm between the antenna and the sample leads to a current of 10 A in order to generate 200 mT using the formula $I = 2\pi r B/\mu_0$. From measurements with a vector network analyzer we know that the insertion loss of the antenna is 35.5 dBm at 2.45 GHz. Considering a microwave power of 0 dBm and an impedance of 50 Ω the current can be estimated to 75 μA, which is 10^5 times smaller than the required current to obtain 200 mT. Moreover, the aluminum bonding wire has an approximate fuse current of 300 mA.

(II) The magnetic field could have been created by the tunnel current through the molecule. This time we can assume a distance of 0.5 nm between the electronic spin and the tunnel current. Using the formula $I = 2\pi r B/\mu_0$ as a rough estimate, results in a required tunnel current of 500 μA. However, the current through the molecule is in the order of 1 nA, which is 5×10^5 times smaller. Even the maximal current through a single molecule, which can be as large as 100 nA, is not sufficient to explain such high magnetic fields.

(III) The magnetic field could have been created by the hyperfine Stark effect, which describes the influence of the electric field on the hyperfine interaction. The hyperfine interaction can be seen as an interaction with an effective magnetic field, which is generated by the electronic spin at the center of the nucleus. Manipulating the interaction constant A by means of an oscillating electric field results in an alternating magnetic field. In order to achieve a magnitude of 200 mT at 0 dBm, the relative variation of the hyperfine constant $\Delta A/A$ should be in the order of the ratio of the corresponding Rabi oscillation to the hyperfine splitting which is 1 MHz/ 2.45 GHz $\approx 10^{-3}$ which would require electric field fluctuations in the order of 1 mV/nm.

The first step towards a verification of the third possibility was to quantify the amplitude of the pulsed oscillating electric field, used to perform the Rabi oscillations. To do so, the full width at half maximum (FWHM) of the dip at the right side of the charge degeneracy point of sample C ($V_g = 2.2$ V in Fig. 5.2.1) was measured as a function of the applied microwave power (see Fig. 7.14a). The observed dip is a signature created by a transition from the inelastic cotunneling between the singlet/triplet state to elastic cotunneling through the singlet state only. In a first approximation, the amplitude of the induced AC voltage is directly proportional to the broadening of the dip. Since the microwave power had to be applied continuously, we could measure only up to a injection power of -20 dBm in order to avoid any damage of the sample. Figure 7.14a shows the evolution of the FWHM from -40 to -20 dBm and an extrapolation up to 0 dBm. From this measurement we see that the induced voltage drop across the molecule is about 2 mV at 0 dBm. Given the size of the molecule to be 1 nm, the generated electric field is estimated to be 2 mV/nm. We will use this value in Eq. 7.4.16 to estimate a relative change of the hyperfine constant to $\Delta A/A = 2 \times 10^{-3}$. This value is in the same order of magnitude than the required value of the consideration given above.

This result emphasized the possibility to use the hyperfine Stark effect to manipulate a single nuclear spin by means of an electric field only. The estimated effective magnetic field in the order of 200 mT and about two orders of magnitude higher

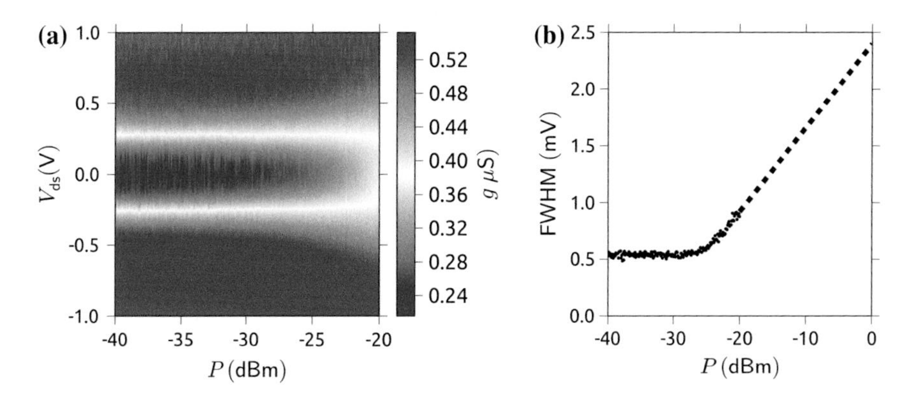

Fig. 7.14 **a** Conductance through the read-out dot as a function of the source-drain voltage V_{ds} and the applied microwave power P. **b** Evolution of the full width half maximum (FWHM) of the dip in (**a**) as a function of the microwave power P

than the fields generated by on-chip micro-coils, which leads to an increase of the clock-speed of the coherent manipulation.

7.4 Theoretical Discussion of the Hyperfine Stark Effect

The model presented in this section was elaborated in cooperation with Rafik Ballou from the Néel institute and is aimed to give an order of magnitude explanation of the experimental data in Sect. 7.3.2. To keep the derivation as intuitive as possible, rather complicated algebraic calculation were cut out, and only the result will be given.
To determine the magnitude of the hyperfine Stark effect, we used to the following strategy. Starting from the isolated terbium ion, we consider the effect of the ligand field as a perturbation on the electronic configurations. Subsequently, the Stark effect is treated as perturbation on the ligand field ground states. In this way, we derive an expression which connects the mixing of the ground state wavefunctions with the electric field. Afterward, we will evaluate hyperfine interaction with the mixed ground states within first order perturbation theory. Thus, we are able obtain an expression correlating the electric field $\mathcal{E}$ with the change of the hyperfine constant A.
The isolated Tb^{3+} ion possesses a ground state configuration of $4f^8$ and an excited state configuration of $4f^7 5d^1$ (see Fig. 7.15a). The latter arises from an excitation of one 4f electron into the 5d orbital and is about 5.5 eV higher in energy . Moreover, the lowest energy states of each configuration (states having $S = max$ and $L = max$) are split into levels of different J due to the spin-orbit interaction (compare Fig. 3.5.2). To

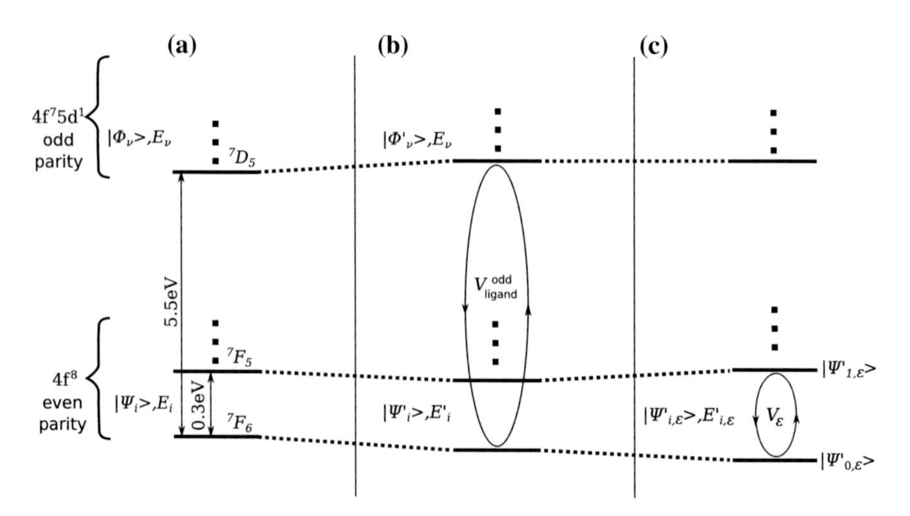

Fig. 7.15 **a** Illustration of the isolated Tb^{3+} electronic ground state configurations $4f^8$, containing the states $\Psi_i\rangle$, and the first excited configuration $4f^7 5d^1$, composed of the states Φ_ν. **b** Under the influence of a symmetry breaking ligand field $V^{\text{odd}}_{\text{ligand}}$, the ground state and excited state configurations are mixed along with their parities. **c** If an additional electric field, described by the operator $V_\mathcal{E}$, is applied, states within ground state configuration are being mixed

distinguish these states from each other, we will use the spectroscopy nomenclature ${}^{2S+1}X_J$, with $S = \sum_i s_i$, $L = \sum_i l_i$, $J = |L - S|...L + S$, and $X = S, P, D, F$ for $L = 0, 1, 2, 3$. For the Tb^{3+} the two lowest energy multiplets are ${}^7F_6$ and ${}^7F_5$, which correspond to states with $S = 3$, $L = 3$, and $J = 6$ or $J = 5$ respectively.

At this point, we want to recall that the parity P of the wavefunction is defined as $P = (-1)^{\sum_i l_i}$ with $l_i = 0, 1, 2, 3, \ldots$ for $s, p, d, f, \ldots$ electrons. Thus, $P = (-1)^{8*3} = 1$ for all the states of the ground configuration $4f^8$ of the Tb^{3+} free ion, whereas $P = (-1)^{7*3+2} = -1$ for all the states of its first excited configuration $4f^7 5d^1$.

If the isolated terbium ion is placed into the electrostatic environment of the molecule, all electronic levels are modified by the ligand field operator $V^{\text{odd}}_{\text{ligand}}$. Since the molecule lacks an inversion symmetry, the operator contains contributions of odd parity, which is able to mix the states $|\Psi_i\rangle$ of the ground configuration $4f^8$ with states $|\Phi_\nu\rangle$ of the excited configuration $4f^7 5d^1$ of opposite parity. In first order perturbation theory, the modified ground state multiplets $|\Psi'_i\rangle$ are calculated as:

$$|\Psi'_i\rangle = |\Psi_i\rangle + \sum_\nu \frac{\langle\Phi_\nu|V^{\text{odd}}_{\text{ligand}}|\Psi_i\rangle}{E_i - E_\nu}|\Phi_\nu\rangle = |\Psi_i\rangle + \sum_\nu \alpha_i^\nu|\Phi_\nu\rangle \tag{7.4.1}$$

where $E_i - E_\nu$ is the energy difference between the states $|\Phi_\nu\rangle$ of the $4f^7 5d^1$ configuration and the state $|\Psi_i\rangle$ of the $4f^8$ ground configuration. Note that without parity breaking, the ligand field operator would have been of even parity and the term $\langle\Phi_\nu|V_{\text{ligand}}|\Psi_i\rangle = 0$.

If, furthermore, an external electric field $\mathcal{E}$ is applied, the lowest energy levels of the ground state configuration $4f^8$ (all terms beginning with ${}^7F$) are themselves mixed due to the Stark interaction $V_{\mathcal{E}} = -\boldsymbol{d}\boldsymbol{\mathcal{E}}$. In first order of perturbation, the in this way altered wavefunctions $|\Psi'_{i\mathcal{E}}\rangle$ are determined as:

$$\begin{aligned}|\Psi'_{i\mathcal{E}}\rangle &= |\Psi'_i\rangle + \sum_j \frac{\langle\Psi'_j|V_{\mathcal{E}}|\Psi'_i\rangle}{E'_i - E'_j}|\Psi'_j\rangle = |\Psi'_i\rangle + \sum_j \beta_i^j|\Psi'_j\rangle \\ &= |\Psi_i\rangle + \sum_\nu \alpha_i^\nu|\Phi_\nu\rangle + \sum_j \beta_i^j|\Psi_j\rangle + \sum_j \beta_i^j \sum_\nu \alpha_j^\nu|\Phi_\nu\rangle \end{aligned} \tag{7.4.2}$$

At this point we have successfully established the correlation of the electric field $\mathcal{E}$ with the mixing of the ground state wavefunctions. All what remains is evaluation of the hyperfine splitting using the perturbed ground states $|\Psi'_{i\mathcal{E}}\rangle$. To do so, we have to determine the expression of the hyperfine Hamiltonian first. Generally, the hyperfine interaction can be considered as a change of the potential energy of the nuclear magnetic moment $\boldsymbol{\mu}_{\text{I}}$, exposed to the magnetic field $\boldsymbol{B}_{\text{elec}}$, which is created by the ensemble of the electrons in the 4f shell. Therefore, the hyperfine Hamiltonian can be written as:

$$H_{\text{hf}} = -\boldsymbol{\mu}_{\text{I}}\boldsymbol{B}_{\text{elec}} \tag{7.4.3}$$

The magnetic field operator $\boldsymbol{B}_{\text{elec}}$ consists of two independent contributions, an orbital contribution $\boldsymbol{B}_{\text{orbit}}$ coming from the motion of the electrons around the core, and a spin contribution $\boldsymbol{B}_{\text{spin}}$ resulting from the magnetic dipole field of the electron's spin. Since the probability density of 4f electrons is zero at the core, there is no contact interaction. To cut down the problem, we are going to consider the orbital contribution first. The magnetic field $\boldsymbol{B}_i$, created by a moving electron i at velocity $\boldsymbol{v}_i$ and distance $\boldsymbol{r}_i$ of the atomic core, is given by the law of Biot-Savart:

$$\boldsymbol{B}_i = \frac{\mu_0}{4\pi} e\boldsymbol{v}_i \times \frac{\boldsymbol{r}_i}{r_i^3} \tag{7.4.4}$$

Since $e\boldsymbol{v}_i \times \boldsymbol{r}_i = -2\frac{e}{2m}\boldsymbol{r}_i \times m\boldsymbol{v}_i = -2\mu_{\text{B}}\boldsymbol{l}_i$ and thus the orbital contribution becomes:

$$\boldsymbol{B}_{\text{orbit}} = -\frac{\mu_0}{4\pi} 2\mu_{\text{B}} \sum_i \frac{\boldsymbol{l}_i}{r_i^3} \tag{7.4.5}$$

Now we turn to the spin contribution $\boldsymbol{B}_{\text{spin}}$. We assume that the spin is localized on each electron, so that the magnetic field seen by the nucleus is just the sum of the magnetic field created by each magnetic moment $\boldsymbol{\mu}_{\text{s}}^i$ at the distance r_i.

$$\boldsymbol{B}_{\text{spin}} = -\frac{\mu_0}{4\pi} \sum_i \frac{\boldsymbol{\mu}_{\text{s}}^i}{r_i^3} - \frac{3\boldsymbol{r}_i(\boldsymbol{\mu}_{\text{s}}^i \boldsymbol{r}_{\text{i}})}{r_i^5} \tag{7.4.6}$$

Substituting $\boldsymbol{\mu}_{\text{s}}^i = -2\mu_{\text{B}}\boldsymbol{s}_i$ and $\boldsymbol{\mu}_{\text{I}} = g_{\text{N}}\mu_{\text{N}}\boldsymbol{I}$, we obtain following hyperfine Hamiltonian

$$H_{\text{hf}} = -\boldsymbol{\mu}_{\text{I}} \left(\boldsymbol{B}_{\text{orbit}} + \boldsymbol{B}_{\text{spin}}\right) \tag{7.4.7}$$

$$= a \sum_i (\boldsymbol{N}_i / r_i^3) \cdot \boldsymbol{I} \tag{7.4.8}$$

where $a = \frac{\mu_0}{4\pi} 2g_{\text{N}}\mu_{\text{N}}\mu_{\text{B}}$ is a constant, $\boldsymbol{I}$ is the nuclear spin and $\boldsymbol{N}_i = \boldsymbol{l}_i - \boldsymbol{s}_i + 3\boldsymbol{r}_i(\boldsymbol{s}_i \cdot \boldsymbol{r}_i)/r_i^2$ is the operator accounting for the interaction with the ith electron having the spin $\boldsymbol{s}_i$ and the angular momentum $\boldsymbol{l}_i$ at a distance $\boldsymbol{r}_i$. On the quantum states from which the electronic degrees of freedom can be factored out into a state $|\Psi_0\rangle$, the electronic part of the hyperfine interaction is given as $\langle\Psi_0|\boldsymbol{N}|\Psi_0\rangle\langle 1/r^3\rangle$, where $\boldsymbol{N} = \sum_i \boldsymbol{N}_i$ and where the radial integral $\langle 1/r^3\rangle$ is a constant within the same electronic configuration. In the absence of any parity breaking interaction the electronic state $|\Psi_0\rangle$ has a well defined parity P. We recall that $P = (-1)^{\sum_i l_i}$ with $l_i = 0, 1, 2, 3, \ldots$ for $s, p, d, f, \ldots$ electrons. Thus, $P = 1$ for all the states of the ground configuration $4f^8$ of the Tb^{3+} free ion, whereas $P = -1$ for all the states of its first excited configuration $4f^7 5d^1$. It is also crucial to recall that the matrix elements of an operator O of even (resp. odd) parity, i.e. invariant (resp. reversed) under the space inversion, are non zero solely between states with the same (resp. opposite) parity. The position vector $\boldsymbol{r}$ is reversed by space inversion whereas the orbital $\boldsymbol{l}$

and spin $\boldsymbol{s}$ moment operators are invariant, which implies that the dipole electric moment operator $\boldsymbol{d}$ is of odd parity but that the operator $\boldsymbol{N}$ is of even parity. It is a matter of standard use of the Racah algebra [16] to compute the matrix element of the spherical components N_q ($q = -1, 0, 1$) of the operator $\boldsymbol{N}$ between any two states of an electron shell. Within the Russel-Saunders coupling scheme and by making use of the Wigner-Eckart theorem one computes

$$\langle 4f^8 \xi SLJM | N_q | 4f^8 \xi' S'L'J'M' \rangle = (-1)^{J-M} \begin{pmatrix} J & 1 & J' \\ -M & q & M' \end{pmatrix} \times$$
$$\times \, (4f^8 \xi SLJ \| L - (10)^{\frac{1}{2}} \sum_i (s^{(1)} C^{(2)})_i^{(1)} \| 4f^8 \xi' S'L'J') \tag{7.4.9}$$

with

$$(\cdots \|L\| \cdots) = \delta(\xi, \xi')\delta(S, S')\delta(L, L') \times$$
$$\times \, (-1)^{S+L+J+1} ([J][J'])^{\frac{1}{2}} (L(L+1)(2L+1))^{\frac{1}{2}} \begin{Bmatrix} S & L & J \\ 1 & J' & L' \end{Bmatrix}$$

and

$$(\cdots \| \sum_i (s^{(1)} C^{(2)})_i^{(1)} \| \cdots) = ([J][1][J'])^{\frac{1}{2}} ([1][2])^{-\frac{1}{2}} \begin{Bmatrix} S & S' & 1 \\ L & L' & 2 \\ J & J' & 1 \end{Bmatrix} (s\|s\|s)(l\|C^{(2)}\|l) \times$$
$$\times \, (4f^8 \xi SL \| W^{(12)} \| 4f^8 \xi' S'L')$$

where $[x] = 2x + 1$, $\delta(X, X') = 1$ if and only if $X = X'$ and $= 0$ otherwise, (:::), {:::} and {⋮⋮⋮} stand for the 3j, 6j and 9j symbols, and the reduced matrix elements of the tensor operator $W^{(12)}$ are tabulated [17] or can be computed by making use of the coefficients of fractional parentages [16].

We shall now consider that the electronic wavefunction is exposed to the ligand field and an external electric field $\boldsymbol{E}$ resulting in the Stark interaction $V_{\mathsf{E}} = -\boldsymbol{d} \cdot \boldsymbol{E}$. Since the molecule lacks an inversion symmetry the electrostatic interactions with the ligand field contains contributions of odd parity $V^{\mathsf{odd}}_{\mathsf{ligand}}$, which mixes the states $|\Psi_i\rangle$ of the ground configuration $4f^8$ with states $|\Phi_\nu\rangle$ of the excited configuration $4f^7 5d^1$ of opposite parity. In first order perturbation theory the new wavefunction $|\Psi_i'\rangle$ is approximated as

$$|\Psi_i'\rangle = |\Psi_i\rangle + \sum_\nu \frac{\langle \Phi_\nu | V^{\mathsf{odd}}_{\mathsf{ligand}} | \Psi_i \rangle}{E_i - E_\nu} |\Phi_\nu\rangle = |\Psi_i\rangle + \sum_\nu \alpha_i^\nu |\Phi_\nu\rangle, \tag{7.4.10}$$

where $E_i - E_\nu$ is the energy difference between the states $|\Phi_\nu\rangle$ of the $4f^7 5d^1$ configuration and the state $|\Psi_i\rangle$ of the $4f^8$ ground configuration. Owing to this admixture, the states of the ground configuration $4f^8$ are themselves mixed under an applied electric field as

$$|\Psi'_{iE}\rangle = |\Psi'_i\rangle + \sum_j \frac{\langle\Psi'_j|V_{\mathrm{E}}|\Psi'_i\rangle}{E'_i - E'_j}|\Psi'_j\rangle = |\Psi'_i\rangle + \sum_j \beta_i^j|\Psi'_j\rangle$$
$$= |\Psi_i\rangle + \sum_\nu \alpha_i^\nu|\Phi_\nu\rangle + \sum_j \beta_i^j|\Psi_j\rangle + \sum_j \beta_i^j \sum_\nu \alpha_j^\nu|\Phi_\nu\rangle, \quad (7.4.11)$$

now to first order in perturbation in V_{E} with respect to $V_{\mathrm{ligand}}^{\mathrm{odd}}$. The influence of the Stark effect on the hyperfine coupling can be evaluated by calculating the matrix element of the operator $\boldsymbol{N}$ on the perturbed state $|\Psi'_{0E}\rangle = |\Psi'_0\rangle + \sum_j \beta_0^j|\Psi'_j\rangle = |\Psi_0\rangle + \sum_\nu \alpha_0^\nu|\Phi_\nu\rangle + \sum_j \beta_0^j|\Psi_j\rangle + \sum_j \beta_0^j \sum_\nu \alpha_j^\nu|\Phi_\nu\rangle$:

$$\langle\Psi'_{0E}|\boldsymbol{N}|\Psi'_{0E}\rangle = \langle\Psi_0|\boldsymbol{N}|\Psi_0\rangle + \sum_{j\neq 0}(\beta_0^j\langle\Psi_0|\boldsymbol{N}|\Psi_j\rangle + \beta_0^{j\star}\langle\Psi_j|\boldsymbol{N}|\Psi_0\rangle) + \cdots, \quad (7.4.12)$$

where contributions involving products of the coefficients α_i^ν and β_i^j are ignored as being negligible. It is emphasized that $\sum_\nu \alpha_0^\nu\langle\Psi_0|\sum_i(\boldsymbol{N}_i/r_i^3)|\Phi_\nu\rangle +$ complex conjugate $= 0$, because $|\Psi_0\rangle$ and $|\Phi_\nu\rangle$ are of opposite parity and $\sum_i(\boldsymbol{N}_i/r_i^3)$ is of even parity. Assuming that $E'_0 - E'_j \approx E_0 - E_j$ and $E_0 - E_\nu \approx \Delta E_{4f^8\to 4f^7 5d^1}$ then using the closure relation $\sum_\nu |\Phi_\nu\rangle\langle\Phi_\nu| = 1$, the coefficient β_0^j can be approximated as

$$\beta_0^j = \frac{\langle\Psi'_j|V_{\mathrm{E}}|\Psi'_0\rangle}{E'_0 - E'_j} \quad (7.4.13)$$
$$= \frac{\{\langle\Psi_j| + \sum_\tau\langle\Phi_\tau|\frac{\langle\Psi_j|V_{\mathrm{ligand}}^{\mathrm{odd}}|\Phi_\tau\rangle}{E_0 - E_\tau}\}V_{\mathrm{E}}\{|\Psi_0\rangle + \sum_\nu \frac{\langle\Phi_\nu|V_{\mathrm{ligand}}^{\mathrm{odd}}|\Psi_0\rangle}{E_0 - E_\nu}|\Phi_\nu\rangle\}}{E'_0 - E'_j}$$
$$\approx 2\frac{\langle\Psi_j|V_{\mathrm{E}}V_{\mathrm{ligand}}^{\mathrm{odd}}|\Psi_0\rangle}{(E_0 - E_j)\Delta E_{4f^8\to 4f^7 5d^1}},$$

The change in the hyperfine interaction may finally be written as

$$\langle\Psi'_{0E}|A\boldsymbol{J}\cdot\boldsymbol{I}|\Psi'_{0E}\rangle = (1 + \Delta A/A)\langle\Psi_0|A\boldsymbol{J}\cdot\boldsymbol{I}|\Psi_0\rangle \quad (7.4.14)$$

with

$$\Delta A/A \approx 4\sum_j \frac{\langle\Psi_j|V_{\mathrm{E}}V_{\mathrm{ligand}}^{\mathrm{odd}}|\Psi_0\rangle}{(E_0 - E_j)\Delta E_{4f^8\to 4f^7 5d^1}} \frac{\langle\Psi_0|\boldsymbol{N}|\Psi_j\rangle}{\langle\Psi_0|\boldsymbol{N}|\Psi_0\rangle} \quad (7.4.15)$$

In general the crystal field experienced by the excited configuration $4f^7 5d^1$ is about ten times larger [18] than the one experienced by the electrons of the ground configuration $4f^8$. It is then reasonable to expect that the effect of $V_{\mathrm{ligand}}^{\mathrm{odd}}$ amounts to around 1–2 eV in energy. On the other hand, given the size of the electronic orbits, which is within the range 0.1–0.2 nm, and the expression of the dipole operator $\boldsymbol{d} = -e\boldsymbol{r}$, the strength of V_{E} under an electric field E measured in mV/nm is estimated in eV to $(1-2)\cdot 10^{-4}$ E. The excited configuration ($4f^7 5d^1$) is separated from the ground configuration ($4f^8$) by about $\Delta E_{4f^8\to 4f^7 5d^1} = 5.5$ eV. The quantity $4\langle\Psi_j|V_{\mathrm{E}}V_{\mathrm{ligand}}^{\mathrm{odd}}|\Psi_0\rangle/\Delta E_{4f^8\to 4f^7 5d^1}$ thus is estimated to $(1.8\pm 1.1)10^{-4}$ (eV) E with E given in mV/nm. If furthermore, we consider only the states of the ground multiplet $^7F_6$ and those of the first excited $^7F_5$ multiplet then only two excited states are mixed by the electric field with the ground state, with $E_0 - E_{j=1}\approx -0.06$ eV and $\langle\Psi_0|\boldsymbol{N}|\Psi_{j=1}\rangle/\langle\Psi_0|\boldsymbol{N}|\Psi_0\rangle = -1/\sqrt{6}$ for the first and $E_0 - E_{j=2}\approx -0.3$ eV and, making use of the Eq. 7.4.9, $\langle\Psi_0|\boldsymbol{N}|\Psi_{j=2}\rangle/\langle\Psi_0|\boldsymbol{N}|\Psi_0\rangle = -0.41576$ for the second. With all these numbers we may reasonably expect a change in the hyperfine constant in the order of

$$\frac{\Delta A}{A}\approx 10^{-3}\ E(\mathrm{mV/nm}) \tag{7.4.16}$$

The result is in the same order of magnitude than the experimental value and shows that the observed nuclear spin response to the electric field is explainable by the hyperfine Stark effect.

7.5 Dephasing Time T_2^*

7.5.1 *Introduction*

In this section, we are going to present measurements of the dephasing time T_2^* of the nuclear spin qubit. The dephasing time is equal to the duration over which the time average coherence of the quantum superposition is preserved. But before turning to the discussion of the experimental results, a brief review about the dephasing of an effective spin 1/2 and the experimental access to this quantity is given. To do so, we will follow the common approach by starting from the time evolution of the 2×2 density matrix ρ:

$$i\hbar\frac{d\rho}{dt} = [H,\rho] = H\rho - \rho H = \frac{\hbar}{2}\left[\begin{pmatrix}\Delta & \Omega\\ \Omega e^{-i\phi} & -\Delta\end{pmatrix}\rho - \rho\begin{pmatrix}\Delta & \Omega\\ \Omega e^{-i\phi} & -\Delta\end{pmatrix}\right] \tag{7.5.1}$$

where H is the Hamiltonian described in Eq. 7.1.8. Expanding this matrix equation and substituting

$$\langle\sigma_{\mathrm{x}}\rangle = (\rho_{21}+\rho_{12}) \tag{7.5.2}$$

$$\langle\sigma_{\mathrm{y}}\rangle = i\,(\rho_{21}-\rho_{12}) \tag{7.5.3}$$

$$\langle\sigma_{\mathrm{z}}\rangle = \rho_{22}-\rho_{11} \tag{7.5.4}$$

we get the equations of motion in the rotation frame:

$$\langle \sigma_\mathrm{x} \rangle = \Delta \langle \sigma_\mathrm{y} \rangle \tag{7.5.5}$$
$$\langle \sigma_\mathrm{y} \rangle = -\Delta \langle \sigma_\mathrm{x} \rangle + \Omega \langle \sigma_\mathrm{z} \rangle \tag{7.5.6}$$
$$\langle \sigma_\mathrm{z} \rangle = -\Omega \langle \sigma_\mathrm{y} \rangle \tag{7.5.7}$$

These equations describe the motion of the spin exposed to an alternating field, however, the effects of relaxation and decoherence are still missing. In 1946, Felix Bloch extended this set of equations by empirical terms to allow for the relaxation to equilibrium. He assumed that the relaxations along the z-axis and in the $x - y$ plane happen at different rates, which are designated as $1/T_1$ and $1/T_2$ for the z-axis and the $x - y$ plane respectively. Including these terms results in the Bloch equations:

$$\langle \sigma_\mathrm{x} \rangle = -\Delta \langle \sigma_\mathrm{y} \rangle - \frac{\langle \sigma_\mathrm{x} \rangle}{T_2} \tag{7.5.8}$$

$$\langle \sigma_\mathrm{y} \rangle = \Delta \langle \sigma_\mathrm{x} \rangle + \Omega \langle \sigma_\mathrm{z} \rangle - \frac{\langle \sigma_\mathrm{y} \rangle}{T_2} \tag{7.5.9}$$

$$\langle \sigma_\mathrm{z} \rangle = -\Omega \langle \sigma_\mathrm{y} \rangle - \frac{\langle \sigma_\mathrm{z} \rangle}{T_1} \tag{7.5.10}$$

In case of no alternating field $\Omega = 0$ one can show that the solution to these equations is:

$$\langle \sigma_\mathrm{x} \rangle = \langle \sigma_\mathrm{x} \rangle_{t=0}\ cos(\Delta t) e^{-t/T_2} \tag{7.5.11}$$
$$\langle \sigma_\mathrm{y} \rangle = \langle \sigma_\mathrm{y} \rangle_{t=0}\ sin(\Delta t) e^{-t/T_2} \tag{7.5.12}$$
$$\langle \sigma_\mathrm{z} \rangle = \langle \sigma_\mathrm{z} \rangle_{t=0}\ (1 - e^{-t/T_1}) \tag{7.5.13}$$

Equations 7.5.11–7.5.13 describe the precession of a spin 1/2 with the detuning Δ around the z-axis. This precession is damped at a rate $1/T_2$ in the $x - y$-plane and with the rate $1/T_1$ along the z-axis. To measure the relaxation in the $x - y$-plane (free induction decay) a series of operations is performed. First, the spin is prepared along the $+z$-axis in the Bloch sphere at $t = 0$. Subsequently, we turn the spin into the equatorial plane using a MW pulse and thus create a superposition between the two spin states. The duration of this pulse was adjusted to perform a 90° rotation around the x-axis, which is why this type of pulse is referred to as a $(\pi/2)$ pulse (Fig. 7.16a). Afterward, we are waiting for the time τ, leading to the precession of the spin according to Eqs. 7.5.11–7.5.12 around the z-axis at the frequency Δ. Notice that $\langle \sigma_\mathrm{z} \rangle$ remains zero and only $\langle \sigma_\mathrm{x} \rangle$ and $\langle \sigma_\mathrm{y} \rangle$ are changing. Then a second $\pi/2$ pulse is rotating the spin back onto the z-axis. This operation transforms the former value of $\langle \sigma_\mathrm{y} \rangle$ to $\langle \sigma_\mathrm{z} \rangle$, which is measured subsequently. Repeating this pulse sequence (Fig. 7.16b) for different values of τ and measuring the resulting value of $\langle \sigma_\mathrm{z} \rangle$ leads to oscillations with a period of $1/\Delta$ — the so called Ramsey fringes (Fig. 7.16c). In the case of a single spin, many measurements are averaged to obtain the expectation value $\langle \sigma_\mathrm{z} \rangle$. Due to the changing environmental influence in each measurement, the spin performs rotations with slightly different angles at a given time τ between the two $\pi/2$ pulses. This dephasing mechanism between subsequent

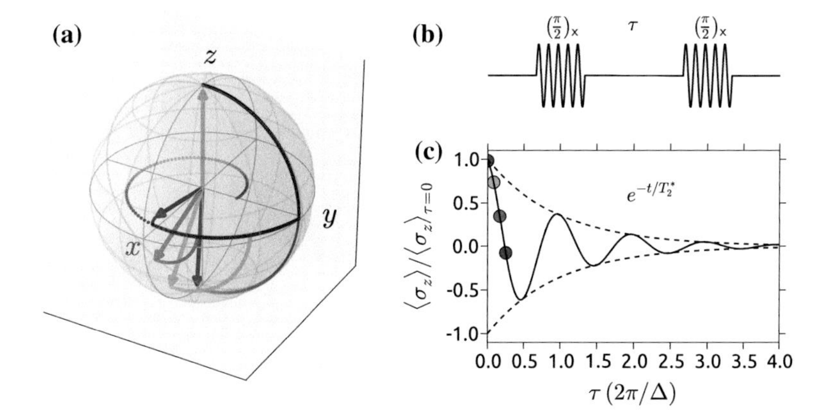

Fig. 7.16 **a** Bloch sphere trajectory of the spin wavefunction during the experiment. The Ramsey fringes are measured by applying a sequence of two MW pulses (**b**) to a spin, which was initially oriented along the z-axis. The first MW pulse rotates the vector by 90° into the $x-y$-plane, which is equivalent to a superposition of the two spin states. Waiting for a time τ, causes a damped precession of the Bloch vector with the frequency $2\pi/\Delta$ and at the rate T_2^*. The second MW pulse rotates the spin again by 90° around the x-axis, thus, mapping $\langle\sigma_y\rangle$ on $\langle\sigma_z\rangle$. **c** Repeating this sequence for different τ and measuring the resulting expectation value $\langle\sigma_z\rangle$ leads to oscillations decaying with e^{-t/T_2^*}

measurements leads to a decay faster than the decoherence time. The envelope of the oscillation is modeled by the function $exp(-t/T_2^*)$, where T_2^* is the dephasing time.

7.5.2 Experimental Results

From the previous section we know that the oscillation frequency of the Ramsey fringes is equal to the detuning $\Delta/2\pi$. Therefore, in order to adjust the oscillation period, the precise position of the resonance frequency ν_0 had first to be obtained. This was done by measuring the visibility of the Rabi oscillations as function of the frequency at a microwave power of 0 dBm (see Fig. 7.17a). By fitting a Lorentzian to the obtained data points, we found the maximum at 2449 MHz for $V_g = 2.205$ V. Afterward, we detuned the microwave source by 100 kHz in order to see Ramsey fringes with an oscillation period of 10 μs. In the next step we measured a full Rabi oscillation at $\nu = 2448.9$ MHz to determine the duration of the $\pi/2$ pulse (see Fig. 7.17b). By fitting the Rabi oscillation to a sine function, the duration of the $\pi/2$ pulse can be obtained and was $\simeq$284 ns.
Having calibrated the $\pi/2$ pulse at the microwave frequency of 2448.9 MHz, we measured the Ramsey fringes following the sequence presented in Fig. 7.18a. First the nuclear spin qubit was initialized by sweeping the magnetic field back and forth until the nuclear spin was in the $|+3/2\rangle$ state. Subsequently, two $\pi/2$ MW pulses were generated with the inter-pulse delay τ. At last, the final state was probed sweeping the magnetic field back to its initial value, while checking for a QTM transition.

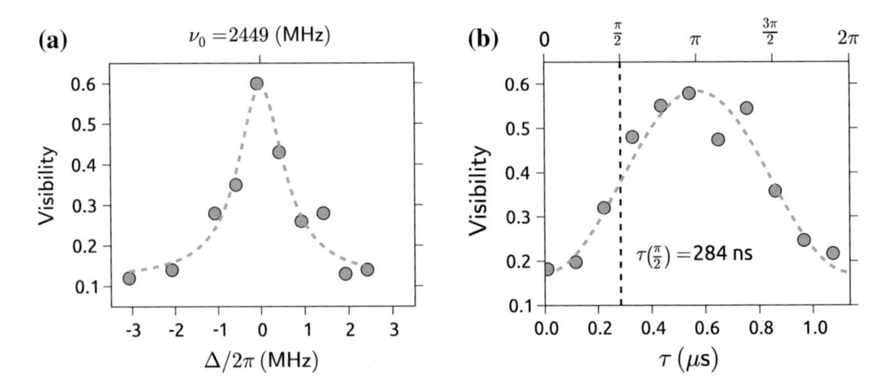

Fig. 7.17 **a** To calibrate the detuning, we measured the visibility of the Rabi oscillations as a function of the frequency at 0 dBm microwave power. The maximum of the visibility corresponds to zero detuning and was found at 2449 MHz. **b** In order to obtain Rabi oscillations with a period of 10 μs, we detuned the microwave source by 100 kHz to 2448.9 MHz and recorded a full period of a Rabi oscillation. Fitting the data to a sine function gave rise to a $\pi/2$ pulse length of 284 ns

If no QTM event was observed, the measurement was rejected. To obtain a good approximation of the expectation value this procedure was repeated 100 times for each inter-pulse delay τ, resulting in the Ramsey fringes as shown in Fig. 7.18b. The measurements exhibit an exponentially decaying cosine function. By fitting the data to $y = cos((\Delta/2\pi)t)exp(-t/T_2^*)$, we extracted a dephasing time $T_2^* \approx 64\ \mu$s.

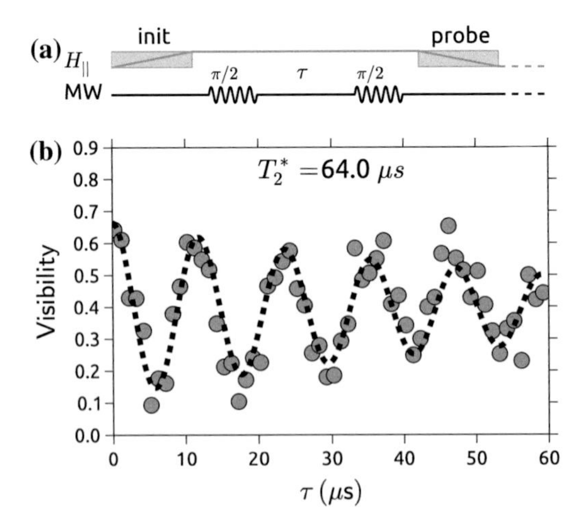

Fig. 7.18 **a** Time dependent external magnetic field $H_{||}$ and pulse sequence generated to measure the Ramsey fringes. Initialization and probe of the nuclear spin qubit are performed using the identical protocol explained in Fig. 3a. The MW sequence consists of two $\pi/2$ pulses, with an increasing inter-pulse delay τ. **b** Ramsey interference fringes obtained by repeating the procedure of (**a**) 800 times. V_g = 2.205 V, corresponding to a Rabi frequency Ω_R = 1.136 MHz and a resonant frequency ν_0 = 2.449 MHz of the nuclear spin qubit. The measured coherence time $T_2^* \approx 64\ \mu$s

Detailed studies suggest that the mayor contribution to the dephasing was caused by charge noise of the oxide and bit noise of the digital to analog converter at the gate terminal.The amplitude of the latter is estimated to be ± 1 bit resulting in gate voltage fluctuations of $\Delta V_{\mathrm{g}}(\text{bit noise}) = \pm 153\ \mu\text{V} \rightarrow$ Fig. 7.13 $\rightarrow \pm 26.2$ kHz $\rightarrow g_{\mathrm{N}}\mu_{\mathrm{N}}B_{\mathrm{eff}}/h \rightarrow \pm 2.6$ mT. Now we turn to the estimation of the noise generated by charges trapped in the gate oxide. If charges are trapped far away from the molecule those fluctuations are small, their frequency, however, is larger due to the multitude of available trapping sites. From our measurements we extracted that within the time scale of averaging over 1 data point we observed an effective gate voltage fluctuation of $\Delta V_{\mathrm{g}}(\text{bit noise}) = \pm 500\ \mu\text{V} \rightarrow$ Fig. 7.13 $\rightarrow \pm 85.9$ kHz $\rightarrow g_{\mathrm{N}}\mu_{\mathrm{N}}B_{\mathrm{eff}}/h \rightarrow$ ± 8.6 mT. Moreover, a charge can be trapped in the close vicinity of the molecule, leading to a gate voltage shift so large that nuclear spin is completely shifted out of resonance. Since the available sites in the close vicinity of the molecular are very few, this event happens in average every 1 to 2 days. Those events will not necessarily increase the decoherence since the changes are so drastic that we recalibrate the resonance frequency every time they occurred. However, they make the measurement of a complete series of Rabi and Ramsey oscillations, which took about 4 days, extremely difficult and time consuming.
In future devices we wil make use of more stabe gate oxides and well stabilized DA converters, which should increase the dephasing time by at least 2 orders of magnitude.

7.5.3 Outlook

In order to enhance the dephasing time T_2^*, the coupling to the environment must be attenuated. This can be achieve by actively controlling the time evolution of the spin precession. This so called dynamical decoupling relies on a series of MW pulses, which is periodically turning the spin [19]. One can show that environmental interactions, which happen on a time scale longer than the pulse series, are canceled out. The most common of these pulse sequences to dynamically decouple the spin from the environment was presented by Hahn [20] and involves as series of three pulses as shown in Fig. 7.19b. To visualize the experiment, we make again use of the Bloch sphere representation (see Fig. 7.19a). The first step, prior to the pulse sequence, is the initialization of the spin. In this example we will use the vector pointing along the $+z$ direction as initial state (grey vector in Fig. 7.19b). By applying a $\pi/2$ pulse, the Bloch vector will be rotated by 90° around the x-axis, thus creating a linear superposition state (blue curve). This state is left to a free evolution during the time τ and decoheres at a rate of T_2. After a waiting time τ, the vector has rotated by an angle ϕ in the equatorial plane, which is different every time we perform the experiment due to the fluctuations of the magnetic field along z. The second MW pulse (red) rotates the vector by an angle of 180° around x. This operation compensates any difference in ϕ, since vectors which were delayed, due to a slightly smaller magnetic field in z direction (dark red arrow), are now in advance. Followed

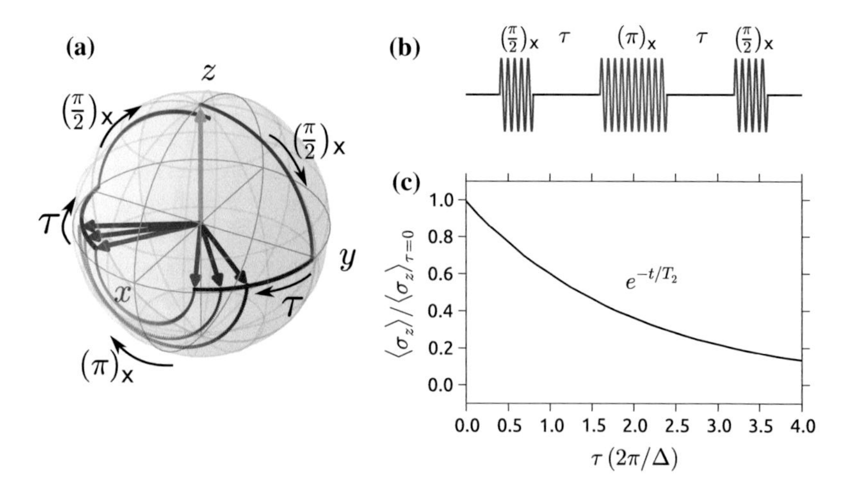

Fig. 7.19 **a** Trajectory of the Bloch vector in the Hahn spin echo experiment, in which a sequence of three MW pulses (**b**) is applied to a spin, initially aligned along the $+z$-axis. First, the vector is rotated by 90° into the equatorial plane using a $\pi/2$ pulse, thus, creating a linear superposition of the two spin states. Afterward, the vector performs a free precession around z for an interval τ, while being damped at the rate e^{-t/T_2}. During this time, magnetic field fluctuations along z result in slightly different precession angles from one measurement to another. A second MW pulse rotates the spin by an angle 180° around x, hence, ending up again in the x-y plane. Due to this operation, vectors which were formerly retarded to to a slightly smaller magnetic field are now in advance. Thus, after waiting for the same period τ, the vector will be aligned along the $-y$-axis. Finally, we project it back to the z-axis using a second $\pi/2$ pulse. **c** Repeating this sequence for different τ leads to an exponentially decaying spin echo signal of e^{-t/T_2}

by a second free precession of duration τ, the Bloch vector will arrive at $-y$ no matter what the local magnetic field was, as long as it remained constant on the time scale of the pulse series. Hence, all magnetic field fluctuations, which were much slower than the pulse sequence, are eliminated. Finally, the vector is rotated by 90° around x, which brings it back to its original position. Yet, the size of the vector is reduced due to decoherence within the $x - y$ plane, resulting in an exponentially decaying spin echo signal (see Fig. 7.19c). However, the characteristic time of the decay is the decoherence time T_2, which is much longer than T_2^* and can theoretically be extended to its fundamental limit $T_2 \leq 2T_1$.
The measurement protocol will be similar to the Ramsey experiment, but with an altered pulse sequence. We expect to eliminate the rather slow gate voltage fluctuations, which were transformed into magnetic field fluctuations by the hyperfine Stark effect and, hence, we should observe a T_2, which is much larger than T_2^*. However, the experimental realization of this experiment has not been performed yet, and could not be presented in my manuscript.

7.6 Summary

In this chapter, we presented the first quantum manipulation of a single nuclear spin qubit in a single-molecular magnet. To overcome the technical problem of generating high magnetic field amplitudes, we proposed and demonstrated the possibility to use the Stark shift of the hyperfine coupling to not only tune the level splitting of our nuclear spin qubit, but also to generate a large effective AC magnetic field at the nucleus. Using local AC electric fields, we performed electrical quantum manipulations of a single nuclear spin qubit at MHz frequencies with a coherence time $T_2^* \simeq 64$ μs. These results open the way to a fast coherent manipulation of a nuclear spin qubit as well as the opportunity to control the entanglement between different single nuclear spin qubits by tuning their resonance frequency using AC and DC gate voltages, by means of the Stark shift of the hyperfine coupling. Since this was only possible due to the unique electrostatic environment of a single molecule magnet, these results will hopefully make molecular based qubits serious candidates for quantum information processing.

References

1. T. Obata, M. Pioro-Ladrière, T. Kubo, K. Yoshida, Y. Tokura, S. Tarucha, Microwave band on-chip coil technique for single electron spin resonance in a quantum dot. Rev. Sci. Instrum. **78**, 104704 (2007)
2. J.J. Pla, K.Y. Tan, J.P. Dehollain, W.H. Lim, J.J.L. Morton, F.A. Zwanenburg, D.N. Jamieson, A.S. Dzurak, A. Morello, High-fidelity readout and control of a nuclear spin qubit in silicon. Nature **496**, 334–338 (2013)
3. K.C. Nowack, F.H.L. Koppens, Y.V. Nazarov, L.M.K. Vandersypen, Coherent control of a single electron spin with electric fields. Science (New York, N.Y.) **318**, 1430–1433 (2007)
4. L. Meier, G. Salis, I. Shorubalko, E. Gini, S. Schön, K. Ensslin, Measurement of Rashba and Dresselhaus spinâĂŞorbit magnetic fields. Nat. Phys. **3**, 650–654 (2007)
5. Y. Kato, R.C. Myers, D.C. Driscoll, A.C. Gossard, J. Levy, D.D. Awschalom, Gigahertz electron spin manipulation using voltage-controlled g-tensor modulation. Science (New York, N.Y.) **299**, 1201–1204 (2003)
6. E. Laird, C. Barthel, E. Rashba, C. Marcus, M. Hanson, A. Gossard, Hyperfine-mediated gate-driven electron spin resonance. Phys. Rev. Lett. **99**, 246601 (2007)
7. C.P. Slichter, *Principles of Magnetic Resonance* (Springer, Berlin, 1978)
8. J. Johansson, P. Nation, F. Nori, QuTiP: An open-source Python framework for the dynamics of open quantum systems. Comput. Phys. Commun. **183**, 1760–1772 (2012)
9. J.R. Johansson, P.D. Nation, F. Nori, QuTiP 2: A Python framework for the dynamics of open quantum systems. Comput. Phys. Commun. **184**, 1234 (2013)
10. R. Haun, J. Zacharias, Stark effect on cesium-133 hyperfine structure. Phys. Rev. **107**, 107–109 (1957)
11. B.E. Kane, A silicon-based nuclear spin quantum computer. Nature **393**, 133–137 (1998)
12. R. Rahman, C. Wellard, F. Bradbury, M. Prada, J. Cole, G. Klimeck, L. Hollenberg, High precision quantum control of single donor spins in silicon. Phys. Rev. Lett. **99**, 036403 (2007)
13. N. Ishikawa, M. Sugita, W. Wernsdorfer, Quantum tunneling of magnetization in lanthanide single-molecule magnets: bis(phthalocyaninato)terbium and bis(phthalocyaninato)dysprosium anions. Angew. Chem. (International ed. in English) **44**, 2931–2935 (2005)

14. C. Hutchison, E. Wong, Paramagnetic resonance in rare earth trichlorides. J. Chem. Phys. **29**, 754 (1958)
15. J.M. Baker, J.R. Chadwick, G. Garton, J.P. Hurrell, E.p.r. and Endor of TbFormula in thoria. Proc. R. Soc. A Math. Phys. Eng. Sci. **286**, 352–365 (1965)
16. B. Judd, *Operator Techniques in Atomic Spectroscopy* (McGraw-Hill Book Compny, Inc., New York, 1963)
17. J.A. Tuszynski, *Spherical Tensor Operators: Tables of Matrix Elements and Symmetries* (World Scientific, Singapore, 1990). ISBN 9810202830
18. R.M. Macfarlane, Optical Stark spectroscopy of solids. J. Lumin. **125**, 156–174 (2007)
19. L. Viola, S. Lloyd, Dynamical suppression of decoherence in two-state quantum systems. Phys. Rev. A **58**, 2733–2744 (1998)
20. E. Hahn, Spin echoes. Phys. Rev. **80**, 580–594 (1950)

Chapter 8
Conclusion and Outlook

In this thesis I developed an entire experimental setup to measure and manipulate quantum properties of single molecule magnets. Equipped with a new dilution refrigerator to meet the needs of ultra sensitive mesoscopic experiments, I designed and constructed innovative equipments such as miniaturized three dimensional vector magnets, capable of generating static magnetic fields in the order of a Tesla and allowing for sweep-rates larger than one Tesla per second.[1] Furthermore, I optimized the noise-filtering system and the signal amplifiers in order to suppress as much as possible the electronic noise pick-up.

After a thorough testing and approving of every part in the measurement chain, I turned to the fabrication of a single-molecule magnet spin-transistor, being probably one of the smallest devices presented in the field of organic spintronics. However, its tiny size of only 1 nm and the technological limitations of state of the art nanofabrication made it extremely challenging to build such a device. Despite this difficult conditions, I was able to perform experiments on three different samples, which demonstrated the feasibility and reproducibility of those cutting edge devices, even though the yield was rather small.

The extreme sensitivity of the molecular spin transistor, opened a path to access pristine quantum properties of an isolated single-molecule magnet, such as read-out of its quantized magnetic moment and the detection of the quantum tunneling of magnetization.

But most important, we were able to detect the four different quantum states of an isolated ^{159}Tb nuclear spin. By reading out the nuclear spin states much faster than the relaxation time T_1, we were able to measure the nuclear spin trajectory, revealing quantum jumps between the four different nuclear qubit states at a timescale of seconds. Finally, a post-treatment and statistical averaging of this data yielded the

[1] The maximum sweep-rate was tested in liquid helium.

S. Thiele, *Read-Out and Coherent Manipulation of an Isolated Nuclear Spin*,
Springer Theses, DOI 10.1007/978-3-319-24058-9_8

relaxation times T_1 of a few tens of seconds, which were resolved for each nuclear spin state individually.
However, the underlying physics, leading to the relaxation of the nuclear spin, remained hidden in the statistical average. In order to extract this very information, we developed a quantum Monte-Carlo code taking the specific experimental conditions into account. By fitting the statistical average of the simulations to the experimental data, we could deduce that the mechanism, dominating the relaxation process, was established over a coupling between the nuclear spin to the electrons tunneling through the read-out quantum dot. Using this knowledge, we could demonstrate that the experimental relaxation times could be modified just by changing the amount of tunnel electrons per unit time.
After having thoroughly investigated the quantum properties of an isolated nuclear spin using a passive read-out only, we wanted to take our experiments to the next level by actively manipulating the nuclear spin in a coherent manner. To overcome the technical problem of generating high magnetic field amplitudes, we proposed and demonstrated the possibility to exploit the Stark shift of the hyperfine coupling to accomplish this task. Not only could we tune the level splitting of our nuclear spin qubit, but also the generation of large effective AC magnetic fields at the nucleus was possible. In this way we performed the first electrical manipulation of a single nuclear spin. In combination with the tunability of the resonance frequency by means of a local gate voltage, the addressability of individual nuclear spins in the spirit of Kane's proposal [1] becomes possible.
During my thesis I could demonstrate that single-molecule magnets are potential candidates for quantum bits as defined by DiVincenzo [2] (see Chap. 1 for further explanation):

- **Information storage on qubits**: I could show that information could be stored on the single nuclear spin of an isolated $TbPc_2$ SMM (Chap. 6).
- **Initial state preparation**: the initial state preparation is up to now rather passive, however, due to the quantum nondestructive nature of the measurement scheme, the initial state can be prepared by the measurement itself.
- **Isolation**: since we used a nuclear spin qubit, isolation is one of its intrinsic properties. I could show in Chap. 7 that the dephasing time T_2^* of our nuclear spin qubit is about 64 μs.
- **Gate implementation**: due to a very large effective magnetic field, created by the hyperfine Stark effect, a coherent manipulation of the nuclear spin could be performed within 300 ns, which was 200 times faster than the coherence time T_2.
- **Read-out**: I demonstrated in Chap. 6 that the read-out of the nuclear spin qubit state was performed with fidelities better than 87 % (sample C). Note that this is no intrinsic limitation of the system and will be improved in future experiments.

However, to be competitive with existing qubit systems, the read-out of the nuclear spin state needs to be speed up by at least three orders of magnitude. To achieve this acceleration of the read-out cycle, a different detection scheme is necessary. On of the most established methods in nuclear spin based qubits takes advantage of the

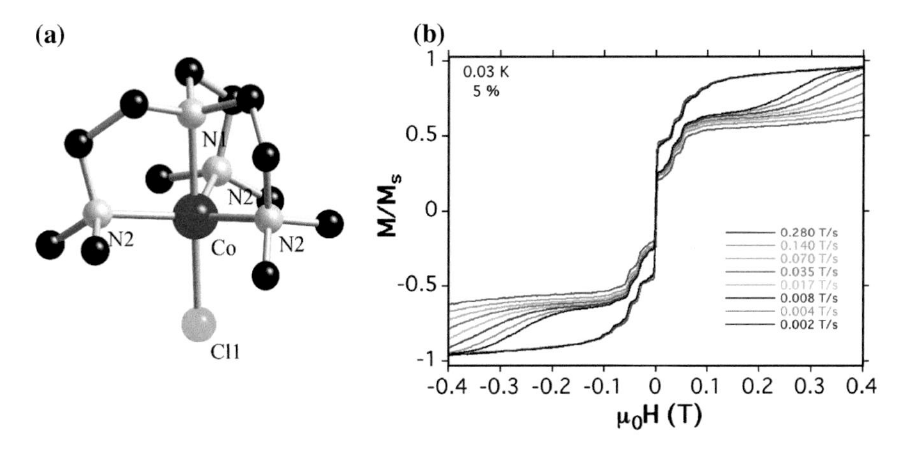

Fig. 8.1 **a** Structure of the Co(Me6tren)Cl SMM. **b** Hysteresis loop of a diluted Co(Me6tren)Cl crystal obtained with a microsquid. Steps in the hysteresis indicate the nuclear spin resolved quantum tunneling of magnetization

nuclear spin resolved electron spin resonance (ESR) as a read-out tool. As already pointed out in Chap. 7, this technique seems to be incompatible $TbPc_2$ SMMs due to the spin ground state of $m_J = \pm 6$ and an excited state separation of about 12.5 THz. However, different kinds of single-molecule magnets with ESR compatible properties could be thought of. Such a molecule is for example the Co(Me6tren)Cl SMM (see Fig. 8.1). It is one of the first mononuclear single-molecule magnets based on transition metal ion. Those molecules are chemically more stable than their polynuclear counterparts, which allows the manipulation and study of its magnetic properties on the single molecule level. It has an Ising type spin ground state of $S = \pm 3/2$, which makes a ESR transition more likely. Furthermore, ^{59}Co is among the 22 existing elements having only one natural isotopic abundance, which is of major importance for its use as a nuclear spin qubit. First, preliminary measurements show, that the hyperfine interaction of the Co^{2+} ion is comparable to the Tb^{3+}, as the steps in the hysteresis curve coming from the hyperfine coupling are well distinguished (see Fig. 8.1(b)). However, its compatibility to the molecular spin-transistor design and the ESR transition between the $S = \pm 3/2$ ground states still needs to be proven. Another important point, which has not been shown yet, is the scalability of our qubits. In contrary to common top-down approaches, like coupling different qubits in a cavity, we want to exploit the potential of organic chemistry in designing molecules including more than one qubit. Starting from the mononuclear terbium double-decker SMM (see Fig. 8.2a), a two qubit system could be made by using a triple-decker SMM with two terbium ions (see Fig. 8.2b). The coupling between the terbium ions is established via the exchange interaction, mediated using an unbound electron of the phthalocyanine ligands. In the single-molecule spin-transistor layout the electron can be easily removed or added by means of the gate voltage. Thus, the coupling between the terbium ions can be switched on or off, allowing for the control of entanglement.

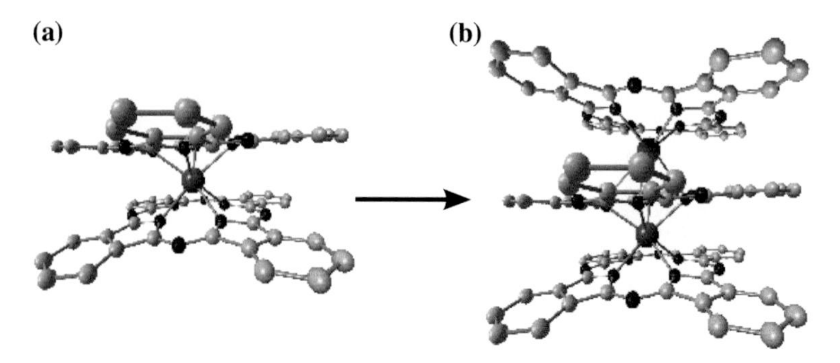

Fig. 8.2 a Structure of the $TbPc_2$ SMM. **b** Structure of the Tb_2Pc_3 SMM

The results presented in this thesis, extent the potential of molecular spintronics beyond classical data storage. We demonstrated the first experimental evidence of a coherent nuclear spin manipulation inside a single-molecule magnet, and therefore build the foundation for the first molecular quantum bits. Their great versatility holds a lot of promises for a variety of future applications and, maybe one day, a molecular quantum computer.

References

1. B.E. Kane, A silicon-based nuclear spin quantum computer. Nature **393**, 133–137 (1998)
2. D.P. DiVincenzo, *Topics in Quantum Computers*, vol. 345. NATO Advanced Study Institute. Ser. E Appl. Sci. vol. 345 (1996)

Appendix A
Spin

A.1 Charged Particle in a Magnetic Field

If an atom is exposed to an external magnetic field it will experience an interaction which can be quantified by the following Hamiltonian.

$$H = \sum_{i=1}^{Z} \frac{1}{2m_{\mathrm{e}}} \left(\boldsymbol{p}_i + e\boldsymbol{A}(\boldsymbol{r}_i)\right)^2 + V(\boldsymbol{r}_i), \tag{A.1.1}$$

where $\boldsymbol{p}$ is the momentum operator, $(-e)$ the charge of an electron and $\boldsymbol{A}$ the vector potential. To simplify the calculation, the Coulomb gauge $div\boldsymbol{A} = 0$ can be used, which makes the operators $\boldsymbol{p}$ and $\boldsymbol{A}$ commute. Moreover, the vector potential is chosen to be $\boldsymbol{A} = \frac{1}{2}(\boldsymbol{B} \times \boldsymbol{r})$. Inserting this into Eq. A.1.1 and expanding the canonical momentum gives:

$$H = \sum_{i=1}^{Z} \frac{\boldsymbol{p}_i^2}{2m_{\mathrm{e}}} + \frac{e}{2m_{\mathrm{e}}} \boldsymbol{p}_i (\boldsymbol{B} \times \boldsymbol{r}_i) + \frac{e^2}{2m_{\mathrm{e}}} \left(\frac{\boldsymbol{B} \times \boldsymbol{r}_i}{2}\right)^2 + V(\boldsymbol{r}_i) \tag{A.1.2}$$

Applying the rules for triple products: $\boldsymbol{p}_i (\boldsymbol{B} \times \boldsymbol{r}_i) = \boldsymbol{B}\left(\boldsymbol{r}_i \times \boldsymbol{p}_i\right)$, and inserting the electron orbital angular momentum $\boldsymbol{l}_i = \boldsymbol{r}_i \times \boldsymbol{p}_i$ we get:

$$H = \sum_{i=1}^{Z} \frac{\boldsymbol{p}_i^2}{2m_{\mathrm{e}}} + V(\boldsymbol{r}_i) + \frac{e}{2m_{\mathrm{e}}} \boldsymbol{B} \sum_{i=1}^{Z} \boldsymbol{L}_i + \sum_{i=1}^{Z} \frac{e^2 \boldsymbol{B}^2}{8m_{\mathrm{e}}} \boldsymbol{r}_i^2 sin^2(\theta_i) \tag{A.1.3}$$

Substituting $\sum_{i=1}^{Z} \frac{p_i^2}{2m_{\mathrm{e}}} + V(\boldsymbol{r}_i)$ with the Hamiltonian in absence of a magnetic field H_0, $\frac{e\hbar}{2m_{\mathrm{e}}}$ with the Bohr magneton μ_{B} and $\sum_{i=1}^{Z} \boldsymbol{l}_i$ with total orbital momentum $\boldsymbol{L}$, gives rise to final Hamiltonian:

S. Thiele, *Read-Out and Coherent Manipulation of an Isolated Nuclear Spin*, Springer Theses, DOI 10.1007/978-3-319-24058-9

$$H = H_0 + \underbrace{\mu_{\mathrm{B}} \frac{\boldsymbol{L}}{\hbar} \boldsymbol{B}}_{H_{\mathrm{para}}} + \underbrace{\sum_{i=1}^{Z} \frac{e^2 \boldsymbol{B}^2}{8 m_{\mathrm{e}}} \boldsymbol{r}_i^2 sin^2(\theta_i)}_{H_{\mathrm{dia}}>0} \tag{A.1.4}$$

This Hamiltonian is divided into three parts. The first one, H_0, describes the atom without a magnetic field. In the second term is a scalar product of $\boldsymbol{L}$ and $\boldsymbol{B}$, which will align two vectors anti-parallel in order to minimize the energy and is responsible for the paramagnetism. The last part of the equation is always positive and therefore increasing the energy of the atom. It describes the diamagnetic response to an applied field. If we put some numbers into the equation, e.g. $|\boldsymbol{L}|/\hbar = \sqrt{2}$, $|\boldsymbol{B}| = 1\mathrm{T}$, $\boldsymbol{r}_i^2 = 0.3\mathrm{nm}$, we get $H_{\mathrm{para}} \approx 100\ \mu\mathrm{eV}$ and $H_{\mathrm{dia}} \approx 1$ neV. Thus, the diamagnetism is much smaller than the paramagnetism and is only of significance in systems with closed or half filled shells. The total magnetic moment $\boldsymbol{\mu}$ is calculated by taking the first derivative of the Hamilton operator with respect to the magnetic field. Using Eq. A.1.4 results in:

$$\boldsymbol{\mu} = \frac{\partial H}{\partial B} = \underbrace{\mu_{\mathrm{B}} \frac{\boldsymbol{L}}{\hbar}}_{\mu_{\mathrm{L}}} + \underbrace{\sum_{i=1}^{Z} \frac{e^2 \boldsymbol{B}}{4 m_{\mathrm{e}}} \boldsymbol{r}_i^2 sin^2(\theta_i)}_{\mu_{\mathrm{ind}}}$$

The last term arises only for finite B, and describes the induced magnetic moment. The first term, however, is present also at zero magnetic field and represents a permanent magnetic moment due to the orbital motion:

$$\boldsymbol{\mu}_{\mathrm{L}} = \mu_{\mathrm{B}} \frac{\boldsymbol{L}}{\hbar} \tag{A.1.5}$$

where $\boldsymbol{L}$ is the total orbital angular momentum with its quantum number L. Its modulus is obtained by:

$$|\boldsymbol{L}| = \hbar\sqrt{L(L+1)} \tag{A.1.6}$$

and its projection on the z axis is given as:

$$L_{\mathrm{z}} = \hbar m_{\mathrm{L}} \tag{A.1.7}$$

with m_{L} being the magnetic orbital quantum number ranging from $-L$ to L, and having therefore $2L+1$ possible values.

A.2 Electron Spin

When Stern and Gerlach did their famous experiment in 1922, they discovered that electrons posses an internal permanent magnetic moment, which is independent of its orbital motion and takes only two quantized values. In analogy to the orbital angular momentum $\boldsymbol{L}$ it is assumed that an additional intrinsic angular momentum $\boldsymbol{S}$, which is called spin, gives rise to this permanent magnetic moment. Similar to Eqs. A.1.5–A.1.7 we define a magnetic moment:

$$\boldsymbol{\mu}_{\mathrm{S}} = g_{\mathrm{S}}\mu_{\mathrm{B}}\frac{\boldsymbol{S}}{\hbar} \tag{A.2.1}$$

where g is the Landée factor and $\boldsymbol{S}$ is the total spin with its quantum number S. The modulus of $\boldsymbol{S}$ is calculated as:

$$|\boldsymbol{S}| = \hbar\sqrt{S(S+1)} \tag{A.2.2}$$

and its projection on the z-axis is given by:

$$S_{\mathrm{z}} = \hbar m_{\mathrm{S}} \tag{A.2.3}$$

In contrary to the orbital angular momentum, the magnetic moment is increased by the Landée factor and S can take half integer values.
The Hamiltonian of a charge particle with spin S modifies to:

$$H = H_0 + \underbrace{\mu_{\mathrm{B}}\frac{\boldsymbol{L} + g_{\mathrm{S}}\boldsymbol{S}}{\hbar}\boldsymbol{B}}_{H_{\mathrm{para}}} + \underbrace{\sum_{i=1}^{Z}\frac{e^2\boldsymbol{B}^2}{8m_{\mathrm{e}}}\boldsymbol{r}_i^2 sin^2(\theta_i)}_{H_{\mathrm{dia}}>0} \tag{A.2.4}$$

A.3 Spin Matrices

Let us first consider a system with only two spin values: $s = 1/2$ and $m_{\mathrm{S}} = -1/2, +1/2$. This is a very simple case, but helps understanding more difficult spin system. When calculating energy levels of spin 1/2 systems it is convenient to work with the matrix representation, where the wave function Ψ is a vector with the spin up and down amplitude and the operator $\boldsymbol{S}$ is a vector of two by two matrices $((2S+1)\times(2S+1))$:

$$\boldsymbol{S} = \frac{\hbar}{2}\boldsymbol{\sigma} \tag{A.3.1}$$

where $\boldsymbol{\sigma}$ are the so-called Pauli matrices:

$$\sigma_{\mathrm{x}} = \begin{pmatrix} 0 & 1 \\ 1 & 0 \end{pmatrix}, \quad \sigma_{\mathrm{y}} = \begin{pmatrix} 0 & -i \\ i & 0 \end{pmatrix}, \quad \sigma_z = \begin{pmatrix} 1 & 0 \\ 0 & -1 \end{pmatrix}$$

Often, instead of σ_{x} and σ_{y}, their linear combinations $\sigma_+ = (\sigma_{\mathrm{x}} + i\sigma_{\mathrm{y}})$ and $\sigma_- = \sigma_{\mathrm{x}} - i\sigma_{\mathrm{y}}$ are used since they are more adapted to the spin up and spin down basis.

$$\sigma_+ = \begin{pmatrix} 0 & 1 \\ 0 & 0 \end{pmatrix}, \quad \sigma_- = \begin{pmatrix} 0 & 0 \\ 1 & 0 \end{pmatrix} \tag{A.3.2}$$

In this representation the Zeeman energy is calculated by diagonalizing the following Hamiltonian:

$$H_{\mathrm{Zeeman}} = \frac{1}{2}\mu_{\mathrm{B}} g_{\mathrm{S}} \left[B_{\mathrm{x}} \begin{pmatrix} 0 & 1 \\ 1 & 0 \end{pmatrix} + B_{\mathrm{y}} \begin{pmatrix} 0 & -i \\ i & 0 \end{pmatrix} + B_z \begin{pmatrix} 1 & 0 \\ 0 & -1 \end{pmatrix} \right]$$

Analogue to the spin 1/2 system those matrices can be calculated for spin systems of order N, where σ_{z} is a $(2N+1) \times (2N+1)$ matrix with only diagonal elements.

$$\sigma_{\mathrm{z}}(N) = \begin{pmatrix} -N & & 0 \\ & \ddots & \\ 0 & & N \end{pmatrix} \tag{A.3.3}$$

The matrices $\sigma_{\mathrm{x}}(N)$ and $\sigma_{\mathrm{y}}(N)$ are obtained via $\sigma_+(N)$ and $\sigma_-(N)$:

$$\sigma_\pm(N) = \sqrt{N(N+1) - m_{\mathrm{N}}(m_{\mathrm{N}} \pm 1)}\ \delta_{i\pm1,j} \tag{A.3.4}$$

A.4 Dirac Equation and Spin-Orbit Coupling

The origin of spin and therefore spin-orbit interaction lies in the relativistic nature of electrons. Relativity theory teaches us that the energy of an electron is calculated by: $E = \sqrt{c^2 \boldsymbol{p}^2 + m_{\mathrm{e}}^2 c^4}$, where c is the speed of light, m_{e} the free electron mass and $\boldsymbol{p}$ the relativistic, classical momentum: $\boldsymbol{p} = \left(1 - v^2/c^2\right)^{-1/2} m_{\mathrm{e}} \boldsymbol{v}$. Due to its non-linearity it is not so easy to translate this equation using the correspondence principle of quantum mechanics into an operator. The only way to solve this problem is to linearize the above equation. It can be shown, that this is only possible by rewriting the standard representation of the Schrödinger equation in the matrix representation. The idea is to find a matrix which multiplied by itself, gives the energy eigenvalues squared. The solution to this problem was found by Paul Dirac in 1928 and has the following form:

$$H_{\mathrm{D}} = \begin{pmatrix} m_{\mathrm{e}}c^2 & 0 & cp_{\mathrm{z}} & c(p_{\mathrm{x}} - ip_{\mathrm{y}}) \\ 0 & m_{\mathrm{e}}c^2 & c(p_{\mathrm{x}} - ip_{\mathrm{y}}) & -cp_{\mathrm{z}} \\ cp_{\mathrm{z}} & c(p_{\mathrm{x}} - ip_{\mathrm{y}}) & -m_{\mathrm{e}}c^2 & 0 \\ c(p_{\mathrm{x}} - ip_{\mathrm{y}}) & -cp_{\mathrm{z}} & 0 & -m_{\mathrm{e}}c^2 \end{pmatrix}$$

which multiplied by itself gives:

$$H_{\mathrm{D}}^2 = \begin{pmatrix} c^2\boldsymbol{p}^2 + m_{\mathrm{e}}^2c^4 & 0 & 0 & 0 \\ 0 & c^2\boldsymbol{p}^2 + m_{\mathrm{e}}^2c^4 & 0 & 0 \\ 0 & 0 & c^2\boldsymbol{p}^2 + m_{\mathrm{e}}^2c^4 & 0 \\ 0 & 0 & 0 & c^2\boldsymbol{p}^2 + m_{\mathrm{e}}^2c^4 \end{pmatrix}$$

The energy eigenvalues of the Dirac Hamiltonian are:

$$E = \pm\sqrt{c^2\boldsymbol{p}^2 + m_{\mathrm{e}}^2c^4}$$

Where each eigenvalue is twice degenerate. The positive energies are describing electrons, whereas the negative energies are for positrons. The time independent Dirac equation is then:

$$\begin{pmatrix} m_{\mathrm{e}}c^2 & 0 & cp_{\mathrm{z}} & c(p_{\mathrm{x}} - ip_{\mathrm{y}}) \\ 0 & m_{\mathrm{e}}c^2 & c(p_{\mathrm{x}} - ip_{\mathrm{y}}) & -cp_{\mathrm{z}} \\ cp_{\mathrm{z}} & c(p_{\mathrm{x}} - ip_{\mathrm{y}}) & -m_{\mathrm{e}}c^2 & 0 \\ c(p_{\mathrm{x}} - ip_{\mathrm{y}}) & -cp_{\mathrm{z}} & 0 & -m_{\mathrm{e}}c^2 \end{pmatrix} \begin{pmatrix} \Psi_{\mathrm{e}}^{\uparrow} \\ \Psi_{\mathrm{e}}^{\downarrow} \\ \chi_p^{\uparrow} \\ \chi_p^{\downarrow} \end{pmatrix} = E \begin{pmatrix} \Psi_{\mathrm{e}}^{\uparrow} \\ \Psi_{\mathrm{e}}^{\downarrow} \\ \chi_p^{\uparrow} \\ \chi_p^{\downarrow} \end{pmatrix}$$

where $\Psi_{\mathrm{e}}^{\uparrow}, \Psi_{\mathrm{e}}^{\downarrow}$ is the up-spin or down spin electron wave function and $\chi_p^{\uparrow}, \chi_p^{\downarrow}$ is the up-spin or down-spin positron wave function, respectively. To describe an relativistic electron in an electro-magnetic field the following substitutions are usually made: $\boldsymbol{p} \rightarrow \boldsymbol{p} + e\boldsymbol{A}$ and $E = E + e\phi$. Where $\boldsymbol{A}$ and ϕ are the magnetic vector potential and the electric scalar potential, respectively. In the following we want combine the up-spin and down-spin component to get smaller expressions. It can be shown easily that: $c\boldsymbol{p\sigma} = c(p_{\mathrm{x}}\sigma_{\mathrm{x}} + p_{\mathrm{y}}\sigma_{\mathrm{y}} + p_{\mathrm{z}}\sigma_{\mathrm{z}}) = \begin{pmatrix} cp_{\mathrm{z}} & c(p_{\mathrm{x}} - ip_{\mathrm{y}}) \\ c(p_{\mathrm{x}} - ip_{\mathrm{y}}) & -cp_{\mathrm{z}} \end{pmatrix}$. Thus we get:

$$\begin{pmatrix} m_{\mathrm{e}}c^2 & c(\boldsymbol{p} + e\boldsymbol{A})\boldsymbol{\sigma} \\ c(\boldsymbol{p} + e\boldsymbol{A})\boldsymbol{\sigma} & -m_{\mathrm{e}}c^2 \end{pmatrix} \begin{pmatrix} \Psi \\ \phi \end{pmatrix} = (E + e\phi) \begin{pmatrix} \Psi \\ \phi \end{pmatrix}$$

This is a coupled equation of Ψ and ϕ. Expanding this matrix equation results in:

$$\left(E - m_{\mathrm{e}}c^2 + e\phi\right)|\Psi> = c(\boldsymbol{p} + e\boldsymbol{A})\boldsymbol{\sigma}|\chi> \tag{A.4.1}$$

$$\left(E + m_{\mathrm{e}}c^2 + e\phi\right)|\chi> = c(\boldsymbol{p} + e\boldsymbol{A})\boldsymbol{\sigma}|\Psi> \tag{A.4.2}$$

We can therefore express $|\chi>$ in terms of $|\Psi>$:

$$|\chi> = \frac{c}{\left(E + m_{\mathrm{e}}c^2 + e\phi\right)}(\boldsymbol{p} + e\boldsymbol{A})\boldsymbol{\sigma}|\Psi>$$

Until now everything is exact. To simplify this equation we use the Taylor series expansion.

$$|\chi > \approx \frac{1}{2m_{\mathrm{e}}c}\left(1 - \frac{E - m_{\mathrm{e}}c^2 + e\phi}{2m_{\mathrm{e}}c^2}\right)(\boldsymbol{p} + e\boldsymbol{A})\boldsymbol{\sigma}|\Psi >$$

Substituting this result into Eq. A.4.1 gives:

$$\begin{aligned}\left(E - m_{\mathrm{e}}c^2 + e\phi\right)|\Psi\rangle &\approx \frac{1}{2m_{\mathrm{e}}}(\boldsymbol{p} + e\boldsymbol{A})\boldsymbol{\sigma}\left(1 - \frac{E - m_{\mathrm{e}}c^2 + e\phi}{2m_{\mathrm{e}}c^2}\right)(\boldsymbol{p} + e\boldsymbol{A})\boldsymbol{\sigma}|\Psi\rangle \\ &= \left[\frac{\left[(\boldsymbol{p} + e\boldsymbol{A})\boldsymbol{\sigma}\right]^2}{2m_{\mathrm{e}}}\left(1 - \frac{E - m_{\mathrm{e}}c^2}{2m_{\mathrm{e}}c^2}\right)\right. \\ &\quad \left. - \frac{e}{4m_{\mathrm{e}}^2c^2}(\boldsymbol{p} + e\boldsymbol{A})\boldsymbol{\sigma}\,(\phi)\,(\boldsymbol{p} + e\boldsymbol{A})\boldsymbol{\sigma}\right]|\Psi\rangle\end{aligned}$$

We used the fact that the operator $(\boldsymbol{p} + e\boldsymbol{A})\boldsymbol{\sigma}$ is not acting on $E - m_{\mathrm{e}}c^2$. Now we want to expand the second term. To do so we recall that the momentum operator $\boldsymbol{p} = -i\hbar\nabla$ and that $\boldsymbol{p}\phi = \phi\boldsymbol{p} - i\hbar\nabla\phi$. Inserting this into the above equation gives:

$$(\boldsymbol{p} + e\boldsymbol{A})\boldsymbol{\sigma}\,(\phi)\,(\boldsymbol{p} + e\boldsymbol{A})\boldsymbol{\sigma} = \phi\left[(\boldsymbol{p} + e\boldsymbol{A})\boldsymbol{\sigma}\right]^2 - i\hbar\left[(\nabla\phi)\,\boldsymbol{\sigma}\right]\left[(\boldsymbol{p} + e\boldsymbol{A})\,\boldsymbol{\sigma}\right]$$

using the equation: $(\boldsymbol{X}\boldsymbol{\sigma})\,(\boldsymbol{Y}\boldsymbol{\sigma}) = \boldsymbol{X}\boldsymbol{Y} + i\boldsymbol{\sigma}\,(\boldsymbol{X} \times \boldsymbol{Y})$ we end up with:

$$-i\hbar\,(\nabla\phi)\,\boldsymbol{\sigma}\left[(\boldsymbol{p} + e\boldsymbol{A})\,\boldsymbol{\sigma}\right] = -i\hbar\,(\nabla\phi)\,(\boldsymbol{p} + e\boldsymbol{A}) + \hbar\boldsymbol{\sigma}\left[(\nabla\phi) \times (\boldsymbol{p} + e\boldsymbol{A})\right]$$

$$\begin{aligned}\left(E - m_{\mathrm{e}}c^2 + e\phi\right)|\Psi > = &\left[\underbrace{\frac{[(\boldsymbol{p} + e\boldsymbol{A})\boldsymbol{\sigma}]^2}{2m_{\mathrm{e}}}\left(1 - \frac{E - m_{\mathrm{e}}c^2 + e\phi}{2m_{\mathrm{e}}c^2}\right)}_{\text{Pauli equation + relativistic correction}}\right]|\Psi\rangle \\ &+ \left[\underbrace{\frac{e}{4m_{\mathrm{e}}^2c^2}\nabla\phi\,(\boldsymbol{p} + e\boldsymbol{A})}_{\text{Darwin--term}} - \underbrace{\frac{\hbar e}{4m_{\mathrm{e}}^2c^2}\boldsymbol{\sigma}\,[\nabla\phi \times (\boldsymbol{p} + e\boldsymbol{A})]}_{\text{spin--orbit--term}}\right]|\psi\rangle\end{aligned}$$

We are now concentrating only on the last term, since it is the most interesting for our purposes.
By changing to the spherical coordinate system:

$$\nabla\phi = \frac{1}{r}\frac{d\phi}{dr}\boldsymbol{r}$$

resulting in:

$$\frac{\hbar e}{4m_e^2c^2}\boldsymbol{\sigma}\left[\nabla\phi \times (\boldsymbol{p}+e\boldsymbol{A})\right] = \frac{\hbar e}{4m_e^2c^2}\boldsymbol{\sigma}\left[\frac{1}{r}\frac{d\phi}{dr}\boldsymbol{r}\times(\boldsymbol{p}+e\boldsymbol{A})\right]$$

Since $\boldsymbol{p}+e\boldsymbol{A}$ is the canonical momentum the expression $\boldsymbol{r}\times(\boldsymbol{p}+e\boldsymbol{A})$ gives us the orbital momentum $\boldsymbol{l}$. The term $\frac{1}{r}\frac{d\phi}{dr}$ is just a scalar and can be combined with the pre-factor to the spin-orbit coupling constant $\xi = \frac{\hbar e}{4m_e^2c^2}\left(\frac{1}{r}\frac{d\phi}{dr}\right)$. Since $\boldsymbol{\sigma}$ is the operator for the spin $\boldsymbol{s}$ we result in the final one electron spin-orbit Hamiltonian:

$$H_{\mathrm{so}} = \xi\,\boldsymbol{l}\boldsymbol{s}$$

If we are now considering systems with more than one electron, there are two possibility of how the spin-orbit coupling effects the orbital energies. The first and for us less interesting case is a system where the spin-orbit coupling is larger than the electron-electron interaction. There each electrons spin $\boldsymbol{s}_i$ couples with its orbit $\boldsymbol{l}_i$ to form an total momentum $\boldsymbol{j}_i = \boldsymbol{l}_i + \boldsymbol{s}_i$. The coupling energy is than given by $H_{l_i,s_i} = c_{ii}\boldsymbol{l}_i\boldsymbol{s}_i$. In the second case the electron-electron interaction, or in other words the coupling between different orbital momenta $H_{l_il_j} = a_{ij}\boldsymbol{l}_i\boldsymbol{l}_j$ and spins $H_{s_is_j} = b_{ij}\boldsymbol{s}_i\boldsymbol{s}_j$ is larger than the spin-orbit coupling. Now the different orbital momenta couple to a total orbital momentum $\boldsymbol{L} = \sum_i \boldsymbol{l}_i$ and the different spins couple to a total spin $\boldsymbol{S} = \sum_i \boldsymbol{s}_i$ before coupling the the total momentum $\boldsymbol{J} = \boldsymbol{L}+\boldsymbol{S}$. The spin-orbit coupling energy is than given by: $H_{\mathrm{so}} = \lambda\,\boldsymbol{L}\boldsymbol{S}$. With this knowledge we can also try to understand the 3. Hunds rule. Therefore we are relating the one electron spin-orbit coupling constant ξ with λ:

$$H_{\mathrm{so}} = \xi\sum_i \boldsymbol{l}_i \sum_i \boldsymbol{s}_i = \lambda\boldsymbol{L}\boldsymbol{S}$$

Therefore

$$\lambda = \frac{\xi\sum_i \boldsymbol{l}_i\boldsymbol{l}_i}{\boldsymbol{L}\boldsymbol{S}}$$

for less than half filled shells $\boldsymbol{s}_i$ is always $\frac{1}{2}$ and can be put in from of the sum. Thus λ becomes positive for less than half filled shells and the ground state is $J = |L-S|$.

$$\lambda = \frac{\frac{1}{2}\xi\sum_i \boldsymbol{L}_i}{\boldsymbol{L}\boldsymbol{S}} = \frac{\xi}{2\boldsymbol{S}} > 0$$

For more than half filled shells $\boldsymbol{s}_i$ has values of $+\frac{1}{2}$ and $-\frac{1}{2}$ and the sum is split in two:

$$\lambda = \frac{\frac{1}{2}\xi \overbrace{\sum_{i}^{half} \boldsymbol{L}_i}^{0}}{\boldsymbol{LS}} - \frac{\frac{1}{2}\xi \overbrace{\sum_{half}^{n} \boldsymbol{L}_i}^{\boldsymbol{L}}}{\boldsymbol{LS}} = -\frac{\xi}{2\boldsymbol{S}} < 0$$

Now λ becomes negative and $J = L + S$ is the new ground state, since it has the smallest energy.

Appendix B
Stevens Operators

$$
\begin{aligned}
O_2^0 &= 3J_z^2 - J(J+1) \\
O_4^0 &= 35J_z^4 - 30J(J+1)J_z^2 + 25J_z^2 - 6J(J+1) + 3J^2(J+1)^2 \\
O_4^4 &= \frac{1}{2}(J_+^4 + J_-^4) \\
O_6^0 &= 231J_z^6 - 315J(J+1)J_z^4 + 735J_z^4 + 105J^2(J+1)^2J_z^2 - 525J(J+1)J_z^2 + \\
&\quad +294J_z^2 - 5J^3(J+1)^3 + 40J^2(J+1)^2 - 60J(J+1) \\
O_6^4 &= \frac{1}{4}\Big[(11J_z^2 - J(J+1) - 38)(J_+^4 + J_-^4) + (J_+^4 + J_-^4)(11J_z^2 - J(J+1) - 38)\Big]
\end{aligned}
$$

where J_z, J_+ and J_- are the generalized Pauli operators of order N.

S. Thiele, *Read-Out and Coherent Manipulation of an Isolated Nuclear Spin*,
Springer Theses, DOI 10.1007/978-3-319-24058-9

Appendix C
Quantum Monte Carlo Code

The following python code is based on a quantum Monte Carlo algorithm and was used to simulate the nuclear spin trajectory.

```
from pylab import *
from scipy.optimize.minpack import curve_fit
import pickle
class QMC:
    def __init__(self,T,Gamma,dt,delta_t,p_LZ):

        self.Psi0 = array([0,0,0,1]) # initial state
        self.H0   = array([0.0,121.0,270.0,448.0]) # in mK
        self.dt   = dt # QMC time step
        self.delta_t = delta_t # measurement interval
        self.Gam= Gamma
        self.T    = T
        self.p_LZ = p_LZ
        self.d_omega = diff(self.H0)
        self.n_T = array([1./(exp((self.d_omega[0])/T)-1),
                          1./(exp((self.d_omega[1])/T)-1),
                          1./(exp((self.d_omega[2])/T)-1)])
        self.scl = array([1.0,2.0,2.2])
        self.C1C1 = self.Gam*array([0,
                                    self.scl[0]*(1+self.n_T[0]),
                                    self.scl[1]*(1+self.n_T[1]),
                                    self.scl[2]*(1+self.n_T[2])])
        self.C2C2 = self.Gam*array([self.scl[0]*self.n_T[0],
                                    self.scl[1]*self.n_T[1],
                                    self.scl[2]*self.n_T[2],
                                    0])
        self.H1 = ones(4)-0.5*dt*(self.C1C1+self.C2C2)
```


S. Thiele, *Read-Out and Coherent Manipulation of an Isolated Nuclear Spin*,
Springer Theses, DOI 10.1007/978-3-319-24058-9

```
        self.res  = []
        self.data = []
        self.miss = []
        self.err  = 0


    def run(self,steps=2**22,tr_rt=True):

        res  = [] #data are stored every quantum jump
        data = [] #data are stored in discrete time intervals
        miss = []
        Psi  = self.Psi0          #initial state

        t = 0      #lifetime of the current state
        rev = 1    #-1: e-spin down
                   #+1: e-spin up

        time = 0 #discrete time
        #number of  cycles for each delta_t
        N = int(delta_t/self.dt)
        #cycle is devided into 5 sections
        t1 = 1.0*N/5.0

        t_arr = round(t1)*ones(10)
        #section numbering
        itv    = array([4,0,1,2,3,4,3,2,1,0])
        #itv[0] current section
        LZ_event = t_arr[0]

        #Monte Carlo Loop
        for ii in range(steps):
            # status report
            if (ii%int(steps/100)==0):
                print str(int(1.*ii/steps*100))+'% completed'

            # increase lifetime of the current state
            t = t + self.dt

            #Thermal contribution
            #--> evolve the population continously
            Psi1 = self.H1*Psi
            dp_rel = dot(Psi1,self.C1C1*Psi1)*self.dt
            dp_exc = dot(Psi1,self.C2C2*Psi1)*self.dt
            dp = dp_rel +dp_exc

            eps = rand()
```

```
        #eps > dp: # nothing happens
        if eps < dp: # quantum jump
            #store population and lifetime every quantum
            #jump for testing
            res.append(Psi[::-rev]*t)
            t = 0 #reset lifetime after quantum jump

            if eps < dp_rel : # relaxation
                Psi = array([Psi[1], Psi[2], Psi[3],0])

            else: #excitation
                Psi = array([0,Psi[0], Psi[1], Psi[2]])


        #Landau Zener contribution
        #every section: possible Landau-Zener transition
        #                if LZT --> inverse and store population
        #Am I at the anticrossing?
        if ((ii == LZ_event) and (tr_rt == True)):
            #cycle through anticrossings
            itv = itv[[1,2,3,4,5,6,7,8,9,0]]
            t_arr = t_arr[[1,2,3,4,5,6,7,8,9,0]]
            # set counter to next anticrossing
            LZ_event += t_arr[0]
            #Am I not at the border?
            if itv[0] < 4:
                #is nuclear spin in the right state?
                if (Psi[::-rev][itv[0]]==1):
                    eps_LZ = rand()
                    # if rand < Landau-Zener probability
                    # -->QTM
                    # --> inverse population
                    # --> store nuclear spin state

                    if (eps_LZ < self.p_LZ):
                        # flip e-spin
                        rev = rev * -1
                        # inverse n-spin population
                        Psi = Psi[::-1]
                        # determine n-spin
                        mj = 1.5-itv[0]
                        #store data
                        data.append([time,mj])
                    elif(eps_LZ > self.p_LZ):
                        mj = 1.5-itv[0]
                        miss.append([time,mj])
```

```
                elif (tr_rt == False):
                    mj = nonzero(Psi==1)[0][0]-1.5  # Psi[0]=1 --> mj=-3/2
                    data.append([time,mj*rev])
            #increase time at the border
            else: time += delta_t

        self.res = array(res)
        self.data = array(data)
        self.miss = array(miss)

    def Pop(self,save=False):
        """
        plot histogram of the time average population
        """
        f = plt.figure()
        ax = f.add_subplot(111)
        ax.set_ylabel('pop total');
        ax.set_title('T = '+str(self.T)+' mK');


        p0 = size(nonzero(self.data[:,1]== 1.5))
        p1 = size(nonzero(self.data[:,1]== 0.5))
        p2 = size(nonzero(self.data[:,1]== -0.5))
        p3 = size(nonzero(self.data[:,1]== -1.5))
        norm = 1.*(p0+p1+p2+p3)
        ax.bar([-1.5,-0.5,0.5,1.5],array([p0,p1,p2,p3])/norm,
               width=0.4,align='center',alpha=1,
               color=['grey','blue','green','red'])
        ax.set_xticks((-1.5,-0.5,0.5,1.5))
        #ax.invert_xaxis()
        ax.set_xticklabels((r'$|+\frac{3}{2}\rangle$',
                            r'$|+\frac{1}{2}\rangle$',
                            r'$|-\frac{1}{2}\rangle$',
                            r'$|-\frac{3}{2}\rangle$'))

        f.tight_layout()
        if save == True:
            f.savefig('Histogram.png',dpi=300)
            f.savefig('Histogram.pdf')


    def DeltaMI(self,save=False):
        temp = []
        for ii in range(size(self.data[:,0])-1):
            temp.append([self.data[ii+1,1]-self.data[ii,1]])
        f1 = plt.figure()
```

```
        ax1 = f1.add_subplot(111)
        self.dmi = array(temp)
        ax1.hist(self.dmi, bins=(-3,-2,-1,0,1.0,2,3),
                 align='left',rwidth=0.5,normed=True, color='r')
        ax1.set_xlabel(r'$\Delta m_{\mathsf{I}}$')
        ax1.set_ylabel('probability')
        ax1.set_xlim((-3,3))
        if save == True:
            f1.savefig('delta_m.png',dpi=300)
            f1.savefig('delta_m.pdf')



    def Lifetime(self,xmax,save = False):
        """
        extract T1
        """
        delta_t = self.delta_t
        temp1 = []
        temp2 = []
        temp3 = []
        temp4 = []

        tau = 0
        for ii in range(size(self.data[:,0])-1):
            if (self.data[ii,1] == self.data[ii+1,1]):
                tau += self.data[ii+1,0]-self.data[ii,0]
            else:
                if (self.data[ii,1] == +1.5): temp1.append(tau)
                if (self.data[ii,1] == +0.5): temp2.append(tau)
                if (self.data[ii,1] == -0.5): temp3.append(tau)
                if (self.data[ii,1] == -1.5): temp4.append(tau)
                tau = 0

        temp = [temp1,temp2,temp3,temp4]

        exp_fit = lambda t,tau,a : a*exp(-t/tau)
        time = linspace(0,120,100)

        f1 = plt.figure(figsize=(12,8))

        for i in range(4):
            ax = f1.add_subplot(2,2,i+1)
            ax.set_yscale('log')
            ax.set_ylim(0.005,1)
            ax.set_yticks((0.01,0.1,1))
```

```
            ax.set_xlabel(r'$t \ (\mathsf{s})$')
            ax.set_ylabel(r'$\langle m_{\mathsf{I}} \ = \ $'
                          +str(1.5-i)+r"$\rangle $");
            ax.set_xlim((0,xmax));
            ax.set_xticks((0,20,40,60,80,100,120))
            #create Histogramm of lifetime distribution
            H1 = histogram(array(temp[i]),
                           bins=linspace(delta_t,120,
                           int(120/delta_t)))

            #extract all nonzero elements #
            lft = (reshape(concatenate((H1[1][nonzero(H1[0]!=0)],
                  H1[0][nonzero(H1[0]!=0)])),
                  (size(H1[1][nonzero(H1[0]!=0)]),2),order='F'))

            params, cov = curve_fit(exp_fit,lft[:,0],
                                    lft[:,1],[10,100])

            fit = exp(-time/params[0])

            #normalize
            lft[:,1] = lft[:,1]/params[1]
            ax.scatter(lft[:,0],lft[:,1],c='k')
            ax.plot(time,fit,'r--',linewidth=3)
            ax.text(0.6,0.8,r'$\tau$'+' = '
                    +str(round(params[0]*100)/100)+'s',
                    fontsize='x-large',transform=ax.transAxes)

        f1.tight_layout()
        if save == True:
            f1.savefig('lifetime.png',dpi=300,format='png')
            f1.savefig('lifetime.pdf',dpi=300,format='pdf')

    def Write(self, outfile):
        """
        save simulated data
        """
        f = open(outfile, "w+b")
        pickle.dump(self.data, f)
        f.close()


if __name__ == '__main__':

    Gam  = 1./41    # 1/s
    T    = 150.0  # temperature in mK
```

```
delta_t = 2.5 # s
dt    = delta_t/60 # s
P_LZ = 0.515  # Landau-Zener probability

sim = QMC(T,Gam,dt,delta_t,P_LZ)
sim.run(2**24)
sim.Pop(True)
sim.DeltaMI(True)
sim.Lifetime(120,True)
#sim.Write("sim_data.file")
```

Appendix D
Qutip Code

The following python program was used to simulate the trajectory of the Bloch vector.

```
from qutip import *
from pylab import *
from numpy import real
from mpl_toolkits.mplot3d import Axes3D
import mpl_toolkits.mplot3d.axes3d as p3
def run():
    #
    # problem parameters:
    #
    delta = 0 * 2 * pi      # qubit sigma_x coefficient
    omega = 1.0 * 2 * pi    # qubit sigma_z coefficient
    A = 0.25 * 2 * pi       # driving amplitude
    w = 1.0 * 2 * pi        # driving frequency
    gamma1 = 0.0            # relaxation rate
    n_th = 0.0              # average number of excitations
    psi0 = basis(2, 0)      # initial state
    #
    # Hamiltonian
    #
    sx = sigmax(); sy = sigmay(); sz = sigmaz();
    sm = destroy(2);
    H0 = - (delta + omega) / 2.0 * sz
    H1 = - A * sx
    #
    # define the time-dependence of the Hamiltonian
    #
    args = {'w': w}
    Ht = [H0, [H1, 'sin(w*t)']]
```


S. Thiele, *Read-Out and Coherent Manipulation of an Isolated Nuclear Spin*,
Springer Theses, DOI 10.1007/978-3-319-24058-9

```
#
# collapse operators
#
c_op_list = []

rate = gamma1 * (1 + n_th)
if rate > 0.0:
    c_op_list.append(sqrt(rate) * sm)       # relaxation

rate = gamma1 * n_th
if rate > 0.0:
    c_op_list.append(sqrt(rate) * sm.dag()) # excitation

#
# evolve and system subject to the time-dependent hamiltonian
#
tlist = linspace(0, 0.70 * pi / A, 100)
output1x = mesolve(Ht, psi0, tlist, c_op_list, [sx], args)
output1y = mesolve(Ht, psi0, tlist, c_op_list, [sy], args)
output1z = mesolve(Ht, psi0, tlist, c_op_list, [sz], args)

#
# Alternative: write the Hamiltonian in a rotating frame,
# and neglect the high frequency component (RWA) so that
# the resulting Hamiltonian is time-independent.
#
H_rwa = - delta / 2.0 * sz - A * sx / 2
output2x = mesolve(H_rwa, psi0, tlist, c_op_list, [sx])
output2y = mesolve(H_rwa, psi0, tlist, c_op_list, [sy])
output2z = mesolve(H_rwa, psi0, tlist, c_op_list, [sz])

#
# Plot the solution
#
fig = figure(figsize=(14,7))
rec1 = [0,0,0.5,1]; rec2 = [0.5,0,0.5,1]
rec3 = [0.,0.9,1,0.1]
ax = Axes3D(fig, rec1, azim=-60, elev=30)
ax2 = Axes3D(fig, rec2, azim=-60, elev=30)
ax3 = fig.add_axes(rec3)
ax3.axis("off")
ax3.text(0.05,0,r"$(\sf{a})$",fontsize=35)
ax3.text(0.55,0,r"$(\sf{b})$",fontsize=35)

b1 = Bloch(fig,ax)
```

```
    b1.add_text(0,0,1.2,"(a)",100)
    b1.font_size = 35; b1.zlabel = [r'$z$','']
    b1.xlpos = [1.3,-1.3]; b1.ylpos = [1.2,-1.2]
    b1.zlpos = [1.2,-1.2]
    b1.vector_color=([0.5,0.5,0.5],[0,0,0])
    b1.vector_mutation = 20
    b1.add_vectors([0,0,1])
    b1.add_vectors([real(output1x.expect[0])[-1],
                    real(output1y.expect[0])[-1],
                    real(output1z.expect[0])[-1]])
    b1.point_color = ("blue")
    b1.add_points([real(output1x.expect[0]),
                   real(output1y.expect[0]),
                   real(output1z.expect[0])])
    b1.draw()
    b2 = Bloch(fig,ax2)
    b2.font_size = 35
    b2.zlabel = [r'$z$','']; b2.xlpos = [1.3,-1.3];
    b2.ylpos = [1.2,-1.2]; b2.zlpos = [1.2,-1.2];

    b2.vector_color=([0.5,0.5,0.5],[0,0,0])
    b2.vector_mutation = 20
    b2.add_vectors([0,0,1])
    b2.add_vectors([real(output2x.expect[0])[-1],
                    real(output2y.expect[0])[-1],
                    real(output2z.expect[0])[-1]])
    b2.point_color = ("blue")
    b2.add_points([real(output2x.expect[0]),
                   real(output2y.expect[0]),
                   real(output2z.expect[0])])
    b2.draw()
    return output1x

if __name__ == '__main__':

    out = run()
```

Curriculum Vitae

Dr. Stefan Thiele

Physicist

Date of Birth
25 Feb 1986

Nationality
German

Current Affiliation
Sensirion AG (Switzerland)

Contact Information
Address: Laubisruetistrasse 50
8712 Staefa

Email: stefan.thiele@sensirion.com

Phone: 0041 / 44 306 40 22

[research interest]

- solid state sensors
- MEMs devices
- quantum physics
- spintronics
- graphene based devices
- semiconductor physics

[education]

2010 – 2014 **PhD Thesis:** Nano Physics — **CNRS / Grenoble / France**
Title: "*Read-out and coherent manipulationof an isolated nuclear spin using single-molecule magnet spin-transistor*"
focus on: - nano-physics
- molecular magnets & spintronics
- quantum electronics & quantum computation
- low-temperature physics & cryogenics

2004 – 2010 **Master Thesis:** Applied Physics — **TU Ilmenau / Germany**, ***MIT / Cambridge / USA***
Title: "*Modeling of the AC and DC characteristics of large-area graphene field-effect transistors*"
major in: - semiconductor/ micro and nano-electronics
- laser physics and measurement technology
- solid state and surface physics
minor in: - electronics, mechanics, business studies

1996 – 2004 **High school graduation** — **Zeulenroda / Germany**
graduated with distinction (1.1)

[awards]

2015 Springer thesis award
Thesis award from the Fondation NanoScience
2010 TU Ilmenau - award for being one of the best graduates of the year
2007 – 2009 Scholarship from the German national academic foundation (available to the best 0.5% of all students)
2004 High school – award for being the best graduate of the year

[publications]

2014 Science **344**, 6188
2013 Physical Review Letters **111**, 037203
2012 Physical Review Letters **109**, 264301
2011 J. of Applied Physics **110**, *034506*
2010 IEEE Trans. on electron Devices **57**, 3231
J. of Applied Physics **107**, *094505*
Applied Physics Letters **96**, *123506*
Nanotechnology **21**, 015601
2009 Nano Research **2**, 509

[Web] LinkedIn.com

S. Thiele, *Read-Out and Coherent Manipulation of an Isolated Nuclear Spin*,
Springer Theses, DOI 10.1007/978-3-319-24058-9

Printed in the United States
By Bookmasters